During the Quaternary Period, which comprises the last 2.3 million years, large areas of the continents were repeatedly glaciated as a result of great global climatic changes. One such area was Scandinavia, whose mountains were the source for repeated glaciation that covered much of eastern, central, and western Europe. With a particular emphasis on the four countries of Denmark, Norway, Sweden, and Finland, this text describes how these glaciations, and their intervening warmer stages, affected Scandinavia and its surrounding areas. The book begins with an investigation of Pre-Quaternary substratum and then proceeds to examine northern Europe in the Quaternary Pre-Saalian, Saalian, and Eeemian stages. In particular, this account focuses on the last cold stage, the Weichselian, with its extensive Late Weichselian glaciation and the subsequent deglaciation, and on the last 10,000 years, the Holocene, with its well-documented environmental changes.

The Quaternary history of Scandinavia provides a cross-frontier synthesis of how the glaciation affected this vast region, and will be invaluable to students and researchers of Quaternary science.

The Quaternary history
of Scandinavia

The Quaternary history of Scandinavia

JOAKIM DONNER
University of Helsinki

CAMBRIDGE UNIVERSITY PRESS
Cambridge, New York, Melbourne, Madrid, Cape Town, Singapore, São Paulo

Cambridge University Press
The Edinburgh Building, Cambridge CB2 2RU, UK

Published in the United States of America by Cambridge University Press, New York

www.cambridge.org
Information on this title: www.cambridge.org/9780521417303

© Cambridge University Press 1995

First published 1995
This digitally printed first paperback version 2005

A catalogue record for this publication is available from the British Library

ISBN-13 978-0-521-41730-3 hardback
ISBN-10 0-521-41730-9 hardback

ISBN-13 978-0-521-01831-9 paperback
ISBN-10 0-521-01831-5 paperback

Contents

Preface

An outline of the Quaternary history of Scandinavia cannot be treated without taking into account the development of northern continental Europe generally, including the whole area affected by the repeated glaciations originating in the Scandinavian mountains. The emphasis in the present account is, however, on the four Nordic countries of Denmark, Norway, Sweden, and Finland. These together are here referred to as Scandinavia, leaving out Iceland, which geologically is a separate entity. The only fuller account in English on the Quaternary of the Nordic countries (edited by Rankama, 1965) had separate authors and chapters for each country and was written over 30 years ago. In the present work the areas bordering the four Nordic countries are included without a strict limit being put to the area treated, but for practical reasons – the language barriers being the most important – the areas southeast and east of Finland are not given the attention they would geologically deserve. The outline given of the Quaternary history cannot in any way be considered exhaustive. It is a subjective attempt at giving a general outline of a complicated history.

The volume is intended not only for students with a knowledge of the main methods employed in Quaternary studies of formerly glaciated areas, but also as an introduction to those for whom the area is less familiar.

The author wants to acknowledge the following persons with thanks for their invaluable help received in the preparation of the manuscript: above all, Tuija Jantunen for skilfully drawing the figures by computer, made possible by support from the University of Helsinki and the Finnish Society of Sciences and Letters; my wife Ruth for carefully reading through the manuscript; Lauri Pesonen for providing Figure 2, and Bjørn Andersen, Jan Lundqvist, and Kaj Strand Petersen for providing a number of photographs. No one other than the author, however, is responsible for the misunderstandings and errors in interpretation that regrettably occur in a treatise of this kind.

Some of the works referred to in the text are general summaries that give further references to the original papers. But as many references as possible were included to aid those readers who want to aquaint themselves with the original papers in more detail. Many of them were published in Danish, Norwegian, Swedish, or Finnish but have summaries in English, as do most papers published in other languages, such as in Russian and Estonian.

Finally, the author wants to thank Catherine Flack and Katharita Lamoza of the Cambridge University Press for their much appreciated assistance in the preparation of the manuscript for publication.

1   Introduction

When the Scandinavian ice sheets spread out from the mountains of Norway and Sweden during the successive Quarternary glaciations of northern Europe they reached at their furthermost as far as The Netherlands and Germany in Central Europe and far into Poland and the Russian Plain in Eastern Europe (Fig. 1.1). But their spread was restricted in the northwest by the deep sea off the Norwegian shelf. They were, however, at times joined across the North Sea floor with the smaller independent ice sheets of the British Isles (Woldstedt, 1958; Flint, 1971; West, 1977; Bowen, 1978; T. Nilsson, 1983). The repeated Quaternary glaciations of Scandinavia resulted in strong glacier action on the underlying rocks and in the laying down of a variety of glacial deposits. The early emphasis in Quaternary studies, starting in formerly glaciated areas of Europe, was focused on glacial deposits, which were used particularly in the reconstruction of the extension of the glaciations. These studies developed from being mainly descriptive, with the inclusion of morphological studies, into a detailed study of the lithostratigraphy of glacigenic sediments. But these original studies were also extended to include nonglacial sediments. The lithostratigraphical studies of, for instance, till beds have been combined with morphostratigraphical studies of moraines in separating ice advances and units in the last deglaciation with the help of end moraines, even if landforms should not generally be used in stratigraphy (see Mangerud et al., 1974). In addition to tracing the extent of former Quaternary glaciations, especially the last one, much stress has been laid on the study of the last deglaciation in general, combined with the study of end moraines, as well as all aspects of the history of the last 10 ka (10,000 years), the Holocene. The chronological frame established with the help of varved clays by Gerard De Geer in Sweden, and later revised and refined, was a major advance in dating the last deglaciation in Sweden and partly also in Finland. Similarly, the introduction by Lennart von Post (1916, see also 1967) of pollen analysis, also in Sweden, gave a powerful tool for biostratigraphical studies, first of the Holocene but later extended into the Pleistocene and even further back in time. The changeover from a study of macrofossils, mainly of plants, to a study of plant microfossils opened up entirely new possibilities in biostratigraphy, and was

later to include the study of other microscopic remains of organisms, such as diatoms. Unraveling the chronology of the last deglaciation of the Scandinavian ice as based on varved clays combined with the relative chronology of the vegetational history based on pollen analysis was given fresh impetus with the introduction of the radiocarbon dating method by Willard F. Libby in 1946. The use of this method in dating organic material, especially those materials postdating the last glaciation, has since the 1950s been decisive in the study of the history of the vegetation and flora, by establishing the spatial differences that could not be demonstrated with pollen analysis alone. This change also affected the dating of the pollen stratigraphical units, formerly used extensively in studies of the Holocene. Radiocarbon dating has made it possible to study the younger dynamic changes in environment, giving results that can also be applied in the study of older deposits outside the reach of radiocarbon dating.

The study of the uplift of the earth's crust, particularly that which followed the last glaciation, has been closely connected with the development of the above-mentioned methods of dating. Large parts of Finland and Sweden were submerged after deglaciation and all coasts of the countries bordering the Baltic Sea and the coasts of Denmark, Sweden, and Norway facing the ocean, have been affected by the isostatic land uplift and eustatic rise of sea level. The study of the interplay between these two changes, with other possible changes involved, has been a particularly favored subject in Nordic Quaternary research in Sweden since the turn of the century (Munthe, 1892; De Geer, 1896), followed by similar studies in the surrounding countries, in Finland particularly by Ramsay (1900). The different models and interpretations presented reflect the complex nature of these changes, which in the Baltic Sea are further complicated by changes in water level at times when the Baltic was an independent lake.

In contrast to the above-mentioned studies of the glacial deposits, mostly of the last glaciation, and of the subsequent Quaternary history of the area formerly covered by the Scandinavian ice sheet, which are by their nature local studies, the stratigraphical studies of the development preceding the last

Figure 1.1. Scandinavia and surrounding countries.

glaciation are largely dependent on results from the marginal areas of glaciation and from outside it. As the stratigraphical evidence, in spite of an increasing number of new sites, is patchy in the central parts of glaciation – in Finland, Sweden, and Norway – the evidence has to be viewed against the results from elsewhere. As the lithostratigraphical control is often poor, unless well-known till beds can be used, the separation of the Quaternary stages has been based on biostratigraphical evidence. Since the pollen analytical studies of interglacial fresh water and marine deposits started in Denmark, for example, by Jessen and Milthers (1928; S. T. Andersen, 1965; Hansen, 1965), the stratigraphical subdivision of the Quaternary of Scandinavia has further developed and been fitted into the larger context of the subdivision of continental northwestern Europe (Mangerud et al., 1974). In dealing with an area as large as that covered by the Scandinavian ice sheet, the spatial differences in the vegetational history do, however, become, important. The correlation between sites far apart is often difficult, especially when there are incomplete successions of organic deposits. The biostratigraphical results are often interpreted by comparisons

with the development which followed the last glaciation, assuming broad similarities between the successive glaciations and the nonglacial intervals. But there is not the same chronological control as in the study of the younger organic deposits where radiocarbon dating can be used. Further, as the resolution is great in Quaternary studies and time differences in the development can be demonstrated, the use of conventional geological methods and divisions is not always ideal. Major climatic changes caused the consecutive growths and decays of large continental ice sheets, with minor fluctuations, but because of timelags in the response to these changes they cannot be directly correlated with the stratigraphical changes (H. E. Wright, 1984). It is therefore difficult to apply formal chronostratigraphic units for the time-transgressive Quaternary events (Watson & H. E. Wright, 1980). In spite of its drawbacks a chronostratigraphic frame is, however, used here in the presentation of the area dealt with, bearing in mind the above-mentioned difficulties.

Before proceeding to the subdivision of the Quaternary Period used in Scandinavia and discussed in Chapter 3, the division

Table 1.1. *Subdivision of the Cenozoic*

Geological time units Stratigraphical units	Eras	Periods	Epochs	
		Systems	Series	
		Quaternary	Holocene	
				10 ka B.P. (radiocarbon years)
			Pleistocene	
				2.3 Ma
	Cenozoic	Tertiary	Pliocene Miocene Oligocene Eocene Palaeocene	
				65 Ma

Table 1.2. *Palaeomagnetic epochs and events in the Quaternary*

Epochs	Events	
Brunhes normal		
		0.73 Ma
	Jamarillo	0.90
		0.97
Matuyama reversed	Olduvai	1.67
		1.90
	Réunion	2.01
		2.04
		2.12
		2.14
		2.48
Gauss normal		

Note: Ages as dated with the K-Ar method.
Source: Revised time scale by Mankinen and Dalrymple (1979).

of the Cenozoic Era into periods and epochs, as used here, should be defined. The division used is given in Table 1.1, according to the general present usage in northwest Europe, with a subdivision of the Quaternary Period into the Pleistocene and Holocene epochs (Mangerud et al., 1974; T. Nilsson, 1983). The age of lower boundary of the Tertiary is given as 65 Ma (65 million years) according to Hambrey and Harland (1981).

The Pliocene–Pleistocene boundary in The Netherlands is placed between the Reuverian and the Pretiglian (Suc & Zagwijn, 1983; Zagwijn, 1985). The age of the boundary is estimated at about 2.3 Ma, as the top of the Reuverian is in the lowermost part of the palaeomagnetic Matuyama reversed polarity zone with a lower boundary at about 2.4 Ma (Suc & Zagwijn, 1983), in the detailed revised geomagnetic polarity time scale based on K-Ar dating given an age of 2.48 Ma (Mankinen & Dalrymple, 1979), as seen in Table 1.2, which shows the younger palaeomagnetic epochs and events relevant to the division of the Quaternary. A modification of the timescale given in Table 1.2 has, however, been suggested by Shackleton, Berger, and Peltier (1990), which increases the ages for Matuyama–Brunhes and Gauss–Matuyama boundaries, as well as for the Olduvai and Jamarillo events, by between 5 and 7 percent. The Pliocene–Pleistocene boundary in The Netherlands has been correlated with the boundary between the Corraline Crag and the Red Crag representing the Waltonian in East Anglia in Britain (Shotton, 1973; West, 1977). The Waltonian is, however, included in the Pliocene in a correlation by Zagwijn (1979), who also demonstrated the existence of great gaps in the preglacial Pleistocene sequence of East Anglia. Through palynological evidence, the base of the Praetiglian in The Netherlands has been correlated with the base of pollen zone P III in the northwestern Mediterranean region, which also there has an age of about 2.3 Ma (Suc & Zagwijn, 1983). The Pliocene–Pleistocene boundary in the marine sequence in Italy as defined

in the Vrica section is younger, with an age of 1.65 Ma (Richmond & Fullerton, 1986), being above the Olduvai palaeomagnetic subzone (Mankinen & Dalrymple, 1979). The suitability of the boundary used in the Vrica section has, however, later been questioned (Jenkins, 1987). If the Pliocene–Pleistocene boundary is placed at the base of the Calabrian, it is also at the base of the Olduvai subzone and has an age of 1.9 Ma (Suc & Zagwijn, 1983). It is in any case above the Pliocene–Pleistocene boundary as defined in The Netherlands and used here. As no deposits found in Scandinavia represent the boundary discussed above, it has so far no practical stratigraphical importance.

In accordance with the recommendation of the Holocene Commission of INQUA the Pleistocene–Holocene boundary has been placed at 10,000 radiocarbon years B.P., using the Libby half life of 5570 ± 30 (or 5568) years with A.D. 1950 as the standard year of reference (see Shotton, 1973, 1986). It was suggested that this boundary be used in Scandinavia (Mangerud et al., 1974) and it is also used here. An acceptable stratotype locality for this boundary fulfilling all requirements has not been found in the selected area in southwestern Sweden (Olausson, 1982).

In an area like Scandinavia, which was overrun several times by large inland ice, the glacial sequence is based on the stratigraphy of deposits left by the ice, in addition to nonglacial deposits. The glacial deposits were formed of material eroded by the ice from the underlying bedrock. The effect of this erosion is treated in the next chapter, with an outline of the pre-Quaternary history of the area and the general changes in morphology that have taken place as a result of repeated glaciations. Of particular interest is the question of the intensity and amount of glacial erosion.

2　Pre-Quaternary substratum

2.1. Pre-Quaternary bedrock

If one uses a geomorphological classification the macro- and mesoreliefcomplexes, which take 100 Ma and 1 Ma respectively to develop (Fogelberg, 1986), are in Scandinavia and surrounding areas dependent on the history of the bedrock, its petrological composition, and the tectonic structures. A contrast to the pre-Quaternary morphological features is found in the Quaternary landforms, which in most areas determine the micro-relief elements formed in 0.1–1 ka. It is only the Scandinavian mountains, in which the Quaternary glaciers were formed and from where they spread, that there were major morphological changes as a result of the abrasion by the glaciers. The Scandinavian mountains provided the "suitable situations" (see Sugden & John, 1976) for the glaciers to develop when the climatic changes during the Quaternary triggered off successive glaciations.

Finland lies in the center of the exposed Precambrian crust forming the Fennoscandian (Baltic) Shield, which in its western part includes most of Sweden and parts of southern and western Norway, in the northeast the Kola peninsula, and in the east reaches the shores of the White Sea and areas beyond Lake Onega (Fig. 2.1). The Fennoscandian Shield, which mainly consists of a resculptured Precambrian erosional surface (Miskovsky, 1985), dips in the east and south under a cover of the Palaeozoic sedimentary rocks of the East European Platform. Some Proterozoic sedimentary rocks, which have not undergone metamorphism, remain within the area of Fennoscandian Shield, such as the "blue clay" of the Karelian Isthmus, formerly placed in the Lower Cambrian (Winterhalter et al., 1981). In Finland, the many-hundred-meter thick Muhos sediments, consisting mainly of sandstones, siltstones, and shales, as well as the Jotnian sediments of Satakunta, mainly sandstones, are also remnants of Proterozoic sediments preserved in tectonic depressions (Simonen, 1980). They both extend into the Gulf of Bothnia. The Jotnian sediments in the southern part, in the Bothnian Sea, underlie Cambrian and Ordovician sedimentary rocks (Winterhalter et al., 1981; Bergman, 1982), but also occur separately in the Åland Sea. Lower Cambrian sediments 300–400 m thick have also been found in the small circular depression of Söderfjärden south of the town of Vaasa (Laurén et al., 1978) and remnants of Ordovician sediments in the depression formed by the bay of Lumparn in the Åland Islands. In Sweden there are also remnants of Proterozoic sedimentary rocks within the Fennoscandian Shield similar to the Jotnian sediments in Finland, the Gävle sandstone at the coast of the Bothnian Sea, and the sandstone in Dalarna further inland (Fig. 2.1). In addition there are remnants of Palaeozoic sedimentary rocks, such as the sediments in Dalarna in the Lake Siljan area and the occurrences of sandstones, siltstones, and limestones further south at Örebro, Motala, Billingen, Kinnekulle, and Halleberg-Hunneberg. At Billingen and Kinnekulle, both in Västergötland, the Cambro–Silurian sediments, including limestones, are as elsewhere erosional remnants of more extensive covers of sedimentary rocks, but they are capped by younger diabases dated to the boundary between the Carboniferous and Permian, such as the Cambrian and Ordovician sediments at Halleberg-Hunneberg, and have therefore remained as hills in contrast to remnants of other sedimentary rocks in depressions elsewhere in the Fennoscandian Shield area. The Palaeozoic igneous rocks of the Oslo graben cut through the area of Precambrian rocks in Norway and are also associated with sedimentary Cambro–Silurian rocks.

The gently dipping beds of the Palaeozoic sedimentary rocks of the East European Platform bordering the Fennoscandian Shield in the east and south, exposed along the shores of the Baltic Sea, Lake Ladoga, and Lake Onega, have been traced across the Baltic Sea southwestwards as far as the Tornquist Line, the major fracture zone cutting through Scania in Sweden thus separating the Fennoscandian Shield and the East European Platform from the Central European Platform (Winterhalter et al., 1981). In the nearly 600 m deep Landsort Trench north of Gotland, bordering the Palaeozoic sedimentary rocks, there are remnants of Jotnian sandstone; elsewhere in the Baltic Sea region there are areas in which these sandstones underlie the Palaeozoic sedimentary rocks (Winterhalter et al., 1981). The outermost zone of Palaeozoic rocks on top of the Precambrian is formed by the narrow zone of Cambrian rocks, with the Ordovician zone going from northern Estonia to the island of Öland, and the Silurian zone including the island of Gotland. Then southeastward follow zones of younger

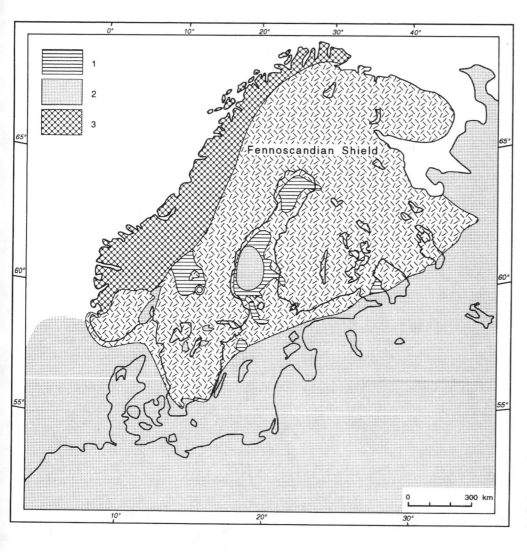

Figure 2.1. Main units of the bedrock of Scandinavia and surrounding areas (after Hjort & Sundquist, 1979; Martinsson, 1980; Simonen, 1980; Winterhalter et al., 1981; Papunen & Gorbunov, 1985). 1, Jotnian (Proterozoic) formations; 2, Paleozoic and Mesozoic rocks; 3, Caledonides.

Palaeozoic rocks, as well as Mesozoic rocks, as far as the Cretaceous sedimentary rocks of the southern areas and the coasts of the Baltic Sea (Martinsson, 1980; Winterhalter et al., 1981).

The Caledonides, the orogenic belt of the Scandinavian mountains northwest of the Fennoscandian Shield, were mainly formed during the Silurian and Devonian as a result of folding and metamorphism of marine geosynclinal Cambro–Silurian sediments and older Precambrian rocks at the edge of the continent of which the Fennoscandian Shield was a part (O. Holtedahl, 1960; Hjort & Sundquist, 1979). The Palaeozoic rocks of the Caledonides have in many places overthrust the Fennoscandian Shield. But within the Caledonides there are outcrops (windows) of Precambrian rocks of younger intrusions, and similar rocks at the coast west of the mountains, particularly in the Lofoten area. The shelf area outside the Norwegian coast consists mainly of Mesozoic sediments. Further south the North Sea is part of the intratectonic Northwest European Basin, with an up to 9 km thick sedimentary sequence of sediments from the Palaeozoic to the Quaternary, the latter sediments reaching a thickness of 1 km (Ziegler & Louwerens, 1979; Jelgersma, 1979).

Denmark and southernmost Sweden are parts of the same basin, the Danish Polish Basin, in which the underlying Precambrian bedrock lies deeper in the south and southwest than at the edge of the platform against the Fennoscandian Shield. In Scania in southern Sweden the outcropping sedimentary rocks represent the Palaeozoic, Mesozoic, and Kenozoic (Tertiary), as they do in Denmark (H. W. Rasmussen, 1966; Hjort & Sundquist, 1979). The Cretaceous limestones in northern Jutland and on the Danish islands to the southeast, mainly on Zealand (Sjaelland), are bordered by zones of Tertiary sediments from Palaeocene to Pliocene in age, getting younger toward the southwest so that most of Jutland is Miocene, with some Pliocene on the west coast. There is, however, no Pliocene sequence merging directly into the Pleistocene (Hansen, 1965). In addition to the Tertiary sediments there are salt domes that have brought Permian and Tertiary sedimentary rocks to the surface of the bedrock (Hansen, 1965). The Danish island of Bornholm in the east, with a Precambrian bedrock, lies at the edge of the Fennoscandian Shield in the same fracture zone as Scania, with younger sedimentary rocks bordering it.

2.2 Pre-Quaternary morphology of bedrock

As mentioned, the bedrock relief in Finland and Scandinavia was largely developed in pre-Quaternary times, and the erosional surface of the Fennoscandian Shield already in the Precambrian. For a surface of this kind the term peneplain can still be applied as a descriptive term (Fogelberg, 1986), without connecting it with the original explanation of its formation as the end product of an erosional cycle of running water in a humid climate (Davis, 1899). Thus Kaitanen (1969), for instance, separated two peneplains in northern Finnish Lapland, formed during different climatic conditions, whereas in earlier works the separation of peneplains was based on the original definition (see Fogelberg, 1986). The Fennoscandian Shield is now considered to be foremost an etchplain, the result of chemical weathering taking place in a humid climate under a thick cover of unconsolidated decomposed bedrock, with the structure and tectonic features governing the intensity of the weathering (Miskovsky, 1985). When the bedrock was subject to intensified erosion, to stripping in other words, the etchplain of the fresh bedrock was exposed. During periods with a dryer climate and therefore less chemical weathering, pediplains were formed when separate erosional surfaces – pediments – merged into one another. The formation of etchplains and pediplans altered during the pre-Quaternary history of the Fennoscandian Shield, depending on its position in relation to the climatic zones and the extent of dry land at different times. The Fennoscandian Shield was, however, as already pointed out, deeply eroded before the end of the Precambrian, with the remains of its youngest sediments, such as the Jotnian sandstone, preserved in down-faults or grabens that were formed during the formation of the joints, faults, and shear zones over a long period of time. The development of the relief of the Fennoscandian Shield after the Precambrian, resulting in the final pre-Quaternary plain, was mainly molded during times with a tropical climate (Fogelberg, 1986). As there is no direct evidence of how intense the weathering was at different times, conclusions are based on what is known about the palaeogeographic position of Scandinavia, and thus its relation to the climatic zones, during the geological periods (Fig. 2.2). During the Cambrian and Ordovician the area was close to the south pole, but later, already during the Silurian but particularly during the Devonian and Carboniferous, it was near the equator and chemical weathering was intense. A renewed time of intense chemical weathering took place from the Jurassic period to the early Tertiary, with a humid marine climate, in spite of the drift northwards of the Fennoscandian Shield (Miskovsky, 1985; Fogelberg, 1986). During the Permian and Triassic periods with a dryer climate, there was stripping and the overburden was largely removed, as it was again later removed by the inland ice during the Quaternary. In addition to the periods of intense weathering alternating with periods of stripping, the relief of the Fennoscandian Shield was affected by crustal movements after the Precambrian, especially during orogenies in adjacent areas. The Caledonian orogeny, culminating in the Devonian, caused an uplift of the Fennoscandian

Shield, and some major faults were still being formed during the Tertiary at the time of the Alpine revolution, which in the northern parts of the shield was also connected with an uplift of the crust and a rejuvenation of the rivers. But Tertiary faults affected other areas as well, such as the Precambrian bedrocks of Scania in southern Sweden.

The morphological features of the pre-Quaternary bedrock of the Fennoscandian Shield were a result of the long evolution beginning in the Precambrian, as described above. It was then that the macro- and mesorelief developed, using the geomorphological classification mentioned in the beginning of the chapter, with landforms that took a long time to develop as compared with those of the Quaternary landforms. Thus the earlier-mentioned even areas with Proterozoic sediments in the tectonic depressions in Satakunta and Muhos in Finland, and the equivalent sandstone areas in Sweden, can be recognized in the landscape, even if they underlie Quaternary sediments. Similarly, the hills of resistant rocks, particularly of quartzite, rise above the surrounding country. The quartzite hills of Tiirismaa near Lahti, Puijo in Kuopio, and Koli at Pielisjärvi in southern, central, and eastern Finland are about 200 m above their surroundings, and even more conspicuous and higher, up to 500 m above their surroundings, are the quartzite hills further north, as Pyhätunturi, Yllästunturi, and Ounastunturi in Finnish Lapland, because of their shape referred to as having been inselbergs that were later reshaped by glaciers (Kaitanen, 1985; Fogelberg, 1986). The earlier-mentioned diabase-capped hills in Sweden are also morphologically important, but represent erosional remnants of Palaeozoic sediments on the Fenno-scandian Shield. Some of the separate quatzite hills in Finland and Sweden are parts of orogenic belts, lying in folded zones like Koli in eastern Finland, which belongs to the Svecokarelidic zone extending from southeastern Finland northwestward to Finnish Lapland (Simonen, 1980). Quartzites are common in this zone, which can now be recognized because of their residual hills. The northern and central parts of Sweden, bordering the Caledonides, were already in later Precambrian times relatively high and have because of later erosion a comparatively high relief, even here reflecting the structure of the underlying bedrock and its resistance to weathering.

The relief of the entire Fennoscandian Shield is, in addition to the lithology and structure of the bedrock, strongly governed by the faults and fracture zones of the bedrock, not only by the larger earlier-mentioned grabens with preserved sedimentary rocks. The broken surface of the crust has resulted in deeper weathering of the fracture zones and the accentuation of smaller landforms. This can particularly be observed in areas where the bedrock is well exposed, as in the large archipelago of southwestern Finland. There the fracture zones can be seen in the grouping and shapes of the islands, and inland by the orientation of the lakes. The later abrasion of the glaciers during the Quaternary often only accentuated and gave the final polish to preexisting landforms of the bedrock.

The present-day relief of the East European Platform bordering the Fennoscandian Shield – both on dry land and on the floor

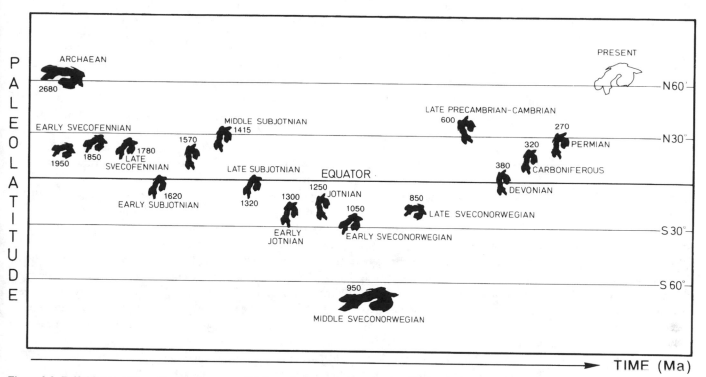

Figure 2.2. Drift history of Scandinavia (Fennoscandia) from Archaean to Permian (modified after Pesonen et al., 1989).

of the Baltic Sea – is dependent on the Palaeozoic sedimentary rocks and further south it is dependent on Mesozoic rocks. They have to some extent been subject to faulting but, as mentioned earlier, they form zones that can be traced throughout the area. Lobate escarpments are often formed, for which the term clint has been used (from the Danish and Swedish word *klint,* Martinsson, 1958, 1980). They are usually marine abrasion cliffs and fluviatile erosion scarps. The escarpments of the coasts of Estonia are such features, and so are the clints of the islands of Gotland and Öland in Sweden, both accentuated by the action of running water. The rivers also dissected the escarpments of this cuesta landscape, that is, by landforms with steep escarpments and gentle dip slopes, which have been preserved even after the repeated glacial abrasion during the Quaternary. The pre-Quaternary drainage pattern of the Baltic Sea, from the Tertiary, can still be traced. The river channels in the Bothnian Bay continue on the sea floor to a depth of about 80 m below present sea level (Tulkki, 1977) and further south, as in Estonia (Tavast, 1987) and the southern parts of the Baltic Sea, there are pre-Quaternary buried channels that influenced the Holocene drainage pattern. The late Tertiary river valleys have also been traced between Öland and the mainland, and between Öland and Gotland (Königsson, 1980). The drainage from almost the whole of the Baltic area passed, however, on the eastern side of Gotland toward the Danzig Depression (Martinsson, 1958) and the Baltic Sea is in its present form a Tertiary feature (Königsson, 1980). During the Tertiary there were generally great changes in the palaeogeography of the area surrounding the Baltic Sea, connected with the Alpine revolution,

particularly at the edge of the East European Platform. It was generally a time with crustal uplift, especially during the Miocene, but in some areas there was subsidence, as in the Gulf of Bothnia and Lake Vänern in Sweden, in addition to Skagerrak and Kattegat further west (Hillefors, 1983, 1985).

The present relief of the Caledonides is also largely a result of late Tertiary uplift. This is in agreement with the finds of reworked marine early Tertiary silicoflagellates, foraminifera, and – particularly – diatoms at several sites in Finnish Lapland (Hirvas & Tynni, 1976; Grönlund, 1977). The Caledonides, formed during the Silurian and Devonian, were eroded and here, as in the erosion of the Fennoscandian Shield, deep weathering in a warm humid climate was an important factor (Gjessing, 1967). An old, paleic, land surface was formed with separate round-shaped residual hills or groups of hills, plains, and basins connected by passes. These paleic landforms rose during the Tertiary to their present altitude asymmetrically, so that the western part, reaching in places 2,000 m above present sea level, is highest. The upland areas (Norwegian *vidde*) are thus remnants of the paleic surface (O. Holtedahl, 1953, 1956). Here, as elsewhere in Scandinavia, the remnants of pre-Quaternary weathered bedrock, to be discussed later, support conclusions about preserved old landforms. Because of the Tertiary uplift of the Caledonides, deep river valleys were cut into the paleic surface, their courses being determined by faults in the bedrock, its structure and lithology, in places weakened by previous weathering.

In connection with the late Tertiary uplift of the Caledonides there were dislocations along major fracture zones along the coast and the continental shelf area of Norway, with long narrow depressions parallel to the coastline inside the more even shelf outside, but also with transverse fracture zones (H. Holtedahl, 1958). Of particular interest is the bedrock plain, the strandflat, in places as much as 40–50 km wide just below, at, or slightly above present sea-level, with a landward cliff (B. Andersen, 1965). The morphologically conspicuous strandflat consists of low islands and shallow sea areas, or low rock platforms. Even if the strandflat lies in the zone affected by major dislocations in connection with the late Tertiary uplift, and therefore has some remnants from the paleic surface (Gjessing, 1967), it is a morphological feature molded by later sea-ice erosion and frost-shattering, as well as by glacial erosion, during the Quaternary, especially during the cold stages (Larsen & H. Holtedahl, 1985). The formation of the strandflat is considered to have taken place over a long period of time, primarily when it was ice-free and in a position several times at sea level.

The continental shelf of the shallow Barents Sea area between North Norway and Svalbard, with thick sedimentary rocks, was dry land during the Tertiary, and the river valleys from that time can still be traced on the shelf (T. O. Vorren, Rønnevik, & Reiersen, 1980).

In contrast to the Tertiary uplift of the Caledonides there was subsidence in the North Sea basin with its thick loads of sediments. The Norwegian trough near the coast, as well as Skagerrak, has been interpreted as having later been scoured out by glaciers (Ziegler & Louwerens, 1979). The irregular surface of the pre-Quaternary land surface in Denmark and in Scania in southernmost Sweden is largely the result of the Tertiary faulting that took place as a result of the Alpine revolution that also affected other parts of Scandinavia. In addition, the previously mentioned salt domes in Jutland continued to rise during the Tertiary, at least still during the Eocene (H. W. Rasmussen, 1966). But there was a general sinking of the basin of which Denmark was a part. It is the early Tertiary marine sediments that are especially well represented; later, during the Tertiary, there were already areas of dry land in which freshwater sediments were formed. It was a time with a changing coastline and therefore varied sedimentary environments (Hjort & Sundquist, 1979).

2.3. Weathering of bedrock

The contact between the Quaternary glacial drift, that is "all rock material in transport by glacier ice, all deposits made by glacier ice, and all deposits predominantly of glacial origin made in the sea or in bodies of glacial meltwaters" and that includes "till, stratified drift, and scattered clasts that lack an enclosing matrix," as defined by Flint (1971, p. 147), and the surface of the underlying sedimentary rocks is not always sharp. In most parts of Fennoscandian Shield and the Caledonides, however, the drift lies on a fresh, often polished surface of the underlying hard, predominantly crystalline, bedrock. But, as noted above, there are in places remnants of a weathered overburden of bedrock, a regolith. It is in places soft and clay-rich as a result of chemical decay and can be classified as saprolite (Strahler, 1965), and is thus

clearly pre-Quaternary, from the earlier-mentioned times of intense chemical weathering in a humid tropical or subtropical climate, when many of the present landforms of the bedrock were formed. But there are also remnants of a regolith that are more difficult to date. If it is till-covered it has been called preglacial, but it cannot always be determined whether it is Quaternary or pre-Quaternary. The remnants of clearly pre-Quaternary, soft, weathered bedrock in exposed parts of the central area of glaciation, as well as of till-covered minerogenic and organic sediments that have been preserved after at least one ice advance, show that they could resist glacial abrasion. Much of the disintegrated bedrock forming angular blocks is, however, primarily a result of incomplete weathering after the last glaciation (Lundqvist, 1985).

There are a number of observations in Scandinavia of till-covered weathered bedrock described as preglacial, which in most cases can be classified as pre-Quaternary, as mentioned earlier (Lidmar-Bergström, 1988). If similar weathered bedrock is in places exposed without being covered by till it can, by analogy, also be called pre-Quaternary. The regolith may have been formed over a long period of time under different climatic conditions and, even if a saprolite is present, it cannot be dated. All occurrences of deep-weathered bedrock that are at least older than the last glaciation are therefore here described together. The preserved regolith in Scandinavia seems to be more frequent in the areas of the ice divide during the glaciations, that is, in Sweden in the northern parts east of the border with Norway (Lundqvist, 1985) and in Finland in central Lapland. Thus Sederholm (1913) described a site with 10–30 m deep-weathered granulite underneath a thick till bed in Laanila, Inari. Later, similar sites were found in Ivalo and Tankavaara in Sodankylä on the lee side of hills in relation to the general movement of the ice, with weathered bedrock up to 100 m thick (Virkkala, 1955). Finds of deep weathering further south are less frequent. At Sulkava in Kiuruvesi, eastern Finland, there is a 1 m thick layer of weathered porphyritic pyroxene granite underneath till, with round unweathered corestones preserved in the gravelly regolith (Brander, 1934). Similar round corestones are found at Pirttikangas, Kauhajoki, in Ostrobothnia in a till-covered, up to 4 m thick layer of weathered granite (Kurkinen & Niemelä, 1979). The weathering has penetrated down into the rock, especially along fractures and around corestones. It can be assumed, as in the granite areas in Jämtland in Sweden (Lundqvist, 1985), that large, well-rounded boulders transported by the ice only a short distance may be corestones from the regolith. The portion of weathered rock material in the drift is generally not known even if, in most cases, it has been assumed that it is small. The drift from the earlier Quaternary glaciations, however, probably contained more reworked weathered rock material. At another site, at Tyrvää in western Finland, a 3–5 m thick till covers bedrock weathered to a depth of 2–8 m (Härme, 1949), and in central Finland there are several sites with 0.5–2 m thick till-covered weathered layers of granitoids, showing both granular disintegration and subsequent alteration as a result of chemical weathering (Lahti, 1985). Even these sites have round corestones. In the central area of the Viipuri rapakivi massif in southeastern Finland the weathered

Figure 2.3. Weathered boulder of Viipuri rapakivi granite with core stones on the Salpausselkä I moraine near Lappeenranta, southeastern Finland. (Photo J. Donner)

rock is 1–3 m thick. The weathered material is usually mechanically and chemically disintegrated material in which the less weathered feldspar ovoids form a gravelly sand (in Finnish locally called *moro*). In addition there is fine-grained clayey material described as saprolite, and as in the other areas described above, corestones (Kejonen, 1985). Many of the saprolites are till-covered and therefore considered preglacial, that is, pre-Quaternary, whereas the gravelly sand was formed primarily after the last glaciation, a process of weathering that still continues. The south-facing sides of rock outcrops and boulders in the rapakivi area often have a weathered layer (Fig. 2.3), and in some areas there is an even layer of weathered rock covering the surface of the rapakivi. It is the only area in Finland in which there has been extensive deep weathering of the bedrock after the last glaciation. In addition to the remnants of weathered rock already described, the fracture zones in Finland have weathered rock material underneath the drift, better preserved in valleys at right angles to the general movement of the ice and thus protected from abrasion, than in those valleys that are in the direction of the ice movement (Niini, 1968).

In Sweden the remains of pre-Quaternary regolith are similar to those in Finland and so are the types of younger weathering of the bedrock. Thus, according to the classification by Lundqvist (1985), the most complete deep-weathering leads to the formation of kaolinite. In places this reaches down to 200 m, and in northern Sweden at Vahtanvaara, east of Kiruna, it is under three till beds and therefore older than the last cold stage with its two glaciations in northern Sweden and Finland. In Sweden the most common form of weathering, however, is granular weathering, a disintegration of the rock into mainly gravel-sized grains, reaching a depth of 10 m in places, around corestones like those found in Finland mentioned previously. This weathering is common in the areas of the former ice divides but is also found elsewhere in Sweden. In northern Sweden and in Ångermanland this form of weathering can also have taken place after the last glaciation, still in Holocene time. The angular fracturing, which is a disintegration of the rock into angular blocks, is a result of frost-shattering and therefore connected with any period or area with a suitably cold climate, either before or just after a glaciation. This frost action is still active in the higher parts of the Scandinavian mountains and in some parts of northern Finland, and will be discussed later in connection with other periglacial features. But similarly fractured bedrock has also been found at several sites in northern Sweden under a drift cover, at Leveäniemi below deposits from the last warm (interglacial) stage (Lundqvist, 1985). In Sweden, as in Finland, the preserved weathered bedrock represents various stages in the development of the regolith, the formation of saprolite being dependent on the earlier chemical weathering in climatic conditions not encountered during the Quaternary. An example is the till-covered Triassic–Tertiary lateritic weathering products with minerals such as smectite, kaolinite, and goethite, found in western Sweden, which are products that have also been mixed into the overlying till (Hillefors, 1985). In addition, there are several sites in the same area with about 1 m thick weathered gneiss or granite underneath till (Hillefors, 1969). In contrast to the weathering of igneous rocks, the karst features in the till-covered Palaeozoic limestones on the island of Öland in the Baltic, with its fissures and cavities, should be mentioned (Königsson, 1977). These features were formed before the end of the glaciation that deposited the till, but when is not known.

In Norway the remains of weathered bedrock are essentially similar to those in Sweden, already described. The presence of deeply weathered bedrock on the summits of the highest coastal mountains or on the outermost islands cannot be used as evidence for the presence of nunataks and icefree areas during the last glaciation (B. Andersen, 1965), as it is evident that remnants of even a pre-Quaternary regolith can survive several glaciations, and that many large block fields were formed in the higher mountains as a result of frost-shattering after the last glaciation (R. Dahl, 1966). That some of the preserved regolith in Norway was also formed in a humid and warm climate is shown, for instance, in a study of sites in western Norway, where the weathered rock had a high content of gibbsite, and by the distribution of smectite and kaolinite. Suitable climatic conditions for the assumed lateritic/bauxitic weathering have not prevailed in the area since late Tertiary times (Roaldset et al., 1982). Even if some regoliths are pre-Quaternary, as shown by mineralogical and chemical analyses, or older than the last glaciation because they are till covered, some later weathering can in places be suspected, as noted in the Lofoten–Vesterålen are in northern Norway, where the weathering of the Precambrian rocks reaches down to 8 m (Peulvast, 1985). Here, as in Finland and Sweden, there are corestones in the weathered rock.

As a result of the erosion of the weathered bedrock, tors, which are essentially a variety of corestones, have sometimes been left exposed, surrounded by slightly displaced corestones and weathered rock, as on top of the hill Lauhanvuori in western Finland, where the main weathering preceded at least the last glaciation and probably even the Quaternary (Fogelberg, 1986). Elsewhere, as in the Narvik area in Norway, active weathering after the last glaciation could have resulted in the formation of torlike forms (R. Dahl, 1966).

In addition to the previously mentioned records of remnants of weathered bedrock from various periods and climatic conditions, there are some observations of micro-weathering after the last glaciation and the removal of the weathering products, called deterration (R. Dahl, 1967). This process, together with exfoliation, a common form of weathering in the northern parts of Scandinavia, as in the mountains, has given those rock outcrops that have been exposed since deglaciation a rough surface from which the finer marks of glacial abrasion have been obliterated. It has been estimated that after the last glaciation the bedrock in southern Finland has weathered to a depth of 1–3 cm and further inland, in an area which has been dry land longer, to a depth of 2–5 cm (Tanner, 1938). By measuring the height of quartz veins above the surrounding rock surface, more exact values can be obtained for the deterration, as has been done in the Narvik area in Norway at different altitudes (R. Dahl, 1967). The values obtained are of the same order of magnitude as those quoted from Finland, with higher values for moss-covered rocks.

2.4. Quaternary erosion of bedrock

In assessing the amount of glacial abrasion, which was the most important form of erosion during the Quaternary, the previously mentioned development of the relief of the bedrock in Scandinavia, and the remnants of pre-Quaternary weathered bedrock, have to be taken into account. The major relief-complexes were already developed before the Quaternary, as noted at the beginning of this chapter. It was only the Scandinavian mountains with a high relief, as well as their marginal areas, that had great morphological changes as a result of the erosion of the glaciers, through abrasion and glacial quarrying or plucking (see Flint, 1971; Sugden & John, 1976). The narrow valleys, especially on the western and northwestern coast of Norway, were first cut by streams into the bedrock as a result of the Tertiary uplift and then deepened and broadened by glaciers into even deeper through-shaped valleys and fjords (O. Holtedahl, 1953; B. Andersen, 1965; B. Andersen & Nesje, 1992). The deepest fjord, Sognefjord, is 1,300 m deep, bounded by mountains 1,000–2,000 m high. Some of the deepened valleys, particularly on the eastern side of the mountains, now have long narrow lakes. Oslofjord, on the other hand, is originally a Palaeozoic graben and the submarine trough a Tertiary graben. But, as mentioned earlier, this trough as well as Skagerrak were scoured out by glaciers during the Quaternary (Ziegler & Louwerens, 1979). The complex formation of the Norwegian strandflat has already been mentioned, where the glacial abrasion only modified the surface while other processes were more important in shaping this wide bedrock plain (Larsen & H. Holtedahl, 1985). In Denmark, with softer sedimentary rocks, the glacial abrasion was selective and resulted in an uneven bedrock surface, as in the area of Copenhagen. Another feature is the presence in Denmark of large "glacial floes," parts of bedrock torn from the underlying bedrock and transported by the ice, the biggest as large as 500 m long and 30–100 m thick (Hansen, 1965). In the clint of the island of Mön, for instance, large floes of white chalk can be seen within the drift.

Of particular interest in the reconstruction of the dynamics of the Scandinavian ice sheets are the mountains sculpted by glaciers along the Norwegian coast and the cirques. During the growth of the ice sheets there was a shift of the center of radial outflow towards the east of the mountains away from the sources of moisture at the coast (Flint, 1971). For the last glaciation this shift was reconstructed with the help of a detailed study of the ice movements by Ljungner (1949). He also demonstrated that the distribution of the cirques on the eastern slope of the mountains shows that there was a lowering of the firn limit resulting in the formation of cirque glaciers before the expansion of the ice sheet. Similarly, cirque glaciers sculpted the sides of the peaks of the coastal mountains of Norway, as the *Syv Søstre* just south of the arctic circle, in an area also subjected to the action of the inland ice. Some of the coastal mountains of Norway were nunataks at some time during the glaciation, as is shown by their sharp features compared with the valleys. Some of these differences were caused by intense frost-shattering of the peaks after the maximum expansion of the ice sheet, when the highest parts of the mountains formed nunataks above the lowering surface of the ice (O. Holtedahl, 1953). Even with the help of studies of weathering and morphology it has not been possible to

Figure 2.4. *Roche moutonnée* with striae, Kaivopuisto, Helsinki. (Photo J. Donner)

show whether or not the highest mountain peaks were ice free through the last glaciation (B. Andersen, 1965); on most of them there is evidence, such as erratics, that at one time during the Quaternary they were covered by ice. The morphological tracing of the upper limit of the receding glaciers has, however, been possible in many parts of the Norwegian coast, combined with a study of the recession of the ice in the fjords and valleys.

The relief of the bedrock of the Fennoscandian Shield as distinguished from the Scandinavian mountains was, as earlier concluded, only slightly altered during the Quaternary. The continental ice sheet eroded the bedrock, removing most but not all of the regolith, and polished the fresh surface of the bedrock, leaving the features that show the direction of the movements of ice. Thus, by repeated sculpting of those parts of the bedrock that were less fractured and weathered than the surrounding parts of the bedrock, the streamlined whaleback forms, including the *roches moutonnées* (ice-smoothed rock mounds), were shaped – forms that occur throughout Finland and Scandinavia (Fig. 2.4). Whereas they reflect the general direction of the abrasion of ice, the striae, as well as grooves, crescentic cracks, and chattermarks, show in great detail the direction and possible changes of ice movement, especially during the last glaciation (Fig. 2.5). The abrasion during their formation affected only the surface of the fresh bedrock.

In using all evidence of glacial abrasion of the bedrock some independent conclusions have been reached, which have then been compared with other evidence in reconstructing the glacial history. Thus, on the basis of the observations of the just mentioned glacial morphology of the Scandinavian mountains, changes in the direction of the striae, and transport of material,

Ljungner (1949) concluded that the last glaciation was twofold, with a shift of the ice divide first from the mountains eastwards and back again and later with another smaller shift, the two expansions of the ice sheet being separated by a nonglacial interval. This conclusion, as will be seen later, gained renewed interest when stratigraphical evidence of the ice advances and nonglacial intervals was obtained, even if later more detailed work on striae has necessitated some modification (Johnsson, 1956).

The movements of the ice during the last glaciation can, as mentioned, be reconstructed chiefly by using striae but also other marks on the bedrock. The youngest movements can be compared with end moraines formed during the recession of the ice. The ice movements also determined the transport directions of the glacial drift. This is clearly seen when all evidence of the glacial action is put together, as will be seen later. In addition to regional studies of the movements of ice, detailed analysis of the glacial rock-sculpture has yielded information of changes in the direction of ice flow and of the properties of the ice that sculpted, in some places plastically, the rock surfaces, as shown in a study of southern Sweden (Johnsson, 1956) and also in a study of an area in northern Nordland in Norway (R. Dahl, 1965). Both examples show the potentialities of the methods used.

During the recession of the ice sheets that covered Scandinavia, erosion by the great quantities of discharged meltwaters could be considerable, particularly subglacially. A channel could be formed under the ice, which on hard rock eroded the underlying bedrock and in places obliterated the marks left by the ice. In the marginal parts of the glaciation, as in Denmark, subglacial rivers eroded long conspicuous tunnel valleys in the drift with irregular longitudinal profiles, the longest valley traced reaching from

Figure 2.5. Striae showing two directions of ice movement. Björkö near Vaasa off west coast of Finland. (Photo J. Donner)

Kattegat to the outermost end moraine of the last glaciation in Jutland (Hansen, 1965). In some areas the pre-Quaternary surface was also eroded, for example, the limestone north and northwest of Copenhagen. There is no regional study of the distribution of the more centrally situated tunnel valleys in the formerly glaciated area. There, however, the presence of round potholes in the bedrock, sometimes many meters deep, shows the concentration of subglacial streams in certain places, often close to glaciofluvial accumulations, such as eskers. Potholes, for instance, are well exposed in southern Finland where the bedrock outcrops are more frequent than further inland (Rosberg, 1925). But, generally, the influence of the subglacial meltwaters on the hard bedrock of the Fennoscandian Shield was small.

Some quantitative estimates have been made of the amount of erosion by the ice during the Quaternary. In Finland the average thickness of the minerogenous sediments is estimated at 8.3 m. Some of the material was transported from areas outside Finland and some of it to areas outside, but it is estimated that the formation of this sediment would have resulted in a lowering of the bedrock by 7 m (Okko, 1964). As some regolith and earlier sediments were redeposited, the lowering would have been less during the last glaciation. Further, the erosion was not even; it was clearly greater in weaker and more weathered parts of the bedrock than elsewhere, even if some pre-Quaternary regolith still remains. In Estonia, south of the Gulf of Finland, the thickness of the drift is 5–10 m (Raukas, 1985), which does not differ from the thickness in Finland, but reaches thicknesses of over 80 m in the Otepää and Haanja regions in south Estonia as a result of exceptional conditions for accumulation (Raukas,

1978). In some other marginal parts of the area of glaciation there are also thicker accumulations of Quaternary deposits, the average for Denmark being 50 m (Hansen, 1965). Some of the material was transported from elsewhere, but the greater thickness may also indicate that the softer pre-Quaternary rocks were more easily eroded than the hard rocks of the Fennoscandian Shield. Denmark, on the other hand, is an area where the accumulation of glacial deposits was comparatively great, as it is close to the Scandinavian mountains and is also an area where sediments older than the last glaciation were preserved. It is therefore not comparable with, for instance, Finland and Sweden. In Norway and in the Scandinavian mountains generally the erosion was, as mentioned earlier, very uneven, with a strong overdeepening of the pre-Quaternary river valleys by glaciers. The total depth of Sognefjord, from the mountaintops to the bottom of the fjord, is over 2 km. Some of the downcutting was done by earlier erosion of water, but a large part was done by subsequent Quaternary glaciers. Elsewhere in Norway the erosion was comparatively small, similar to that in Finland, with some of the pre-Quaternary regolith still remaining. An estimate that the volume of the drift in the marginal parts of the Scandinavian glaciation in central and eastern Europe would fill the basin of the Baltic Sea and the lake basins in Scandinavia and cover the Scandinavian peninsula with a layer 25 m thick may be an exaggeration, but it gives an idea of the amount of material transported from the Scandinavian mountains, particularly from its valleys and fjords (Flint, 1971). The thick Quaternary sediments deposited in the North Sea basin were not included in this estimate.

3 Northern Europe in the Quaternary

3.1. General stratigraphical division

The division of the Quaternary terrestrial stratigraphy in northern Europe is based on criteria particularly suitable for this area. The area was subject to pronounced climatic fluctuations, which at times led to the area being glaciated. At other times the climate was similar to that of today. These climatic fluctuations have been used in defining the Quaternary stages in the Nordic countries (Mangerud et al., 1974), as also in the British Isles (West, 1968, 1977; Shotton, 1973). Those periods that had a climate with a climatic optimum at least as warm as that during the Holocene climatic optimum in the area studied have been called interglacial stages – in northern Europe they are also called temperate stages (West, 1977). The interglacials were separated from cold stages during which glaciers spread over northern Europe and have therefore also been called glacial stages or glacials. The glacial stages were not defined on the basis of glaciations, nor were interglacials consequently bounded by periods of glaciations. The terms are therefore not suitable in a twofold division of climatically controlled stages. As most parts of northern Europe discussed here lie north of the vegetation zone described as temperate, apart from Denmark and the southern coasts of Sweden and Norway (see Ahti, Hämet-Ahti, & Jalas, 1968), the temperate stages are here called warm stages, as opposed to cold stages, even if the authors first defining them called them interglacials or temperate stages. The use of the terms warm and cold is in agreement with that used by T. Nilsson (1983), who considered them to be suitably noncommittal terms. In general correlations with other areas outside Europe only the names of stages are used. During the intervening cold stages there were shorter nonglacial intervals during which the climate ameliorated but was cooler than that of the warm stages. These intervals have been called interstadials (see Table 3.1). Ideally, the term should only be used if, on the basis of pollen diagrams, there was a succession in the vegetation history showing an amelioration of climate followed by a deterioration (Frenzel, 1989). An organic sediment where the pollen diagram only shows what the vegetation was during a part of a cool stage, without any changes, should then be called a nonglacial interval, having no particular

chronostratigraphic significance. The organic sediment may just represent a facies in a cold periglacial environment, in which patches of peats or muds formed in suitable depressions in a bare landscape, similar to that in the present high arctic. The sediments here give information about the surrounding vegetation but do not necessarily represent a climatic oscillation. The intervals defined as interstadials, however, are preceded and followed by stadials – cooler periods – but these do not necessarily have to be connected with evidence of ice advances (Rose, 1989). In some areas the interstadials and stadials are considered to represent substages within cold stages (see Zagwijn, 1973). The glaciations, for which the term glacierizations has also been used, during the cold stages are dealt with separately for each stage and compared with the chronostratigraphy. In this chapter only the definitions of the warm and cold stages are discussed; the more detailed subdivisions will be mentioned later.

In northern Europe the division of the Pleistocene stages preceding the Holocene is based on the identification of warm-stage, fresh-water sediments in Denmark by Jessen and Milthers (1928), and later defined through comparisons with The Netherlands (Zagwijn, 1957, 1963; S. T. Andersen, 1965). With the help of pollen analysis it could be shown that the vegetational sequences of the warm stages differ enough from one another so that they can be identified over large areas, and that pollen analysis can be used in drawing the boundaries between the warm and cold stages. Nevertheless, the use of vegetational changes as a basis for defining the boundaries has its drawbacks. When dealing with an area as large as northern Europe these boundaries are diachronous (Shotton, 1973), as will be discussed later. But pollen analysis has so far been the most useful method in defining and limiting the warm stages, as well as in describing the interstadials.

The climatic sequence for a warm stage in Denmark is, according to Jessen and Milthers (1928): arctic–subarctic–boreal–temperate–boreal–subarctic–arctic. The changeover from an open vegetation to a closed forest, in pollen diagrams from pollen spectra dominated by the nonarboreal (nontree) pollen to arboreal (tree) pollen, has been used as the lower boundary for a warm stage. Similarly, the changeover from forest to an

Table 3.1. *List of chronostratigraphic terms used*

Terms used	Corresponding terms commonly used
Cold stage	Glacial
Warm stage	Interglacial, temperate stage
Within cold stages:	
Stadial	
Interstadial	Substages (in The Netherlands)
Nonglacial interval	

open vegetation has been used as the upper boundary. The temperate climate of a warm stage resulted in the spread of deciduous forests in Denmark during its climatic optimum. The climatic sequence from arctic to temperate and back to arctic did not, however, result in a symmetrical development of the forest history. This history was largely governed by the successive immigrations of the trees, with many trees, for example, *Picea,* reaching many parts of northern Europe only during the latter part of the warm stages. But it was also influenced by soil degeneration. Thus Iversen (1958) divided the warm stage cycle into three parts, beginning with the protocratic period with basic–neutral grassland–woodland, followed by the mesocratic period with climax forest and ending with the telocratic period with acid woodland–moorland. To this three-fold division S. T. Andersen (1965) added a fourth period, the oligocratic, between Iversen's mesocratic and teleocratic periods, to mark the soil degeneration during the period of a warm stage that led to changes in the vegetation before those during the telocratic period. Turner and West (1968) recognized four main warm-stage subperiods of vegetational development closely similar to those described by S. T. Andersen, which in pollen diagrams can be identified as assemblage zones. These are the pretemperate zone, I, the early temperate or mesocratic zone, II, the late temperate or oligocratic zone, III, and the post–temperate zone, IV. Iversen's (1958) terms protocratic and telocratic were not used by Turner and West (1968) because in Iversen's division these two stages extended into the cold stages, if the nonarboreal / arboreal pollen ratio is used in defining the lower and upper boundaries for warm stages.

The fourfold division was used as a frame for more detailed subdivisions into pollen assemblage zones of the warm stages in Denmark (S. T. Andersen, 1965) and used as type localities (or type sites) in the provisional chronostratigraphical subdivision of the Pleistocene in the Nordic countries by Mangerud et al. (1974). The division is based on type localities, which are geographic localities, as it cannot be based on stratotypes, which are specific profiles (Hedberg, 1976; West, 1989). The names for the cold and warm stages in the subdivision shown in Table 3.2 were those used earlier in The Netherlands (Zagwijn, 1957, 1963), but with later additions (Zagwijn, 1985, 1989). The oldest warm stage in Denmark, preceding the Holsteinian, is called the Harreskovian (S. T. Andersen, 1965). It was first correlated with the Cromerian,

but the Cromerian in The Netherlands includes four warm stages ("interglacials"), I, II, III, and IV, separated by three cold stages ("glacials") – A, B, and C – and is therefore called the Cromerian Complex (Zagwijn, 1985, 1986)). Because of certain features in the forest history reflected in the pollen diagram from Harreskov, that is, no *Carpinus* and *Abies,* high *Alnus,* and a *Taxus* zone, a correlation of the Harreskovian with warm stage II in the Cromerian Complex was suggested by Zagwijn, van Montfrans, and Zandstra (1971). The Cromerian warm stage in East Anglia has been correlated with warm stage IV in The Netherlands (Zagwijn, 1975; West, 1977) and thus does not correspond to the Harreskovian in Denmark. From the correlations presented in Table 3.2 it follows that the type localities in Denmark used in the chronostratigraphical subdivision in Scandinavia represent an incomplete warm stage succession and cannot alone be used as a frame for correlations, except for the localities representing the Holsteinian and Eemian. These two warm stages are well represented in Denmark. For correlations of possible, older, warm stages in Scandinavia, other sites in continental northern Europe have to be taken into account, in addition to those representing the Harreskovian. The cold stage preceding the Harreskovian in Denmark cannot with certainty be correlated with any of the known cold stages mentioned in Table 3.2 and has therefore been called Pre-Harreskovian.

In the subdivision of the Pleistocene into Early (Lower), Middle, and Late Pleistocene the boundary between the Middle and Late Pleistocene has, in accordance with the division in The Netherlands, been placed between the Saalian and the Eemian stages (Zagwijn, 1957, 1963, 1985). The boundary between the Early and Middle Pleistocene has in The Netherlands been placed between the Bavelian and the first warm stage in the Cromerian Complex, Interglacial I (Zagwijn, 1985). However, the International Union for Quaternary Research (INQUA) Working Group on Major Subdivision of the Pleistocene has proposed that the Early/Middle Pleistocene boundary should provisionally be placed at the Matuyama/Brunhes paleomagnetic reversal (Richmond, 1990), which is somewhat younger than the boundary above the Bavelian (Table 3.2) in The Netherlands. Wherever the Early/Middle Pleistocene boundary is drawn, the stages so far identified in Scandinavia represent the Middle and Late Pleistocene as well as the Holocene.

The schematical pollen diagrams of the type localities of the three Pleistocene warm stages identified in Denmark – the Harreskovian, the Holsteinian, and the Eemian – are shown in Fig. 3.1, and the localities, all in Jutland, are marked on the different maps of the warm stage sites. In the pollen diagrams, as in all later pollen diagrams, the order of the tree taxa is according to their present climatic tolerance, as shown by their northern limits, starting with *Betula, Pinus,* and *Picea,* which have their northern limits furthest to the north, followed by *Alnus,* etc. (see Chapter 3, section 3).

The Harreskovian warm-stage type locality is situated in western Jutland (S. T. Andersen, 1965, earlier studied by Jessen & Milthers, 1928). The deposits consist of nearly 5 m thick

Table 3.2. *Subdivision of the Quaternary in continental northern Europe*

Periods, systems	Epochs, series	Ages, stages		Type localities for warm stages in Scandinavia
Quaternary	Holocene	Flandrian w		
	Late Pleist.	Weichselian c		
		Eemian w		Hollerup
		Saalian c		
		Holsteinian w		Vejlby
		Elsterian c		
	Middle Pleist.	Cromerian w (East Anglia)	Intergl. IV w	
			Glacial C c	
			Intergl. III w	
		"Cromerian Complex" (The Netherlands)	Glacial B c	
			Intergl. II w	Harreskov
			Glacial A c	
			Intergl. I w	
	Early (Lower) Pleist.	Bavelian c		
		Menapian w		

c = cold

w = warm

freshwater sediments on clay and are overlaid by till, gravel, and sand. After the *Betula* and *Pinus* maxima of the short zone 1, the pollen diagram, divided into six zones, is characterized by the *Ulmus* maximum of zone 2 and the *Taxus* maximum of zone 4. Compared with Flandrian pollen diagrams, the *Corylus* curve starts late and the *Picea* curve early. The upper part of the diagram shows the climatic deterioration that resulted in the increase of *Pinus* and *Betula* in zone 6, in which the relative amount of nonarboreal pollen, NAP, also increases, prior to the changeover to the open vegetation of the next cold stage. *Carpinus* and *Fagus* are both absent in the diagram. The Harreskov pollen diagram is considered to differ enough from diagrams of the next, better-known warm stage, the Holsteinian, to be placed in an older warm stage. There is, however, no stratigraphical control of its age, apart from it being older than the Eemian, and the conclusion about its position in the Pleistocene sequence has been based only on a pollen diagram (T. Nilsson, 1983). This has led, as mentioned earlier, to the reinterpretation of its age by Zagwijn et al. (1971).

The type locality for the Holsteinian stage is Vejlby at Rands Fjord northwest of Frederica in southern Jutland (S. T. Andersen, 1965). The INQUA Subcommission on European Quaternary Stratigraphy has, however, proposed that the Lower Elbe area be chosen as the type area for the Holsteinian warm stage (Jerz & Linke, 1987). The warm-stage freshwater sediment at Vejlby is an about 15 m thick diatomite, with some clay layers, on sand and till and covered by till. The pollen diagram is divided into nine zones and starts with the *Betula* and *Pinus* zones. *Alnus*

has high percentages starting from zone 2 and there is a *Taxus* maximum in zone 4. The early appearance of *Picea*, already in zone 2, the low percentages of *Carpinus* and *Corylus*, and the occurrence of *Abies* and *Buxus* further up in the diagram, and of *Pterocarya* in zone 8, are all features which can be used in identifying the Vejlby diagram as representing the Holsteinian stage as defined in areas further south. Of these tree taxa *Abies*, for instance, did not reach Denmark any more during the Eemian (Zagwijn, 1990). The Vejlby diagram covers most of the warm stage, including the increase of nonarboreal pollen towards the top. Compared with the pollen diagram from Harreskov and with Eemian pollen diagrams, the diagram from Vejlby, as other Holsteinian diagrams, has no pronounced features, such as particularly big maxima of particular tree taxa, and it is this comparative monotony which distinguishes it from the other diagrams and makes the Holsteinian warm stage recognizable. The till stratigraphy at Vejlby does not date the warm stage sediments; the upper till is probably from the last cold stage (T. Nilsson, 1983).

The type locality for the Eemian warm stage is Hollerup west of Randers in northern Jutland (S. T. Andersen, 1965; earlier studied by Jessen & Milthers, 1928). The warm stage freshwater sediment is over 8 m thick and consists of marl, calcareous mud, and diatomite. The lake sediments are in a hollow of the Saalian till and overlaid by about 10 m of sand and a thin cover of Weichselian till, here separating tills of two cold stages. The pollen diagram is divided into seven zones. After the initial maxima of *Betula*, zone 1, and *Pinus*, zone 2, follows a maximum of

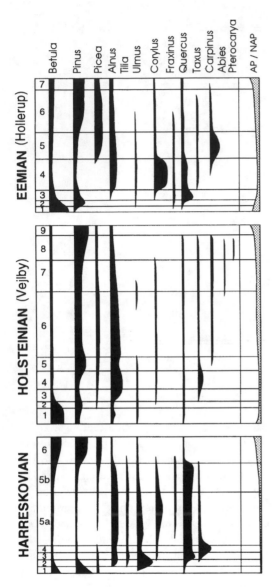

Figure 3.1. Schematic pollen diagrams of the type localities of Harreskovian, Holsteinian, and Eemian warm stages in Denmark (after S. T. Andersen, 1965).

kovian, Holsteinian, and Eemian, all have complete freshwater sedimentary sequences from the beginning to the end of the stages they represent. The Holsteinian and Eemian can be correlated with similar sediments and with coastal marine sediments in the marginal areas covered by the Scandinavian ice sheet during the glaciations of northern Europe, and with sediments in areas bordering the formerly glaciated area. The glacial deposits related to the sediments of the warm stages can therefore also be referred to particular cold stages, even if there are no type localities for these stages. In the more central parts of the area covered by the Scandinavian ice sheet the sequences of warm stage sediments are as a rule incomplete. In addition, the further away they are from the complete sequences the more difficult it is to correlate them with any particular warm stage, because the vegetational history further north, especially outside the area with a temperate climate, differed from that in the area of the type localities, as it has done during the Flandrian. It has therefore been essential, in accordance with the recommendations on stratigraphical usage, to establish local lithostratigraphical units defined in type sections and then make an attempt to correlate them with the general chronostratigraphical division of northern Europe. This was done, for instance, by Gibbard (1979) in a study of the Late Pleistocene stratigraphy of an area around Muthos in northern Finland, by Bouchard, Gibbard, and Salonen (1990) in a study of pre-Weichselian and Weichselian sediments in southern and western Finland and by Berglund and Lagerlund (1981) in a study of the Eemian and Weichselian stratigraphy of south Sweden, where formations, members, and beds were named in lithostratigraphical division. All studies have not adhered to this principle equally rigidly. As the evidence of warm stages is fragmentary and poses difficulties in correlation, the glacial sediments connected with them are often difficult to place in the general chronostratigraphy. Similar difficulties are encountered when correlating interstadial sediments with the glacial sequence, as will be seen later. Thus, regardless of methods and nomenclature used, local successions are difficult to correlate with successions in other regions (West et al., 1974).

As has been seen, the boundaries of the warm stages were pollen analytically defined at the type localities as the horizons at which there was a changeover from an open treeless vegetation to forest and a changeover from forest to a treeless vegetation. But these horizons based on climatically controlled changes in vegetation are, by definition, time transgressive (see Flint, 1971). A warm stage, even when it is considered to represent a chronostratigraphical unit, is thus shorter in the northernmost parts of northern Europe than in Denmark where they were defined, if the boundaries are drawn on the basis of the changes in the pollen diagrams. The time differences are, however, not more than a few hundred years, if the older warm stages were similar to the Flandrian. But there are no means of dating these time differences; only the vegetational history of the Flandrian can be accurately dated. The definition used for the warm stages becomes only seriously unsuitable in those areas which are so far

Quercus, zone 3. The pronounced maximum of *Corylus* in zone 4 and the subsequent strong representation of *Carpinus* in zone 5 are both characteristic features for the Eemian. *Pinus, Picea,* and *Alnus* dominate the upper part of the diagram, and the relative increase of nonarboreal pollen reflects the deterioration of climate before the changeover to the open vegetation in the beginning of the Weichselian cold stage, also included in the diagram. What is unusual about the Eemian, as compared with the two older warm stages and the Flandrian, is that the climatic optimum of the Eemian, as reflected in the vegetation, was warmer and also more oceanic than during the corresponding times of the other warm stages.

The type localities for the three warm stages, the Harres-

north or above the tree limit that they never had a forest vegetation. In these areas warm stage sediments can only be identified through correlations of the glacial sediments, with which they are associated, with corresponding glacial sediments in areas where the sediments are related to a particular warm stage. Further, the criteria in drawing the pollen analytical boundaries for interstadials have not been defined for any part of northern Europe dealt with here. An interstadial is only defined as a nonglacial period either too short or too cold to permit the development of temperate deciduous forest (Jessen & Milthers, 1928; West, 1977, 1984), a definition not well suited to the northern areas that did not, as mentioned, have temperate deciduous forests even during warm stages. Thus, it becomes clear that as the warm stage and interstadial sediments used cannot be dated, except for the Flandrian, the units used are not strictly chronostratigraphical, even if the boundaries are defined at type localities, in an area as large as northern Europe. Further, even if the use of formal assemblage zones in warm-stage pollen diagrams is essential, their use in correlations between different sites is inexact.

In the warm stage–cold stage cycles, the warm stages represent less than 10 percent of the duration of the Quaternary and the maximum glaciations about 10–15 percent of the total time; the rest of the time represents cool periods with interstadials and periods of ice advances (Faure, 1980). Some estimates have been made of the lengths of the warm stages, which can be compared with the length of the Flandrian, which, however, is not a complete warm stage. The length of the Flandrian is 10,000 radiocarbon years, which is the age of the boundary to the Weichselian stage and also, as mentioned earlier, the age of the beginning of the Holocene. This date marks approximately the change from an open vegetation to forest in large parts of northern Europe, as will be seen later. The length of the Eemian has been estimated with the help of annually laminated diatomite at the Eemian site of Bispingen in the Luhe valley in northwestern Germany (Müller, 1974b). The varves were counted in parts of the sedimentary sequence and the time of sedimentation for the remaining parts estimated on the basis of the sedimentation rate. The length of the Eemian was in this way estimated at about 11 ka (Table 3.3). The length of the Holsteinian has similarly been studied at sites in Lower Saxony between Hamburg and Hannover. At Munster-Breloh about 6,900 varves were counted from the sediments of the upper part of the warm stage (Müller, 1974a) and at Hetendorf-Bonstorf 3,000–4,000 varves from the diatomite of the lower part of the stage (Meyer, 1974). The total length of the Holsteinian was estimated at 15 ka to 16 ka. Laminated clay-muds at the Hoxnian (Holsteinian) site of Marks Tey in East Anglia have also been used for an estimation of the length of the warm stage. Assuming that the lamination in the sediments is annual, the total length was first estimated at 30 ka to 50 ka (Shackleton & Turner, 1967), but later revised to 20 ka or at most 25 ka (Turner, 1975; see also Bowen, 1978), which is more in agreement with the estimates from Germany given in Table 3.3.

Table 3.3. *Estimated lengths of the younger Quaternary warm stages in northern Europe and their ages based on correlations with deep-sea sediments*

Stages	Lengths of warm stages	Ages for deep-sea sediments
Flandrian	10 ka (radiocarbon dating)	
Weichselian		
Eemian	11 ka (annually laminated sediments, Bispingen/Luhe) (Müller, 1974b)	Stage 5e, 125 ka (uranium dates for corals)
Saalian		
Holsteinian	15–16 ka (annually laminated sediments, Munster-Breloch and Hetendorf-Bonstorf) (Müller 1974a; Meyer, 1974)	Stage 7, between 190 ka and 250 ka (estimate based on sedimentation rate and other calculations) Likely alternative: Stage 9, about 330 ka (Stage 11, about 400 ka, has also been suggested)

The warm stages defined with the help of pollen analysis of terrestrial sediments have been correlated with marine sediments studied in sections and borings, particularly at the coasts of southern Scandinavia, with the help of foraminifera. In these sediments they mainly represent benthic (benthonic) species. As the distribution of many of these foraminifera is dependent on the sea-water temperature, they can be used in stratigraphical studies of Quaternary marine sediments. In this way marine sediments representing the Holsteinian and Eemian have been defined, as well as sediments covering most of the Weichselian with its interstadials, and Flandrian sediments (Knudsen & Feyling-Hanssen, 1976; Feyling-Hanssen & Knudsen, 1979). Holsteinian and Eemian sediments in northwest Germany have similarly been studied, where the foraminiferal zonation has been compared with the pollen zonation (Knudsen, 1988). Longer sequences with older Quaternary marine sediments have been studied in the central North Sea, where the oldest sediments are clearly older than the terrestrial sediments found in Denmark (Jensen & Knudsen, 1988). In spite of the presence of sediments described as Early Pleistocene, being below the Matuyama–Brunhes palaeomagnetic boundary, the sedimentary sequence is incomplete. Especially the warm-stage deposits are poorly represented, as pointed out by Jensen and Knudsen (1988).

In addition to using foraminifera in the stratigraphical study of marine sediments in northern Europe, comparisons have been made between the incomplete terrestrial sequence and the oxygen–isotope stratigraphy of deep-sea pelagic sediments based on both benthic and planktonic foraminifera. The ages have been based on direct palaeomagnetic studies (Shackleton &

Opdyke, 1973), on comparisons with high positions of sea level with uranium-series dates (Shackleton & Matthews, 1977), and finally on dates based on variations in the earth's orbital geometry, using orbital tuning techniques and comparisons with the previously mentioned dates obtained earlier (Martinson et al., 1987). A close correlation of the terrestrial sequence in northern Europe with the deep-sea chronology has, however, only been established for the Late Pleistocene. The best point for correlation in the deep sea chronology after the Matuyama–Brunhes boundary is stage 5e, during which the sea level was at or slightly above the present-day sea-level and thus has been correlated with the Eemian stage (Shackleton, 1969; Shackleton & Opdyke, 1973). The deep-sea chronology was correlated with tectonically uplifted coral reef terraces on Barbados, of which the Barbados III terrace was correlated with stage 5e, through oxygen-isotope determinations of mollusks from the coral terraces, and dated at 125 ka (Table 3.3) with the method based on ^{230}Th / ^{234}U measurements of corals (Shackleton & Matthews, 1977). The age of the peak of stage 5e was by Martinson et al. (1987) dated at 122.56–125.19 ka and the end of this stage at about 115 ka, but the oxygen-isotope curve cannot be used in dating the boundaries of the Eemian as defined on the basis of terrestrial sediments. In an earlier estimation based on the sedimentation rate of 17.1 cm / 10^4 years since the Matuyama–Brunhes boundary in a Pacific core the age of the peak of stage 5e was estimated at 123 ka (Shackleton & Opdyke, 1973).

The correlation of the Holsteinian stage with the deep-sea chronology depends largely on the interpretation of the evidence in Central Europe of one or even two possible warm stages between the Holsteinian and Eemian stages (Frenzel, 1973), as will be mentioned later. If the chronostratigraphy for northern Europe as presented in Table 3.2 is used, the Holsteinian should be correlated with stage 7 in the deep-sea chronology (Table 3.3), with an age between about 195 ka and 250 ka based on the prior sedimentation rate (Shackleton & Opdyke, 1973) and between 190 ka and 250 ka calculated by Martinson et al. (1987), with the isotope peak around the latter date. These dates agree with the average age of 220 ± 20 ka obtained with the electron spin resonance (ESR) method in dating marine shells from Holsteinian sites in northwest Germany (Linke, Katzenberger, & Grün, 1985). In applying the amino acid racemization dating method on marine shells from warm stage sites in northwest Europe Miller and Mangerud (1985) concluded that stage 7 most likely corresponds to the Holsteinian, but gave as an alternative the correlation of stage 9 with the Holsteinian. This was mainly because there is, as seen later, strong evidence for an additional warm stage between the Holsteinian and the Eemian. These results from terrestrial warm stage sites have therefore led to the conclusion that stage 9 should be correlated with the Holsteinian, with an age of about 330 ka (Miller & Mangerud, 1985). This alternative is also given in Table 3.3. By taking into account ice advances, and nonglacial intervals during the Saalian, which will be discussed in detail later, and correlating them with the deep-sea isotope curve representing changes in ice volume,

Ehlers, Gibbard, and Rose (1991) correlated stage 11 with the Holsteinian, as was already done by Sibrava (1986) in his correlation of European glaciations with the deep-sea record. The Holsteinian would then be about 400 ka old. A correlation of the Holsteinian with deep-sea stage 11 has, however, not been considered possible because the Holsteinian would then be too close to the palaeomagnetic Matuyama–Brunhes boundary (G. H. Miller & Mangerud, 1985).

A correlation of warm stages older than the Holsteinian with deep sea stages is not attempted here, particularly as they are not important in the chronology used for northern Europe. But even if terrestrial sediments from the Early and part of the Middle Pleistocene are not known from this area, the oxygen-isotope stratigraphy in the North Atlantic shows that the first horizon of ice-rafting occurred already at about 2.4 Ma ago, preceded by a minor pulse at about 2.5 Ma, the dates being based on palaeomagnetic measurements (Shackleton et al., 1984). The beginning of ice-rafting coincides with the Pliocene–Pleistocene boundary and shows that there was a glacial interval "with an ice volume similar to the maxima during the middle Pleistocene" (Shackleton et al., 1984). The evidence from deep-sea sediments of glaciation in the Early Pleistocene shows how incomplete the terrestrial sedimentary sequence is, in which no evidence of equally old glaciations has been found in the area of northern Europe dealt with here. In addition to depicting the glaciations, the oxygen-isotope stratigraphy also shows that 7.5 ka would be a long enough time for the build-up of an ice sheet with a half-width of about 500 km, that is, the size of the present Greenland ice sheet (Shackleton, 1969), an estimate important in the discussion of the length of the glaciations during the cold stages.

The deep-sea stratigraphy reflects foremost the changes in the volume of all ice sheets and therefore minor fluctuations in the oxygen-isotope curves are not necessarily contemporaneous with periods of advances or retreats of the comparatively small Scandinavian ice sheet. Some likely correlations can, however, be made between, for instance, the deep-sea stages 5a and 5c with Weichselian interstadials, to be discussed later.

3.2. Extent of glaciations

The extensive glaciations are characteristic for the cold stages in northern Europe, even if the successive glaciations lasted for only a comparatively short time of each cold stage, as mentioned earlier. The till beds that were formed during the glaciations and as already noted can be placed in particular cold stages because of their position in relation to warm-stage sediments, are stratigraphically important as they can often be identified and correlated over wide areas. The till beds provide in many areas a frame for the division of the cold stages; other minerogenic sediments formed during these stages are more difficult to place because of their more local distribution, such as aeolian sands or glaciofluvial sands and gravels. Most tills, however, represent the last Weichselian stage, especially in the

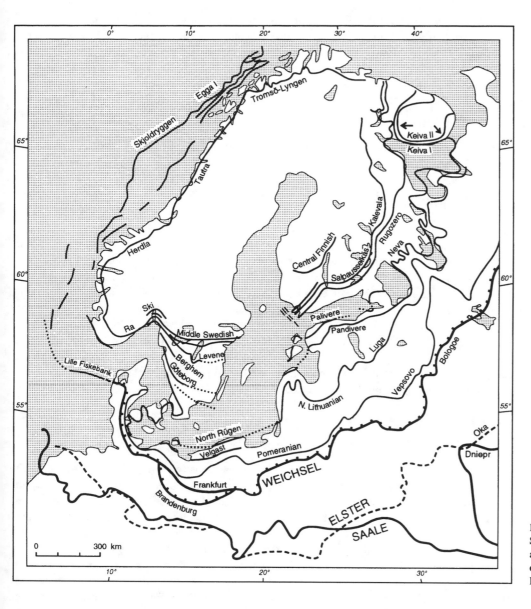

Figure 3.2. Extent of the Elsterian, Saalian, and Weichselian glaciations and the main ice-marginal positions of the Late Weichselian and Early Holocene deglaciation.

more central parts of the glaciation. It is only in the marginal parts and in areas near the former ice divide that tills from older cold stages have beeen preserved to a larger extent. This agrees broadly with the concept of zonal differences in glacial erosion with an intermediate zone of predominant erosion lying between an inner and an outer zone of little erosion, the latter with mainly deposition (Flint, 1971). Tills represent glaciations that, however, can produce more than one till bed, and cold stages can have several glaciations. The cold stages can therefore not stratigraphically be separated by counting the number of till beds downwards, starting from the youngest at the top.

The extent of the Pleistocene glaciations in northern Europe, when the ice sheets spread from the Scandinavian mountains, has been determined on the basis of their marginal formations, which are often clear end moraines. In this way the outer limits of the glaciations during the Elsterian, Saalian and Weichselian

cold stages have been mapped, the Scandinavian ice sheet having been more extensive during the two older stages than during the Weichselian, as seen in Fig. 3.2. The knowledge of older cold stages is too fragmentary to permit reconstructions of the extent of their glaciations.

End moraines have often been given names after the areas in which they occur and linked together when they morphologically can be traced over great distances. These landforms should not, however, generally be used as stratigraphic units. Further, even if they can laterally be traced over great distances, a particular end moraine is not necessarily synchronous throughout the whole area in which it occurs. But locally, sequences of end moraines can be used in describing events in the deglaciation of an area, and the moraines given names used with capital letters as morphostratigraphic units (Mangerud et al., 1974), if the limitations of this approach are kept in mind.

It can be seen that south of Scandinavia the outer limit of the maximum Saalian glaciation is in some ares south of that of the Elsterian glaciation and that in other areas the limit of the Elsterian glaciation is further south. The map in Fig. 3.2 is compiled from various sources. The general, somewhat schematic, map of northern Europe produced by Woldstedt (1969) was modified by using more detailed maps of particular areas, such as maps by Oele and Schüttenhelm (1979) for The Netherlands; by Ehlers, Meyer, and Stephan (1984) for Central Europe from The Netherlands to Poland; by Rózycki (1961) for Poland; and by Gerasssimov, Serebryanni, and Cebotareva (1965) for areas further east as far as just east of Moscow. In The Netherlands the limit of the Saalian glaciation is clearly further southwest than that for the Elsterian, whereas beyond the Harz mountains, where the limits are close to one another, the limit for the Elsterian glaciation is further south than that for the Saalian, the Elsterian limit curving against the mountains of Saxony. It is from this area, from the tributaries Saale and Elster of the river Elbe, the latter flowing through Leipzig, that the two glaciations have got their names. Northeast of the Sudeten mountains the Elsterian limit again joins that of the Saalian, but further east, north of the Carpathian mountains, they again part, the Elsterian limit again being further south than the Saalian. But in Belarus the two limits cross each other, the Saalian limit reaching far southeast in the valley of the river Dniepr into Russia, whereas the irregular Elsterian limit continues northeast to the area south of Moscow, crossing the river Oka, a tributary of the Volga. The two glaciations have therefore in this area been named Dniepr and Oka respectively. South of Moscow the limit of the former (Saalian) glaciation is again close to the latter. In Poland these two glaciations, in a broader sense, have also been given local names. The Elsterian glaciation has been called the Cracovian and the Saalian the Middle Polish glaciation. The subdivision of the Saalian–Middle-Polish–Dniepr cold stage will be discussed later. The limits for the glaciation during the Elsterian and Saalian cold stages, as presented in Fig. 3.2, are interpretations by the authors mentioned; different interpretations have been suggested for the limits, especially in eastern Europe (see T. Nilsson, 1983), but are not discussed here as the limits are well outside the area of northern Europe dealt with in more detail.

The outer limit of the maximum glaciation of the Weichselian cold stage is everywhere clearly inside the limits of the Elsterian and Saalian glaciations. The Weichselian limit can be followed from Jutland in Denmark through Germany to Poland, where the glaciation is called the Vistulian, and to Russia, where it is called the Valdai glaciation. The general trend of the outer limit of the Weichselian was drawn foremost according to Woldstedt (1969), Jansen, van Weering, and Eisma (1979), B. Andersen (1981), and Ehlers, Meyer, and Stephan (1984) in the west and to Gerassimov, Serebryanni, and Cebotareva (1965), Chebotareva (Cebotareva) and Makarycheva (1982), Faustova (1984) and Velichko et al. (1984) in the east. The younger ice-marginal positions of the Weichselian ice sheet will be discussed in

connection with the Late Weichselian deglaciation. It can, however, already here be noted that there is a broad belt with moraines formed during the Weichselian cold stage and that very few moraines were formed during the final retreat and decay of the Scandinavian ice sheet inside the group of moraines formed during the last major standstills or readvances of the ice sheet, before the beginning of the Holocene. There was thus during the Weichselian cold stage, as presumably during the earlier Elsterian and Saalian stages, a considerable shrinkage of the ice sheet long before the beginning of the subsequent warm stages as earlier defined biostratigraphically. In contrast to the marginal areas of glaciations becoming ice-free at the end of cold stages, with a periglacial environment, most of Scandinavia in the central area of glaciations did not become ice-free until the beginning of the warm stages. These regional differences are naturally reflected in the stratigraphical record. Even if the twofold division into cold and warm stages may seem illogical in northern Europe against the evidence of glaciations, the division is useful when viewed against results from larger areas of the northern and southern hemisphere where global changes can be identified and correlated. The qualifying words cold and warm are principally used to make it easier to separate the stages, which have names that do not describe their nature, such as the Eemian and Weichselian.

3.3. Present vegetation zones and tree limits, and their comparison with pollen diagrams

In contrast to the successive glaciations of northern Europe, with expansions and withdrawals of the Scandinavian ice sheets described in the previous chapter, there were advances northwards of plants, followed by their retreat, during the times between the glaciations of the cold stages. In the study of the warm stages the forest history is of particular interest, because the definition of these stages is based on the forest history as depicted in pollen diagrams. In the study of the cold stages, with their interstadials and nonglacial intervals, the interpretation of nonarboreal pollen of the open treeless vegetation is, in addition, important. As a starting point in the interpretation of the tree pollen diagrams the present range of the tree taxa represented in the diagrams can be considered. The tree line, the upper northern limit of trees, is formed in Scandinavia and Finland by the limit of birch, *Betula pubescens,* near its limit by *B. pubescens* ssp. *tortuosa,* except in the easternmost part of northern Finland. The somewhat lower or southern forest line marks the limit for the birch forest belt, north of the coniferous forest with pine, *Pinus sylvestris,* and spruce, *Picea abies.* Their limits are quite close to that for birch, and the forest line is already in the White Sea area formed by *Picea obovata* and further east by the larch *Larix sibirica.* In the eastern parts of the Scandinavian mountains the "birch zone" is often only 50 m broad, but in the central and western parts its vertical width may be several hundred meters (Ahti, Hämet-Ahti, & Jalas, 1968). The forest line for birch varies in height in Scandinavia; it is

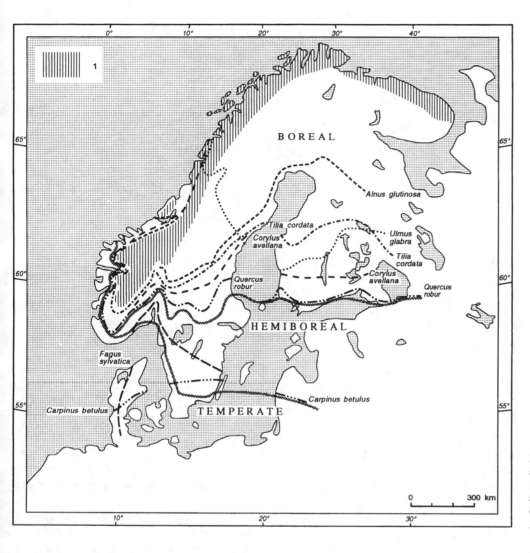

Figure 3.3. Tree lines and boundaries between main vegetation zones; 1 indicates treeless mountains and arctic lowlands of the oroarctic zone (after Hultén, 1950; Hintikka, 1963; Ahti et al., 1968; Alalammi, 1988).

about 500 m above sea level near the coast in southern Norway but 1200 m in the mountains, and rises from about 200 m near the coast to 500 m further inland in northernmost Norway (Treter, 1984). The position of the tree line borders the treeless area shown schematically on the map in Fig. 3.3 and representing the oroarctic vegetation zone (Ahti, Hämet-Ahti, & Jalas, 1968). Of the two conifers *Pinus* and *Picea* close to the forest line, the latter is a continental species and therefore has a more restricted area of distribution along the Norwegian coast than *Pinus* (Hultén, 1950; Hintikka, 1963). *Picea* is not native in Denmark or in southernmost Sweden. In addition to these three common taxa, aspen, *Populus tremula*, also grows in the forests in the mountains and in the north and is common throughout Scandinavia (Hultén, 1950), but is poorly represented in pollen diagrams. Alder, *Alnus glutinosa*, has a more restricted range and its northern limit in Sweden and Finland is clearly south of the forest line (Fig. 3.3). Whereas the alder pollen in the diagrams are interpreted as mainly representing *A. glutinosa*, they may also be partly from the gray alder, *A. incana*, which has a wider range and is common throughout Scandinavia and grows also in

northernmost Finland and Norway, although more sparsely. It is classified as a continental species, as *Picea* (Hintikka, 1963). Of the tree species with their limits further south those of small-leaved lime, *Tilia cordata;* elm, *Ulmus glabra;* hazel, *Corylus avellana;* common oak, *Quercus robur;* beech, *Fagus sylvatica;* and hornbeam, *Carpinus betulus,* are shown in Fig. 3.3, according to Hultén (1950) and Hintikka (1963). Of these, the limit of *Ulmus* differs from the others in that it crosses the Scandinavian mountains from Sweden to Norway; south of the line, there are isolated finds even high in the mountains. But it is common only in the southern parts of Sweden and in Denmark. The northern limit of ash, *Fraxinus excelsior*, not shown in Fig. 3.3. because like *Populus* it is poorly represented in the pollen diagrams, follows roughly that of oak. The limit of yew, *Taxus baccata*, also not shown in the figure, is south of that of oak. It is present on Åland but not on the mainland of southwestern Finland. Of these tree species it may be noted that *Fagus* is absent from western Jutland in Denmark and from the west coast of Norway, whereas the northern limit for *Carpinus* cuts through southernmost Sweden and Jutland. The northern limits of all these deciduous trees get closer to one another toward the west against the Scandinavian mountains, and are very close to one another in southern Norway with its mountains near the coast.

The order in which the tree taxa were drawn in the pollen diagrams is according to their present range, starting with those taxa with their limits furthest to the north, as mentioned earlier (Chapter 3, Section 1). For *Ulmus* the place in this succession was determined by the general trend of its limit in Finland, as the pattern of its range in Scandinavia differs from those of the other tree taxa. The order used in the pollen diagrams is the following: *Betula – Pinus – Picea – Larix – Populus – Alnus – Tilia – Ulmus – Corylus – Fraxinus – Quercus – Taxus – Fagus – Carpinus. Abies* and *Pterocarya* were, when present, given after these taxa.

The tree line, the forest line, and the limits for the tree taxa are all expressions of the present climate, but also the end result of the changes that have taken place during the Flandrian. Thus, the present succession of trees is not the same as the order in which these tree taxa spread northwards in the beginning of the Flandrian or during the earlier warm stages. A good example for this is *Picea*, which is a late arrival compared with other trees. The vegetation zones have similarly undergone changes. The present composition of the vegetation is comparatively young and has no equivalents earlier in the Flandrian. Not only has the climate changed but there has also been soil degeneration, mentioned earlier, in addition to differences in the spread of individual species. A detailed correlation of the present vegetation zones with the changes in the pollen diagrams cannot be made, even if some conclusions can be drawn about displacements of the main vegetation zones, keeping in mind that their floral composition has changed. Some of the main vegetation zones, as defined by Ahti, Hämet-Ahti, and Jalas (1968; see also Alalammi, 1988), were therefore included in Fig. 3.3. The previously mentioned oroarctic zone represents the treeless mountains and the arctic lowlands. The orohemiarctic zone, which is above the continuous forests, with "scattered stands of forests or isolated, low trees, mainly birch but also spruce and pine" (Ahti et al., 1968), lies between the oroarctic and the three boreal zones – the northern, middle and southern boreal zones – not separated in Fig 3.3. The northern limit of the hemiboreal zone follows broadly the northern limit of *Quercus robur* and has therefore earlier by Du Rietz been called the conifer forest region with oak (see comparison with earlier divisions in Ahti et al., 1968). The northern limit of the temperate zone with its deciduous forests follows the south and west coast of Sweden and the south coast of Norway; the whole of Denmark lies within this zone. In the interpretation of pollen diagrams the main division into a treeless vegetation, corresponding to that in the present oroarctic zone, and the forest vegetation of the boreal, hemiboreal, and temperate zones can be used in broad outlines if, as mentioned earlier, it is remembered that the components within these zones have changed. A general usage with more neutral terms is therefore more correct, by naming the vegetation, as interpreted on the basis of pollen diagrams, as being treeless, coniferous forest, or deciduous forest, and then listing the tree taxa identified

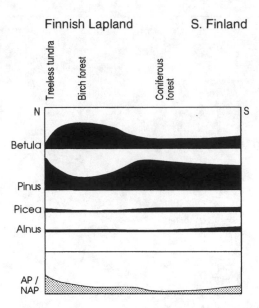

Figure 3.4. Schematic pollen diagram of surface samples from different vegetation zones in Finland (after L. Aario, 1940).

and describing their relative importance. A forest with the taxa *Ulmus, Tilia, Quercus,* and *Fraxinus* has in pollen-analysis traditionally been called mixed oak forest, QM (*Quercetum mixtum*), sometimes in diagrams presented as a separate curve (Faegri & Iversen, 1975). The use of general terms in the description of the vegetation is followed in the present account in order to avoid, if possible, misleading interpretations of the vegetational history.

Pollen percentage diagrams have generally been used in Quaternary stratigraphy, but in addition, various methods have been employed to find out to what extent the pollen diagrams reflect the vegetation which produced the recorded pollen assemblages (for details, see Janssen, 1974; Faegri & Iversen, 1975; Birks & Birks, 1980). In his study of the pollen composition of surface samples from Lapland and also of samples from southern Finland, and a comparison with samples from northwest Germany, L. Aario (1940) demonstrated that in the northern birch forest the birch pollen dominate, and that pine is dominant in the coniferous forest further south, and that there is an increase of pollen of deciduous trees further south in northwest Germany, in the present temperate zone (Fig. 3.4), being more pronounced further south in Germany (L. Aario, 1940). Of special interest is the tree pollen composition of the surface samples from the treeless area, where the relative amount of pine is the same as in the coniferous forest. The pollen rain in the treeless area thus reflects the forest vegetation in a wide area – including the birch forest – in which pine is the dominant tree. The actual amount of tree pollen is, however, low in the surface samples from the treeless area, in L. Aario's study expressed as arboral pollen in 50 mg of peat. The amount of nonarboreal pollen is also low in this area, but its relative

VAKOJÄRVI 1971

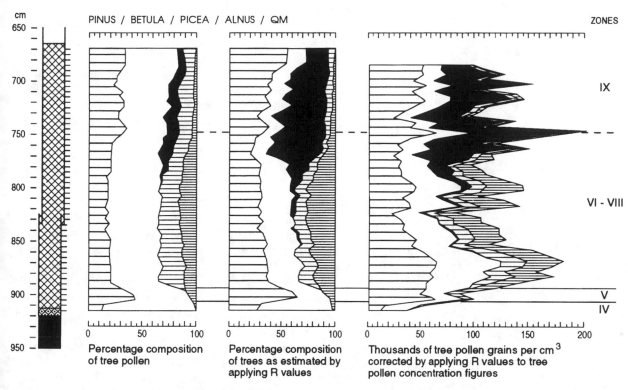

Figure 3.5. Percentage composition in Lake Vakojärvi of tree pollen, percentage composition of trees when applying R values, and frequency curve after application of R values to tree-pollen concentration values (after Donner, 1972).

amount as compared with arboreal pollen is high. It was on this observation that the separation in pollen diagrams of the open treeless vegetation from the forest was first based (see Firbas, 1949), which is of particular importance in defining the boundaries of the warm stages, as noted earlier. When pollen concentration values, that is the number of pollen per cm^3, have been changed to pollen influx values, that is, the annual deposition of pollen per cm^2, by using figures for the annual sedimentation rates obtained from radiocarbon dates, the low production during the initial Flandrian treeless period could be demonstrated in northern Scandinavia (Hyvärinen, 1976), a result in agreement with L. Aario's conclusions.

The pollen influx diagrams have not essentially altered the conclusions about the forest history drawn on the basis of percentage diagrams, either in southern Finland (Donner, 1972) or in northern Scandinavia (Hyvärinen, 1975a, 1976). But as it can be assumed that there are great differences in pollen production between different tree species, attempts have been made to recalculate the percentages for the trees by using values for their relative pollen production, which, however, are likely to have varied through time and also as a result of differences in environment (S. T. Andersen, 1970; Birks & Birks, 1980). The corrections of the pollen diagrams have been based on R values for the tree species obtained by dividing the pollen percentage of

the species with the vegetational percentage of the species (West, 1971). In the determination of the R values for a Flandrian diagram from Vakojärvi in southern Finland (Donner, 1972) the vegetational percentages were based on the first national forest inventory (1921–4), as selective cutting has changed the vegetational percentages of tree species (Ilvessalo, 1956). The R values were then used in recalculating both the percentages and the concentration figures for the trees, as shown in Fig. 3.5. The R values for the four most common trees were for Betula 2.5, Pinus 0.84, Picea 0.25, and Alnus 2.8, which shows that Picea has a relatively low pollen production and Alnus a high production, the former therefore being underrepresented in a percentage diagram and the latter overrepresented. In later calculations based both on surface pollen spectra from lake sediments and on the second forest inventory, results from around the lakes gave the following R values for southern Finland: Betula 1.8, Pinus 1.0, Picea 0.42, and Alnus 3.4, broadly similar to the preceding values (Parsons, Prentice, & Saarnisto, 1980). For northern Finland the values were for Betula 2.6, Pinus 1.0, and Picea 0.38, and – for Lapland as a whole – for Betula 0.28. The R values were also used in correlations between pollen spectra and the vegetation in Finland and north Norway by Prentice (1978). Thus, the studies give an indication of how much the proportions between the tree species are distorted in the pollen diagrams

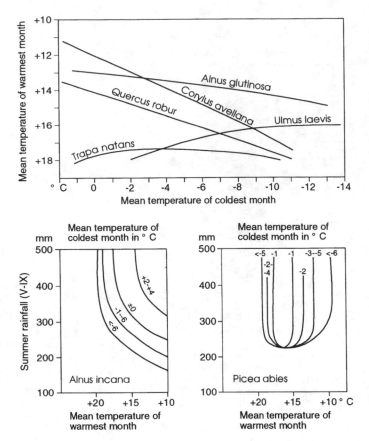

Figure 3.6. Temperature and rainfall requirements for some trees and for water chestnut, *Trapa natans* (after Hintikka, 1963).

Thus, by using as parameters the mean temperature of the warmest month (in C°) and the mean temperature of the coldest month, Iversen (1944) showed with curves in diagrams the limits for the distribution, called thermospheres, of *Ilex aquifolium, Hedera helix,* and *Viscum album* in Denmark. *Ilex* is the most oceanic species and *Viscum* the most continental, demanding relative high summer temperatures but tolerating low winter temperatures (see Birks & Birks, 1980). Later Hintikka (1963) compared the distribution in northern Europe of several species, including the common trees, with meteorological observations from a number of stations and in this way correlated the limits of their distribution with temperature, and for some species, also with rainfall. By using diagrams, as Iversen had done, with the mean temperatures of the coldest month as x-axis and the mean temperature of the warmest month as y-axis, Hintikka determined the limits for several species, of which a few curves are shown in Fig. 3.6, including some trees and the water chestnut, *Trapa natans.* Of these *Ulmus laevis* is classified as a continental species. So is *Picea abies* for which a diagram was constructed in which the summer rainfall in mm (May–September) was included in addition to the warmest and coldest monthly mean temperatures. Another continental species, *Alnus incana,* was presented in a similar diagram.

The diagrams can be used in assessing former climatic changes, when the former distributions of certain species are known. This is particularly the case with plant species for which earlier distributions are based on dated macroscopic remains in lake sediments or in peats of mires. In this way, however, one can get only a rough estimate about former climatic conditions, as it is difficult to determine if the changes in climate were caused by changes in mean summer or mean winter temperatures or both, and to what extent the mean rainfall changed. By combining observations of several species an estimate can, however, be made of, for instance, changes in the mean temperature of the warmest month, as was done by Iversen (1944) in a comparison of the Flandrian climatic optimum with that of today.

Even if these examples show the potential of pollen analysis for detailed interpretations of changes in vegetations and climate, it has not been employed in the general study of Quaternary stratigraphy. There the conclusions have been based on conventional pollen-analysis studies by using percentage diagrams that can easily be compared with one another. The results from the refined methods used in the study of the Flandrian will therefore not be discussed further.

compared with their real vegetational percentages, without giving exact figures applicable for the whole Flandrian forest history or for earlier warm stages.

Further south, in the temperate vegetation zone of Denmark, the relative pollen production was studied in detail (S. T. Andersen, 1970) and R values used in constructing a corrected pollen diagram for the site Eldrup in eastern Jutland (S. T. Andersen, 1973; also described in Birks & Birks, 1980).

In addition to using the changes in pollen assemblages in the pollen diagrams as a stratigraphical tool, or in order to reconstruct the vegetational history, the presence or absence of certain species in the diagrams have been used for conclusions about the climate.

4 Pre-Saalian stages

4.1. Pre-Elsterian stages

In North Germany there is no evidence of glaciations older than the Elsterian (Ehlers, Meyer, & Stephan, 1984), but in Denmark the pollen diagrams from Harreskov, the type locality for the Harreskovian, and from Ølgod, also representing the Harreskovian, show that this warm stage was preceded by a cold stage (S. T. Andersen, 1965). This fact is also indicated by the sediments. At Harreskov the warm stage organic sediments lie on a stone-free clay, already interpreted by Jessen and Milthers (1928) as a glacial sediment (see also Rasmussen, 1966; T. Nilsson, 1983), and it thus represents a pre-Harreskovian cold stage. If the Harreskovian is correlated with interglacial II of the Cromerian Complex in The Netherlands (see Table 3.2), the cold stage is at least as old as the first Middle Pleistocene cold stage in the Cromerian Complex. The warm stage lake sediments at Ølgod similarly rest on clay (S. T. Andersen, 1965). The evidence of terrestrial pre-Harreskovian sediments is, however, too scanty to allow a reconstruction of the development during the pre-Harreskovian cold stage. Further, no sediments representing this stage have been recorded from Norway, Sweden, or Finland.

In contrast to the incomplete terrestrial record, the already mentioned Quaternary marine sediments, up to 1,000 m thick, in the North Sea basin include Early Pleistocene sediments. Of the 14 foraminiferal assemblage zones in boring 81/29 zones 8–14 are below the Matuyama-Brunhes boundary and the others in the Brunhes normal epoch, representing the Middle and Late Pleistocene (Jensen & Knudsen, 1988). Similar sediments have also been recorded from other borings in the North Sea area and Skagerrak (Knudsen & Feyling-Hanssen, 1976; Feyling-Hanssen & Knudsen, 1979). These thick marine sediments show that during the Early and Middle Pleistocene the alternating warm and cold stages affected Scandinavia in the same way as later in the Pleistocene, from which time organic sediments remain. The erosion of the terrestrial sediments has thus been rather effective in the central parts of the glaciations, with only incomplete remnants left in some places.

In addition to the type locality at Harreskov there is in Denmark the Ølgod site, also in Jutland, from which there is a pollen diagram for the Harreskovian warm stage (Table 4.1). Both diagrams cover practically the whole stage and show the same features on the basis of which this stage can be separated from the Holsteinian and Eemian (S. T. Andersen, 1965), and according to Zagwijn, van Montfrans, and Zandstra (1971; Zagwijn, 1990) from interglacial III and IV (Cromerian in a narrow sense), in the Cromerian Complex. The earlier mentioned pronounced maxima of *Ulmus, Quercus*, and *Taxus* are conspicuous in the lower parts of both diagrams. No organic sediments, which on the basis of their stratigraphical position or their pollen content could be referred to the Harreskovian warm stage, have been found in the Scandinavian peninsula or in Finland.

Further south, in Thuringia south of the Harz mountains but north of the outer limit of the Elsterian glaciation, in an area not reached by the Saalian glaciation, there are two sites with organic lake sediments described as representing warm stages in the Cromerian Complex (Erd, 1970). The younger site representing the Voigstedtian is covered by Elsterian till and has been correlated with interglacial IV in the Cromerian Complex, that is, with the Cromerian in a narrow sense (Cepek, 1990). The incomplete pollen diagram of the end of the warm stage has features not found in the diagram from Harreskov in Denmark, thus confirming the conclusion that the two diagrams represent two different warm stages. The diagram from Voigstedt has no *Taxus*, high percentages of *Alnus* and *Ulmus* as well as of *Abies* and *Carpinus*, before a rise of *Pinus* and *Picea*. It lacks Tertiary relics such as *Pterocarya, Carya*, and *Tsuga* and is younger than the Matuyama–Brunhes palaeomagnetic boundary (Cepek, 1986). The pollen diagram from the older site locally representing the Arternian warm stage is also incomplete and is considered by Erd (1970) to be separated from the Voigstedtian by a cold stage. In Bilshausen between Göttingen and the Harz Mountains there are also sediments referred to the Voigstedtian warm stage and an older warm stage, the Rhume stage

Table 4.1. *Pre-Elsterian sites*

Localities	References	Deposits	Biostrati-graphical evidence
Warm stages			
Denmark			
(= Interglacial II in "Cromerian Complex")			
Harreskov, Jutland	S. T. Andersen, 1965	f	p (type loc.)
Ølgod, Jutland	S. T. Andersen, 1965	f	p
Germany			
(= Interglacial IV in "Cromerian Complex")			
Voigstedt (Voigstedtian)	Erd, 1970	f	p
Bilshausen	Müller, 1965	f	p
Early cold stage interstadials			
Denmark			
Ølgod II	S. T. Andersen, 1965	f	p (type loc.)
Ølgod I	S. T. Andersen, 1965	f	p (type loc.)

f = freshwater
p = pollen

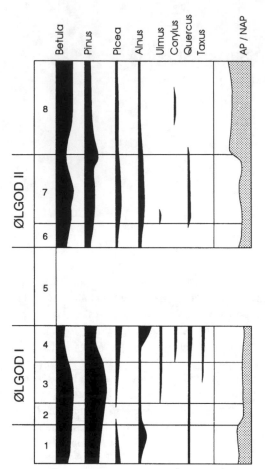

Figure 4.1. Schematic pollen diagrams of early Elsterian(?) Ølgod I and Ølgod II interstadials (after S. T. Andersen, 1965).

(Menke & Behre, 1973). An even older stage, the Osterholz stage, has been separated in Germany (Frenzel, 1973; Menke & Behre, 1973). In the correlation between wider areas, that between Voigstedt and Bilshausen seems to be likely (T. Nilsson, 1983). As can be seen from the examples from Germany, there are sites with sediments of at least two pre-Elsterian warm stages, both probably younger than the Harreskovian in Denmark. Of these, however, there is no terrestrial evidence from Scandinavia.

The earlier marine sediments in the North Sea basin represent in part the time prior to the Matuyama–Brunhes boundary and the beginning of the Brunhes epoch (Jensen & Knudsen, 1988), and are thus older than the warm stages described above. The coastline at the time of the later part of the Cromer Complex, broadly corresponding to the Voigstedtian (interglacial IV), was reconstructed for the southern part of the North Sea area by Zagwijn (1979).

At Ølgod in Denmark the Harreskovian warm-stage freshwater deposits are covered by a sequence of clay, mud, clay, mud, sand, clay-mud, and thick sand. The pollen diagram from this site shows that the two muds on top of the Harreskovian represent two short, early, cold-stage interstadials, named by S. T. Andersen (1965) Ølgod I and Ølgod II, which he preliminarily placed in the Elsterian. On the basis of the results from Germany and The Netherlands they are, however, likely to rep-

resent an older cold stage in the Cromerian Complex (Table 4.1). Ølgod I shows a succession of a *Juniperus* maximum followed by the rise of the *Betula* and *Pinus* curves and later by *Picea*, *Alnus*, *Quercus*, *Ulmus*, *Corylus*, and *Taxus*. Ølgod II has a similar initial succession but the upper part has less pollen of deciduous trees and was therefore cooler than the Ølgod I interstadial (Fig. 4.1). Ølgod II was followed by a rise of non-arboreal pollen. At Ølgod there is no stratigraphical evidence of a glaciation.

4.2. Elsterian stage

The tills in Denmark, which on the basis of their relationship to warm-stage sediments can be referred to the Elsterian, represent two directions of ice advance differentiated with the help of their lithology. The first advance, depositing till A in an older classification (Hansen, 1965), brought in material from Norway, whereas the second advance was more from the east, from the Baltic area, and deposited a sandy till in northern Jutland (Jensen & Knudsen, 1984). In northwest Germany there were similar changes in the lithology of the Elsterian tills (Grube, Christense, & Vollmer, 1986). These tills are considered to represent a single glaciation, during which a change in the direction of ice movement caused the differences in lithology, which can be used in reconstructing the fans for different indicators (Ehlers, 1983; Ehlers, Meyer, & Stephan, 1984).

In eastern Germany two Elsterian tills, I and II, have been separated, but at Voigstedt the warm stage sediments are only covered by one till, the Elster II till according to Cepek (1986), which therefore alone would represent the Elsterian glaciation. But according to Ehlers et al. (1984) the till on top of the Voigstedtian represents the older Elsterian till. What seems likely, however, is that there is only one till unit between the Voigstedtian and the Holsteinian. Similarly, one Elsterian till has been separated in Estonia, representing the Oka glaciation of eastern Europe (Serebryanny & Raukas, 1966). Thus, remnants of the till deposited during the extensive Elsterian glaciation can be traced in the marginal parts of the Scandinavian glaciation. In cores from the Alnarp valley in southern Sweden the lowest till of five till beds was assumed to have been deposited during the Elsterian glaciation, or possibly in the beginning of the Saalian, in which case two till beds would represent this stage (U. Miller, 1977). The age of the lowest till is uncertain, because the mainly lacustrine sediments on top of it either represent the Holsteinian or a Saalian interstadial, the conditions, however, according to the pollen diagrams being more likely similar to those during a warm stage (U. Miller, 1977). In Finnish Lapland six till beds have been differentiated, the three uppermost, I–III, representing the Weichselian and IV the Saalian (Hirvas & Nenonen, 1987). If a compressed peat at Naakenavaara in Kittilä represents the Holsteinian, as suggested by Hirvas and Eriksson (1988; Hirvas, 1991) then there are till beds in Finnish Lapland, V and VI, that are at least as old as the Elsterian.

Deep channels are typical features in the marginal areas of the Elsterian glaciation. In northwest Germany around Hamburg some are 400 m deep, the greatest measured depth being 434 m (Ehlers et al., 1984). They are considered to have been cut by glaciers and particularly by meltwaters (Ehlers & Linke, 1989), but some channels, such as the previously described channels reaching a depth of 60 m in Estonia, have been interpreted as showing the pre-Quaternary drainage pattern (Tavast, 1987, 1992; see also Königsson, 1980). In The Netherlands erosion channels, locally 300 m deep, presumably originated as tunnel valleys (Zagwijn, 1979, 1989). It seems likely that, in many areas, preexisting valleys were deepened and broadened by the Elsterian ice sheet and subglacially eroded by its meltwaters; whether some of the valleys were formed already before the time of the Elsterian is not known. They were, however, filled with glaciolacustrine clays during the retreat of the Elsterian ice that in Schleswig-Holstein and the Hamburg area are called the Lauenburg Clay (Ehlers et al.,

Table 4.2. *Sites showing late Elsterian transgression in Denmark*

Localities	References	Deposits	Biostrati-graphical evidence
Hostrup, N Jutland	Knudsen & Feyling-Hanssen, 1976	m	fa
Gyldendal, Limfjord	Jensen & Knudsen, 1984	m	fa
Skaershøj, Limfjord	Ditlefsen & Knudsen, 1990	m	fa
Esbjerg, S Jutland	Knudsen & Penney, 1987	m	fa

m = marine clay
fa = foraminifera

1984) and in The Netherlands *potklei* (Zagwijn, 1979). The beginning of the marine transgression at the end of the Elsterian is shown by the marine shallow-water sediments with arctic assemblages of foraminifera found in Denmark. These borehole records are reported from Hostrup (Knudsen & Feyling-Hanssen, 1976; Feyling-Hanssen & Knudsen, 1979) and from Gyldendal in the western Limfjord area (Jensen & Knudsen, 1984). Further south in Jutland the sediments are found at Tornskov and at Inder Bjergum south of Esbjerg, and at Esbjerg there are also dislocated clays of the same age (Knudsen & Penney, 1987). In addition, late Elsterian marine sediments were found in a cliff exposure at Skaershøj in the western Limfjord area, where the fauna represents a glacio-marine environment (Ditlefsen & Knudsen, 1990). These sites are listed in Table 4.2. The clays are also reported to occur in Schleswig-Holstein (Feyling-Hanssen & Knudsen, 1979). In addition to the sites mentioned above, there are occurrences described from northern Jutland and the Kattegat area with similar assemblages of foraminifera (Knudsen, 1978).

4.3. Holsteinian stage

The type locality used for the Holsteinian in Scandinavia is Vejlby, northwest of Frederica in Jutland, with diatomite underneath a 10 m thick till bed (S. T. Andersen, 1965). The pollen diagram covers, as noted earlier, most of the warm stage. A similar pollen diagram was obtained from the core of marine sediments at Tornskov on the west coast of Jutland, south of Esbjerg (S. T. Andersen, 1965). It is the same site from which the foraminifera were analyzed and where late Elsterian marine sediments were also found, as stated earlier (Knudsen & Penney, 1987). Here the foraminiferal zones could be directly compared with the pollen zones for the Holsteinian. Pollen zone 8 with *Pterocarya* found in Vejlby is, however, not present in the Tornskov sequence. Further north in Jutland, in the Limfjord

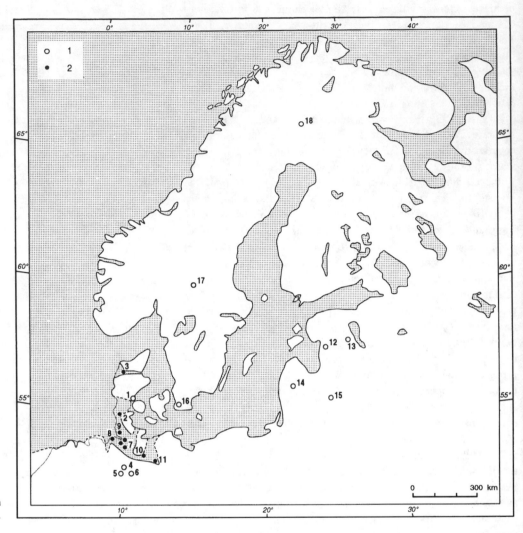

Figure 4.2. Holsteinian sites (listed in Table 4.3). 1, freshwater; 2, marine.

area, the late Elsterian shallow-water marine sediments, with an arctic foraminiferal fauna, were replaced by marine Holsteinian sediments deposited at a greater water depth and with a boreal fauna. These sediments were reported from Kås Hoved below Saalian till (Jensen & Knudsen, 1984).

There are several pollen diagrams from Holsteinian sites in northwest Germany in which the same features in the vegetational history as those found in Denmark (see Fig. 3.1) can be recognized; it was therefore suggested that the lower Elbe area be the type area for the Holsteinian warm stage (Jerz & Linke, 1987). It was also in this area, in Lower Saxony between Hamburg and Hannover, that the length of the Hosteinian was estimated (see Table 3.3) on the basis of annually laminated sediments at Munster-Breloh (Müller, 1974a) and at Hetendorf-Bonstorf (Meyer, 1974), sediments from which there are also detailed pollen diagrams. Müller (1974a) included in his study of Munster-Breloh a previously analyzed pollen diagram from the diatomite at Ober-Ohe in the same area. *Pterocarya* is also found in the uppermost parts of the pollen diagrams from Munster-Breloh and

Hetendorf-Bonstorf, as it is in the diagram from Vejlby in Denmark. In addition to these Holsteinian freshwater sediments, which have been particularly suitable for palynological studies, there are the Holsteinian marine sediments in the Hamburg area. The thick Elsterian glaciolacustrine Lauenburg clay in the deep channels is overlain by limnic, marine, and fluviatile Holsteinian sediments in an area that has tectonically subsided (Linke, 1986). The early Holsteinian transgression can thus be studied in this area, where the sediments have a rich molluskan fauna. The Holsteinian sediments have, for instance, been studied in cores from deep boreholes in Dockenhuden and Billbrook, and in the disturbed sediments of the brickworks at Hummelsbüttel (Linke, 1986). Marine Holsteinian sediments have also been recovered from borings closer to the North Sea coast, in the Cuxhaven area, where their foraminiferal faunas have been studied by Knudsen (1988). At Wacken in Schleswig-Holstein, north of Hamburg, there are well-documented thick Holsteinian marine sediments, from which the pollen diagram shows the familiar features for a good part of this warm stage, even if the sediments have been disturbed by the overriding ice during the Saalian (Menke, 1968). The foraminifera show the changeover from an Elsterian to a Holsteinian fauna (Knudsen, 1989). Southeast of Hamburg, east of the River Elbe, there are also sites with Holsteinian sediments. At Granzin, closer to Hamburg, the marine influence is stronger than at Pritzwalk (Prignitz), where there is only a short episode of brackish conditions in the Holsteinian sequence (Erd, 1970). Both pollen diagrams have pollen of Tertiary relics, the former *Pterocarya* and the latter *Celtis*. The water fern *Azolla filicuoides* was present at both sites (Erd, 1970). Finds of megaspores of massulae of *Azolla* have been reported in Europe only from the Hoxnian/Holsteinian and older warm-stage lake deposits, but not any more from Eemian sediments (Godwin, 1975; West, 1977). It obviously became extinct in Europe but has later been reintroduced from the New World, where it still grows (Godwin, 1975). It may, however, as will be seen later, have survived into the early Saalian.

The occurrences of marine Holsteinian sediments give an idea of the areas submerged during this stage. The coastline drawn on the map in Fig. 4.2 for the Holsteinian shows schematically the areas of northern Germany invaded by the North Sea, for the first time after the Miocene (Zagwijn, 1979; Ehlers, 1990b). In detail the area was more varied than shown on the map. The sea mainly penetrated into the deep buried channels of the area (Ehlers et al., 1984). The Holsteinian coastline in southern Jutland was drawn according to Knudsen and Penney (1987) and in northern Jutland according to Knudsen (1987).

In addition to the Holsteinian sites described in Germany, a few sites may be mentioned from areas further east. In the three Baltic states there are freshwater sediments, including peats, that, on the basis of their pollen diagrams, are placed in an older warm stage than the Eemian, i.e. in the Holsteinian correlated with the Likhvinian of eastern Europe (Liivrand, 1984). In Estonia, as in the Hamburg area, the warm-stage sediments have been preserved in the previously described deep drainage valleys. The pollen diagram of the till-covered organic sediments at Karuküla in southern Estonia lacks the Eemian features but differs also from the pollen diagrams from the Holsteinian sites in Denmark and Germany. In the Karuküla diagram *Picea* and *Abies* are more strongly represented, which is likely to be a result of a more continental climate. The absence of *Taxus* is probably only a result of its pollen not being identified. *Pterocarya* is not recorded but, on the other hand, remains of some plant species found in Likhvinian and older warm-stage sediments in the East European Plain were identified (Liivrand, 1984, 1991). The low percentages of *Corylus*, without a distinct maximum, is a feature in the Karuküla diagram that makes it different from an Eemian diagram. A similar diagram as that from Karuküla was obtained from the till-covered mud from a borehole in Korveküla in southeastern Estonia (Liivrand, 1991). The pollen diagrams from Karuküla and Korveküla are similar to a diagram from Pulvernieki in Latvia and a diagram from Butenai in Lithuania, also consid-

ered to represent the Holsteinian (Liivrand, 1984). If correctly placed in the stratigraphical sequence, the diagrams together give an idea of the Holsteinian vegetational history south of the Gulf of Finland. The correct stratigraphical positions of the sediments, however, are often difficult to determine because of later disturbances caused by ice advances. Thus, for instance, the disturbed warm-stage sediments at Karuküla are covered by only one till representing the Weichselian (Liivrand, 1991). In the stratigraphical division for Poland the Holsteinian warm stage has been correlated with the Masovian I stage (for details, see T. Nilsson, 1983).

In Scania in southernmost Sweden (Fig. 4.2) there is, as mentioned earlier, lacustrine silty fine sand with mud covered by four till beds in the Hyby core from the Alnarp valley (U. Miller, 1977). Miller called the lake sediment below 68 m the Hyby I interglacial and she tentatively correlated it with the Holsteinian or, as an alternative, interpreted it as being a Saalian interstadial, a possibility to be discussed later. The sediments have some reworked older pollen and spore types that are mainly Tertiary, but the pollen diagram from the Hyby core has pollen spectra similar to the Holsteinian diagrams from Denmark and northern Germany, with *Ulmus, Quercus, Fraxinus, Tilia, Carpinus, Fagus, Picea, Abies,* and *Corylus,* in addition to *Betula* and *Pinus.* Furthermore, *Taxus* is rather common and massulae of the *Azolla* type were also present. There are thus many of the characteristic features used in identifying the Holsteinian of older warm stages and the correlation of Hyby I with the Holsteinian (U. Miller, 1977) has been considered the most likely (Lundqvist, 1986a).

At Öje in central Sweden there are till-covered sediments, including laminated silts with organic layers or homogenous silts with organic material. On the basis of their pollen assemblages, dominated by *Picea* and *Pinus* in addition to some *Latrix,* it was concluded that the sediments represent the end of a warm stage with a vegetation of coniferous forest, either the end of the Eemian or possibly the Holsteinian. The presence of spores of *Osmunda* cf. *claytoniana,* however, implied a Holsteinian age (Robertsson, 1988). The study of the macrofossils, with remains of *Abies* spp. and *Picea omorica,* further strengthened the correlation with the Holsteinian (García Ambrosiani, 1990), as, for instance, *Abies* was not present any more in the Eemian of Estonia (Liivrand, 1984). Öje was therefore tentatively included in the list of Holsteinian sites (Table 4.3). Its stratigraphical position, being covered by one till, does not, however, give additional evidence about its age.

The till-covered peat at Naakenavaara in Kittilä, in northern Finland, is far from the Holsteinian sites described (Fig. 4.2). The up to 1.3 m thick, compressed-peat bed at 6.5 m depth is covered by till bed IV of northern Finland, which stratigraphically underlies organic beds interpreted as Eemian (Hirvas, 1991). According to its stratigraphical position the Naakenavaara peat is thus older than the Eemian. The pollen diagram has a dominance of *Pinus,* and smaller relative amounts of *Betula, Picea, Larix,*

Table 4.3. *Holsteinian sites*

Localities	References	Deposits	Biostratigraphical evidence
Denmark			
1. Vejlby, Jutland	S. T. Andersen, 1965	f	p (type loc.)
2. Tornskov, Jutland	S. T. Andersen, 1965	m	p
	Knudsen & Penney, 1987		fa
3. Kås Hoved, Jutland	Jensen & Knudsen, 1984	m	fa
N W Germany			
4. Munster-Breloch	Müller, 1974a	f	p
5. Hetendorf-Bonstorf	Meyer, 1974	f	p
6. Ober-Ohe	Müller, 1974a	f	p
7. Dockenhuden, Hamburg	summary by Linke, 1986	f/m	p, mK, d, fa
Billbrook, Hamburg			
Hummelsbüttel, Hamburg			
8. Cuxhaven area	Knudsen, 1988	m	fa
9. Wacken	Menke, 1968	m	p
10. Granzin	Erd, 1970	m/f	p
11. Pritzwalk (Prignitz)	Erd, 1970	b/f	p
Estonia			
12. Karuküla	Liivrand, 1984	f	p
13. Korveküla	Liivrand, 1991	f	p
Latvia			
14. Pulvernieki	Liivrand, 1984	f	p
Lithuania			
15. Butenai	Liivrand, 1984	f	p
Sweden (Hyby I Igl.)			
16. Hyby I, Alnarp valley	U. Miller, 1977	f	p
17. Öje	Garcia Ambrosiani, 1990	f	p, mf
Finland (Naakenavaara Igl.)			
18. Naakenavaara, Kittilä	Hirvas, 1991	f, peat	p

Deposits: b = brackish   Biostratigr. evidence: d = diatoms
            f = freshwater   fa = foraminifera
            m = marine   mf = macrofossils
                                                             mK = mollusks
                                                             p = pollen

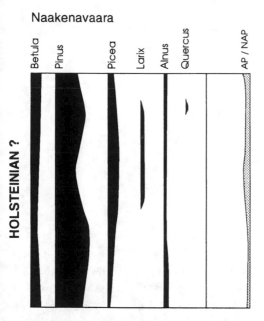

Figure 4.3. Schematic pollen diagram of Naakenavaara, northern Finland, interpreted as Holsteinian (after Hirvas, 1991).

and *Alnus* (Hirvas, 1991). Earlier pollen counts of a few samples from the Naakenavaara peat gave a similar result without, however, records of *Larix* pollen (Hirvas et al., 1977). Compared with Holocene pollen diagrams from the same area, the Naakenavaara pollen assemblages represent a warm stage (Fig. 4.3). In addition to the pollen a number of seeds of *Aracites Johnstrupii* (Hartz) were found – a plant that has been recorded from Pliocene deposits in The Netherlands and from deposits of the Likhvinian (Holsteinian) warm stage in Smolensk and from other areas in the east, but not from younger sediments (Katz, Katz, & Kipiani, 1965). In a later study, the seeds from Naakenavaara were described as those of *Aracites interglacialis* Wieliczk (Aalto, Eriksson, & Hirvas, 1992). On the basis of its stratigraphical position and its pollen diagram, and supported by the presence of *Aracites* seeds, the warm stage represented by the Naakenavaara peat, named the Naakenavaara interglacial, was correlated with the Holsteinian (Hirvas, 1991), even if this conclusion could not be drawn on the basis of the pollen diagram alone.

5   Saalian stage

In continental northern Europe the Holsteinian warm stage was followed by the Saalian cold stage (see Table 3.2), the stage during which the glaciation in many parts of the continent was more extensive than during the Elsterian cold stage and probably any other Pleistocene stage (see Fig. 3.2). Looking at the Saalian cold stage in a wider context it was more complex than the sites in northern Europe, for example, in Denmark, would suggest. It is therefore necessary to include the stratigraphical results from central Europe for additional information about the Saalian in order to get as complete a picture as possible of this stage, as a background scheme for discussion of the development in the central parts of the Saalian glaciation in Scandinavia.

In Denmark the Holsteinian was followed by two interstadials, as with the Harreskovian warm stage. At the Holsteinian type locality at Vejlby the warm-stage freshwater diatom mud was directly overlain by clay, diatom mud, clay, again diatom mud, and then Saalian till (S. T. Andersen, 1965). The clays, containing redeposited pollen and spores, and in some samples up to 40 percent pre-Quaternary microfossils, represent solifluction deposits from cold periods, whereas the diatom mud represent two interstadials, named Vejlby I and Vejlby II (Fig. 5.1) by S. T. Andersen (1965). The pollen diagram shows – for both interstadials – a succession of a treeless vegetation being replaced by forest and then a return to a treeless vegetation. The interstadial forest vegetation had only *Betula* and *Pinus;* the climatically more demanding tree species were absent. Thus the Vejlby interstadials represent a cooler climate than the Ølgod interstadials, which also had pollen of *Picea* and a number of deciduous trees.

The previously addressed Elsterian marine clay at Gyldendal and the Holsteinian marine sediments at Kås Hoved, both in Limfjord in northern Jutland, are overlain by two Saalian tills representing kinetostratigraphic units. They were separated on the basis of an analysis of their fabric and tectonic deformations, showing the directions of ice movement when the tills were deposited (Jensen & Knudsen, 1984). The first Saalian till, a sandy till, was formed by an ice advance from the northeast, with erratics from Norway, and the second clayey till by an advance from the southeast from the Baltic. Jensen and Knudsen called the tills the Drenthe and Warthe tills respectively, using the north German division, or the Saale I and Saale II tills (Table 5.1). A Weichselian ice advance from the northeast laid down and deformed glaciofluvial deposits on top of the two tills. The division of the Saalian in the Limfjord area agrees with the conclusions elsewhere in Denmark, separating a first ice advance from the north, covering the whole of Denmark, and a second advance from the Baltic area reaching central Jutland. The presence of a soil horizon in western Jutland separating Drenthe from Warthe suggests that there was a nonglacial interval between the two ice advances. It is specifically on the basis of a paleosol at Oksbøl near Esbjerg that Sjörring (1981) separated the Saalian glaciations, and the interval was later referred to as the Oksbøl interstadial (see Lundqvist, 1986a), even if its vegetational history or climate has not been established as for other interstadials referred to. Sjörring (1981) also mentioned a third Saalian ice advance from the north, which laid down a flint-conglomerate till as far south as Esbjerg but which was restricted to western Jutland, a late Saalian till earlier identified as having been deposited by an ice advance from Norway (Hansen, 1965).

The twofold glaciation during the Saalian cold stage is also found in northwestern Germany, with the Drenthe (in Schleswig-Holstein, locally called Lippe) ice advance reaching The Netherlands in the west (Zagwijn, 1986) being more extensive than the subsequent Warthe advance. The interval between the two ice advances, in Schleswig-Holstein as in Denmark identified with the help of a paleosol, was called the Treene interstadial, but as it has not biostratigraphically been defined its stratigraphical significance is not yet established (Menke & Behre, 1973; Ehlers, Meyer, & Stephan, 1984; Grube, Christensen, & Vollmer, 1986). Thus the Treene and the Oksbøl interstadials have only been included in Table 5.1 to show their suggested stratigraphical position. Even if two glacial advances have generally been identified in Germany, three tills have been separated lithostratigraphically and their names vary in different federal laender. The names used are given in Table 5.2 according to Ehlers et al. (1984) and Menke and Meyer (in Linke, 1986). It can be seen, for instance, that Drenthe in Lower Saxony includes Older and Middle Saale, and therefore Ehlers et al. suggested

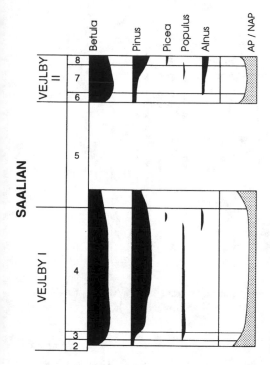

Figure 5.1. Schematic pollen diagrams of Saalian Vejlby I and Vejlby II interstadials (after S. T. Andersen, 1965).

Table 5.1. *Division of the Saalian stage*

Denmark	Schleswig-Holstein in NW Germany[a]	Area east of the River Elbe[b]	The Netherlands[c]
Ice advance in W Jutland, till Warthe ice advance (Saale II), till	Warthe		
Interval (Oksbøl Interstadial), paleosol (Sjörring, 1981)	Interval (Treene Interstadial), paleosol	Saalian 1–3	
Drenthe ice advance, till	Drenthe (= Lippe)		Stadial III
			Bantega Interstadial
			Stadial II
	Wacken* Interstadial, freshwater sed.	Dömnitzian* Interstadial, freshwater sed.	Hoogeveen Interstadial
	Mehlbeck	Fuhne	Stadial I
Vejlby*Interstadial II, fresh water sed (S. T. Andersen, 1965) Vejlby Interstadial I, fresh water sed. (S. T. Andersen, 1965)			
Holsteinian	Holsteinian		Holsteinian

*Shown on map.

Sources: [a]Menke & Behre, 1973
         [b]Erd, 1970
         [c]Zagwijn, 1973, 1986

Table 5.2. *Division of Saalian tills in Germany*

General division	Schleswig-Holstein	Hamburg	Lower Saxony
Younger Saale	Warthe 2	Fuhlsbüttel	Warthe
Middle Saale	Warthe 1	Niendorf	Drenthe 2
Older Saale	Drenthe	Main Drenthe	Drenthe 1

Sources: Ehlers et al., 1984; Menke and Meyer in Linke, 1986.

that Drenthe and Warthe should not be used in correlations between different regions. In Table 5.1 the division of the Saalian stage in Denmark is given according to the division of the neighboring area of Schleswig-Holstein (Table 5.2). As the nature of the possible interval between Drenthe and Warthe is not established, the two ice advances have not been referred to as representing stadials.

In contrast to the stratigraphical scheme of the Saalian presented in Table 5.1 for Denmark, another scheme has been suggested that includes a warm interstadial, by some described as a warm stage, after the Holsteinian but before the Drenthian. This gives an older age for the Holsteinian than that given as the first alternative in Table 3.3, which is in agreement with the previously noted alternative correlation suggested by Zagwijn (1990). At Wacken in Schleswig-Holstein the previously mentioned Holsteinian marine sediments disturbed by overriding ice during the Saalian (Menke, 1968) are overlain by white sands, peats, mud, and glaciofluvial sands. The pollen diagram from the organic fresh water sediments is generally similar to Holsteinian diagrams but with some differences (Fig. 5.2). *Abies* is missing, the *Carpinus* curve starts comparatively early and that of *Picea* late. *Azolla filiculoides* is also present here as in the already discussed Holsteinian diagrams (Menke, 1968; Menke & Behre, 1973). On the basis of the pollen diagram from the freshwater sediments at Wacken the Wackenian warm stage was separated from the Holsteinian by a cold period called the Mehlbeck, during which the white sands but no glacial sediments were formed (Menke & Behre, 1973). Sediments similar to those at Wacken were also recorded from a clay pit at nearby Muldsberg (Menke, 1970). In the upper part of the white sands at Wacken, which are probably partly aeolian, there are convolutions indicating periglacial conditions (Menke in Linke, 1986).

A sequence similar to that at Wacken was recorded from the Holsteinian site at Pritzwalk (Prignitz) east of the River Elbe (Erd, 1970). The partly brackish water sediments, from which the Holsteinian pollen diagram was obtained, are covered in the same section by glaciolacustrine sand formed, according to the pollen diagram, in an arctic cold steppe environment. Above the sands there is a freshwater clay-mud. The pollen diagram, similar to the lower part of the Holsteinian, was interpreted as representing the beginning of a warm stage, named the Dömnitzian (Erd, 1970). *Alnus* and *Quercus* are strongly

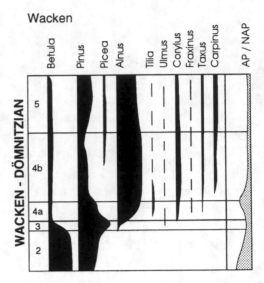

Figure 5.2. Schematic pollen diagram of Wacken of the Wacken–Dömnitzian warm stage (after Menke, 1968).

represented and the continuous curves for *Corylus, Carpinus,* and *Taxus* all begin in the uppermost part of the diagram. *Azolla filiculoides* is also present in this diagram. The freshwater sediments are covered by the till of the first Saalian ice advance (Erd, 1970). In the stratigraphical scheme suggested by Erd, the Holsteinian and Dömnitzian warm stages were separated by the Fuhnian, in Erd's terminology a glacial similar to the three Saalian glacials after the Dömnitzian (Table 5.1).

The similarities between the stratigraphies at Wacken and Pritzwalk and between their pollen diagrams have, as already seen, led to the correlation of Mehlbeck with Fuhne and of Wacken with Dömniz, the latter two collectively included in the Wacken–Dömnitzian warm stage (Menke & Behre, 1973; Frenzel, 1973). Further west in The Netherlands there are two early Saalian interstadials, Hoogeveen and Bantega, which preceded the Drenthe ice advance. Of these interstadials the Hoogeveen probably corresponds to the Wacken-Dömnitzian stage (Zagwijn, 1986). As the vegetational history of the Hoogeveen interstadial is very similar to that of a warm stage it could, according to Zagwijn (1973), also be regarded as a short warm stage. It was preceded by Stadial I of the Saalian in The Netherlands, characterized by a polar desert with permafrost. Stadial I and the Hoogeveen interstadial together represent two Saalian substages, as shown in Table 5.1 where the younger Saalian substages are also included (Zagwijn, 1973). The interstadials Vejlby I and II in Denmark are probably older than the interstadials in The Netherlands and the Wacken-Dömnitzian in Germany, as they followed directly after the Holsteinian (Zagwijn, 1973).

On the basis of the sites already discussed, it seems that in Germany and The Netherlands there was a short, warm, nonglacial interval, at least an interstadial, that can be classified as a warm stage, after the Holsteinian and separated from it by a stadial or cold stage. The Wacken–Dömnitzian stage is older than the first Saalian ice advance, the Drenthe ice advance. The difference between a short, warm stage and an interstadial, however, is not always clear. An interstadial in northwestern Europe can further east in a more continental climate be more similar to a warm stage (West, 1977). This difference is, as will be seen later in dealing with the Weichselian, particularly pronounced during early interstadials of a cold stage, when the sea level had fallen from its warm-stage position and the influence of a continental climate increased in eastern Europe.

If, as concluded, the Wacken-Dömnitzian warm stage is climatically similar to other warm stages, its correlation with stage 7 in the deep-sea chronology seems likely, which means that the Holsteinian corresponds to stage 9, or possibly stage 11, both mentioned as alternatives in Table 3.3.

The results from eastern Europe give no conclusive stratigraphical evidence of a warm stage between the Holsteinian and the Eemian. However, interstadials have been recognized in the Saalian (Middle Polish glaciation) in Poland (Rzechowski, 1986) and also further east in Europe, and in these regions two Saalian ice advances have also been separated, corresponding to those of Drenthe and Warthe further west. Thus, the Odintsovian interstadial separates the Dnieprovian and Moscovian glacial advances in the Russian plain (Lukashov, 1982), correlated with the Odra and Warta stadials of Poland (Rózycki, 1961; see summary by T. Nilsson, 1983).

The stratigraphical division for Denmark presented in Table 5.1 is essentially similar to the schemes for Central Europe. The relationship between the Vejlby I and II interstadials to the Wacken–Dömnitzian is, as noted, not clear. They are all older than the first Saalian ice advance but they cannot be correlated with one another. In the further discussion of the Saalian stratigraphy in Scandinavia the possibility of a comparatively warm Saalian interval, warmer than either Vejlby I and II, has, however, to be taken into account. Thus the previously mentioned Hyby I warm stage in the Alnarp valley in Scania in southern Sweden could be early Saalian instead of Holsteinian or, as suggested by U. Miller (1977), a warm interstadial between the Drenthe and Warthe ice advances. The presence of *Azolla* does not mean that it has to be Holsteinian or older, as *Azolla* was also present in the sediments representing the Wacken–Dömnitzian both at Wacken and at Pritzwalk, as already mentioned. But a warm interstadial between Drenthe and Warthe with *Azolla* seems unlikely against the evidence from elsewhere. Thus, in spite of its uncertain stratigraphical position Hyby I was placed in the Holsteinian (see Table 4.3).

The basis for placing all these sediments, glacial sediments as well as those of interstadials, as well as the relatively warm Wacken–Dömnitzian, stratigraphically into the Saalian (in a broad sense), has been that they are underlain by recognizable Holsteinian sediments and covered by Eemian sediments. In the more central parts of the Saalian glaciation, as in Finland and also in Estonia (Liivrand, 1984), this requirement is not fulfilled.

If glacial sediments are covered by sediments correlated with the Eemian they have been placed in the previous cold stage, the Saalian, even if stratigraphically they can only with certainty be said to be pre-Eemian. One of the exceptions is the already noted site in the Alnarp valley in Scania in southern Sweden, but then it is close to Denmark, where the stratigraphical control of the Saalian is comparatively good. The Hyby I warm-stage sediments, correlated with the Holsteinian, are covered by one till, the Old Baltic till, representing the Saalian, fine-grained sediments partly of Eemian age and three Weichselian tills (U. Miller, 1977). There is thus at this site at least one Saalian till, but its correlation with a particular Saalian stadial is not possible. It was mentioned, in connection with the discussion of the Elsterian stage, that the Hyby I warm-stage sediments are on top of a till, the Old Northern till, which – according to the interpretation favored here – corresponds to the Elsterian. That there are at least one and possibly two Saalian tills in Scania is also shown at Stenberget, where two tills underlie sediments dated as Eemian (Berglund & Lagerlund, 1981). Sometimes Eemian sediments are directly on top of the bedrock, which may be disintegrated and weathered, without any older glacial sediments, as at Leveäniemi in Swedish Lapland (Lundqvist, 1971). On the west coast of Norway, south of Bergen, Eemian sediments at Fjøsanger cover two tills, the Straume till and the Paradis till, both of which have been placed in the Saalian (Mangerud et al., 1981a). Even if there are localities in Sweden and Norway with preserved pre-Weichselian, probably mostly Saalian tills, they are not common near the Scandinavian mountains. Of the three sites mentioned (Table 5.3, Fig. 5.3) two are in Scania and one on the Norwegian west coast, away from the mountains.

In contrast to Sweden and Norway there are a number of sites in Finland with pre-Eemian glacial sediments and some non-glacial sediments. This results from a systematic survey of the Quaternary stratigraphy. But there also seems to be a regional difference between the relatively flat country of Finland and the Scandinavian mountains and their immediate surroundings in Sweden and Norway. Particularly in northern Finland, a till bed from a Saalian ice advance has been described from many different localities, where test pits were used in stratigraphical studies. Lithological studies together with stone orientation measurements were used to separate the previously noted three Weichselian tills and the older pre-Weichselian tills, and to reconstruct the flow patterns of the ice advances which deposited the tills (Hirvas et al., 1977; Hirvas, 1983, 1991). Till bed IV underlies organic sediments of a warm stage, regionally called the Tepsankumpu warm stage after a site in Kittilä and correlated with the Eemian (Hirvas, 1983; see also Donner, Korpela, & Tynni, 1986; Donner, 1988). At the site at Naakenavaara the compressed peat underneath till bed IV, representing the Naakenavaara warm stage, was correlated with the Holsteinian. Even if this correlation is somewhat uncertain, the placing of till bed IV in the Saalian seems likely; its

Table 5.3. *Saalian sites in Sweden, Norway, and Finland*

Localities	References	Stratigraphy	Correlations
Sweden			
1. Alnarp Valley	U. Miller, 1977	Old Baltic till	= Saalian s.l.
		Hyby I freshwater sed.	= Holsteinian
2. Stenberget	Berglund &	Stenberget Formation	= Eemian & E.W.
	Lagerlund, 1981	Ramslid till	= Saalian
		Stallerhult till	
Norway			
3. Fjøsanger	Mangerud et al., 1981a	Fjøsangerian sediments	= Eemian
		Paradis till	= Saalian
		Strauma till	
Finland			
4. Naakenavaara,	Hirvas & B. Eriksson, 1988	Till bed IV	= Saalian
Kittilä		Naakenavaara peat	= Holsteinian
5. Evijärvi	B. Eriksson et al., 1980	Mud	= Eemian
		Till	= Saalian
6. Vesiperä,	Nenonen, 1986; Hirvas &	Organic sediment	= Eemian
Haapavesi	Nenonen, 1987	Till	= Saalian
7. Ollala,	Forsström et al., 1988	Organic sediment	= Eemian
Haapavesi		Till	= Saalian
8. Norinkylä	Hirvas & Niemelä, 1986	Mud	= Eemian
		Till	= Saalian
9. Vimpeli	Aalto et al., 1989	Peat	= Eemian
		Older till	= Saalian
10. Harrinkangas,	Gibbard et al., 1989	Organic sediments	= Eemian
Kauhajoki		Esker sand & gravel	= Saalian

stratigraphical control is fairly well established. The main directions of the Saalian ice advance in northern Finland are included in Fig. 5.3, based on a more detailed map with all sites (Hirvas et al., 1977). Compared with the directions of Weichselian ice advances, the Saalian ice advance over northern Finland is from a more southerly direction, from the southwest. This shows that at the time of the formation of the till the center of the ice sheet and the ice divide were more southerly than during the Weichselian ice advances, probably as a result of the Saalian ice being more extensive than the Weichselian ice sheets.

In Ostrobothnia (Pohjanmaa) on the west coast of Finland there are a few sites with till-covered organic sediments correlated with the Eemian, overlying tills that have been interpreted as Saalian. Such sites are Evijärvi (B. Eriksson, Grönlund, & Kujansuu, 1980), Vesiperä, Haapavesi (Nenonen, 1986; Hirvas & Nenonen, 1987), Ollala, Haapavesi (Forsström et al., 1988), Norinkylä (Niemelä & Tynni, 1979; Hirvas & Niemelä, 1986; Donner, 1988) and Vimpeli (Aalto et al., 1989), all listed in Table 5.3 and shown in Fig. 5.3. At Vimpeli the stone orientation measurements of the upper part of the till, called the Older till, gave a northwest–southeasterly flow direction to the ice, which was interpreted as giving the direction of the ice during the Saalian deglaciation. The till fabric in the lower part

of the till gave a northeast–southwesterly direction, at right angles to the direction in the upper part, and was interpreted as a transverse orientation. Compared with northern Finland (Fig. 5.3) the possibility of a Saalian ice movement from the southwest also in Ostrobothnia cannot be entirely ruled out.

There are a number of sites in Ostrobothnia were eskers are covered by a Weichselian till (Niemelä & Tynni, 1979; Niemelä, 1979), and some sites that will be described later where early Weichselian till-covered interstadial organic sediments lie on top of these eskers. At Harrinkangas in Kauhajoki there are organic sediments which were tentatively correlated with the Eemian (Gibbard et al., 1989). It thus seems as if eskers from the time of the Saalian deglaciation have been preserved in the coastal regions of Ostrobothnia. Till-covered gravels and sands are also common in other parts of Finland in addition to Ostrobothnia, particularly in northern Finland (Niemelä, 1979), but they are not all of Saalian age, many being younger.

Even if sites with a Saalian till have been found at several sites in Finland, the ice advance which deposited the till cannot be correlated with any of the two main Saalian glaciations as separated further south (Table 5.1). The possibility of the presence of two Saalian tills, as in southernmost Sweden and in Denmark, must also be taken into account. The eskers formed

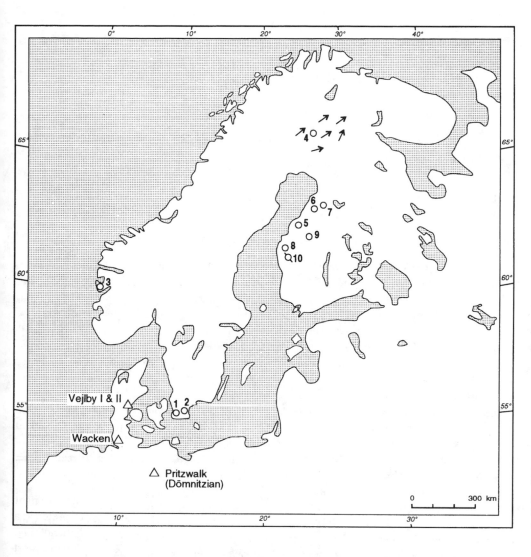

Figure 5.3. Saalian sites (listed in Tables 5.1 and 5.3) and direction of ice movement in Finnish Lapland. Site 10 shows direction of Saalian esker.

during the Saalian deglaciation before the Eemian and preserved underneath a Weichselian till in Ostrobothnia also show that minerogenic sediments other than tills have been preserved from the Saalian stage. The dating of the sands and gravels is, however, only possible when they are connected with organic sediments such as at Harrinkangas.

6 Eemian stage

The Eemian marks the beginning of the Late Pleistocene. It is the last Pleistocene warm stage preceding the Weichselian, and more sediments have been preserved from the Eemian than from any other warm stage. Further, as was pointed out earlier in the description of the type locality at Hollerup in northern Jutland, the Eemian pollen diagrams have characteristic features by which they can be identified and separated from older warm stages, particularly in the areas which at that time had deciduous forests of the temperate zone. As the climate was warmer during the Eemian than during the warm stages previously described and during the Flandrian, the temperate forest vegetation extended relatively far north in Norway, Sweden, and Finland as compared with its present northern limit. Beyond the influence of the temperate forest the identification of the Eemian with the help of pollen diagrams is more problematic, as earlier pointed out. The type locality at Hollerup is unique in the respect that bones of fallow deer (*Dama dama*) found at the site were split by man in order to get the marrow. It is the only pre-Weichselian evidence of human beings in Denmark (Petersen et al., 1987); no similar sites have been recorded in Norway, Sweden, or Finland.

In Jutland there are a number of Eemian freshwater sediments that inside the outer limit of the Weichselian glaciation are covered by glacial sediments but outside this limit are covered by periglacial solifluction sediments (Hansen, 1965). The stratigraphy in the latter area led to the separation of a different type of Eemian deposits called the Herning type after the site in western Jutland studied by Jessen and Milthers (1928). The organic sediments covered by solifluction deposits were originally interpreted as having two warm horizons separated by a middle bed of clay formed during subarctic conditions. Later studies, however, showed that only the lower organic sediments represent the Eemian, whereas the solifluction sediments contain reworked organic material, including pollen, and were formed at the beginning of the Weichselian (S. T. Andersen, 1961). A Herning-type stratigraphy was also recorded at some other sites, with similar Eemian sediments covered by solifluction layers. Nowhere is there, however, evidence of a twofold Eemian vegetational history separated by a cooler climate. At the type

locality of Hollerup the Eemian freshwater sediments of marl and diatom mud are covered by sand and a thin Weichselian till. The pollen diagram is similar to that of the organic sediments at Herning, but at Hollerup there is hardly any influence of reworked pollen in the upper part of the diagram (S. T. Andersen, 1965). The cooling of the climate at the transition to the Weichselian is clearly reflected in the pollen diagram as well as in the sediments by a relative increase in clay in the lake sediments.

In southern Jutland, in the areas which were not covered by ice during the Weichselian, the Eemian sediments overlie Saalian till and marine deposits. Undisturbed marine Eemian sediments are found particularly on the west coast, whereas they were disturbed in the eastern and northern parts by the overriding Weichselian ice, as also on the islands of Fyn, Als, and Aerö to the east of Jutland and at the few sites on Sjaelland (Hansen, 1965).

In Schleswig-Holstein, south of Denmark, there are sites with Eemian freshwater sediments, some of which are covered by sediments of Early Weichselian stadials and interstadials (Fig. 6.1, Table 6.1). One of these sites is Loopstedt, where the Eemian sediments consist of both muds and peat (Schütrumpf, 1967). A similar site, also studied by Schütrumpf, is Geesthacht by the River Elbe east of Hamburg, where the Eemian sediments also consist of muds and peat, as at the site of Oerel between Hamburg and Wilhelmshaven (Behre & Lade, 1986). At Rederstall there is a 32 m thick limnic deposit covering the time from the early Eemian to Early Weichselian (Menke & Tynni, 1984), and at Odderade the Eemian muds and peats represent the upper part of the warm stage (Averdieck, 1967; Averdieck et al., 1976). The site mentioned at Bispingen in the Luhe valley, where the length of the Eemian was estimated (Müller, 1974b), has, on the other hand, a very complete record of the Eemian. There are some local differences between the Eemian sites in Schleswig-Holstein but there is no need, according to Averdieck et al. (1976), to introduce new warm stages because of this variety. The differences in vegetational history between the Eemian sites further apart in Central Europe can be seen in the comparison made by Menke and Tynni (1984) of a number of pollen diagrams, schematically redrawn from the original dia-

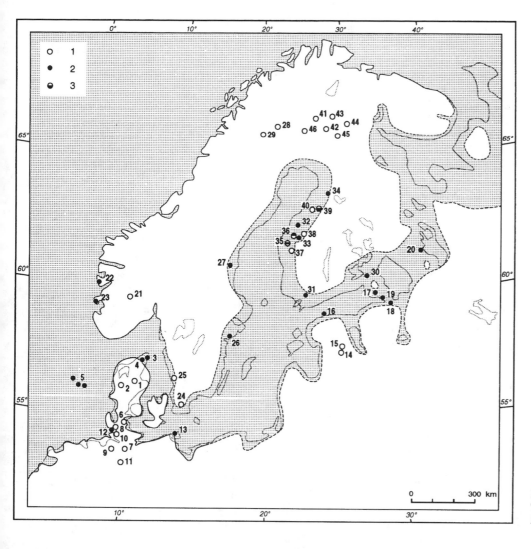

Figure 6.1. Eemian sites (listed in Table 6.1). 1, freshwater; 2, marine; 3, marine and freshwater.

grams. Further east, in Estonia, the pollen diagram from Röngu (Thomson, 1941; Liivrand, 1984, 1991) has comparatively low percentages of *Tilia* and *Abies*. *Picea*, on the other hand, is strongly represented here, as it generally is also in Flandrian pollen diagrams in eastern Europe. At Röngu the Eemian lake sediments are in a secondary position, above a gray till that elsewhere in south Estonia covers Eemian sediments, as at Peedu, another Eemian site (Liivrand, 1991).

The zoning of the Eemian diagrams varies from area to area. Thus, in Estonia the diagrams were divided both according to the zonation by Jessen and Milthers (1928) for the Eemian and by Grichuk (1961) for the Mikulian, the boundaries being identical. Averdieck et al. (1976) similarly used the zoning by Jessen and Milthers but also a more detailed division earlier used in northwestern Germany, and as earlier mentioned S. T. Andersen (1965) divided the Eemian into seven zones, which differ from those used in Germany, as seen in the comparison made by Menke and Tynni (1984). For the often incomplete pollen diagrams from Norway, Sweden, and Finland, local assemblage zones have mostly been used, and some attempts have been

made to apply the same division as that used in diagrams from areas further south. Its usefulness in areas with a vegetational history different from that in the area for which the division was originally intended is, however, limited, even if the sequence of the main changes in the pollen diagrams is similar to that in the diagram from the earlier described type locality of Hollerup in Jutland and in other diagrams in that area.

The Eemian was, as mentioned, comparatively warm but it was also characterized by a relatively high sea level, being at or slightly above the present ocean level. Further in the area of northern Europe dealt with here, the submergence was clearly greater after the Saalian glaciation than after the Weichselian and possibly after the older glaciations. This was a result of a comparatively great downwarping of the earth's crust during the extensive Saalian glaciation. The eustatic transgression of sea level could be determined in The Netherlands, outside the area of isostatic uplift, with the help of pollen-analytically studied submerged organic deposits from eight levels, two of them in the North Sea (Zagwijn, 1983). The sea level rose from about -45 m to about -8 m during a time corresponding to zone f in the

Table 6.1. *Eemian sites*

Localities	References	Deposits	Biostratigraphical evidence
Denmark			
1. Hollerup, Jutland	S. T. Andersen, 1965	f	p (type loc.)
2. Herning, Jutland	S. T. Andersen, 1961	f	p
3. Apholm, Jutland	Knudsen, 1984, 1985c, 1986b	m	fa
4. Skaerumhede I Jutland and surroundings of Fredrikshavn	Knudsen, 1985c	m	fa
5. North Sea, Roar, Skjold, and Dan	Knudsen, 1985a	m	fa
NW Germany			
6. Loopstedt	Schütrumpf, 1967	f	p
7. Geesthacht	Schütrumpf, 1967	f	p
8. Rederstall	Menke & Tynni, 1984	f	p
9. Oerel	Behre & Lade, 1986	f	p
10. Odderade	Averdieck, 1967; Averdieck et al., 1976	f	p
11. Bispingen	Müller, 1974b	f	p
12. Oldenbüttel	Knudsen, 1985b	m	fa
13. Kap Arkona, Rügen	Erd, 1970	f/m	p
Estonia			
14. Rõngu	Thomson, 1941; Liivrand, 1984	f	p
15. Peedu	Liivrand, 1991	f	p
16. Prangli	Liivrand, 1984, 1991	m	p
Russia			
17. Kyyrölä	Sokolova et al., 1972	m/f	p,d,mK
18. Mga	Gross, 1967	m	p,mK
19. Jukki	Maljasova, pers. comm.	m	p,mK
20. Petrozavodsk	Lukashov, 1982	m/f	p,fa,mK
Norway			
21. Hovden	T. O. Vorren & Roaldset, 1977	f	p
22. Fjøsanger	Mangerud et al., 1979b, 1981a	m	p,fa,mK
23. Bø	B. Andersen et al., 1983	m	p,mK
Sweden			
24. Stenberget	Berglund & Lagerlund, 1981	f	p
25. Margreteberg	Påsse et al., 1988	f	p
26. Nyköping	U. Miller & Persson, 1973	m, redep.	p,d
27. Bollnäs	Halden, 1948; Garcia Ambrosiani, 1990	b/m	p,d
28. Leveäniemi	Lundqvist, 1971	f	p
29. Seitevare	Robertsson & Rohde, 1988	f	p
Finland			
30. Rouhiala (now in Russia)	Brander, 1937, 1943	m, redep.	p,d
31. Somero	Donner & Gardemeister, 1971	m, redep.	p,d
32. Evijärvi	B. Eriksson et al., 1980	m, redep.	p,d
33. Alajärvi	Niemelä & Tynni, 1979	m, redep?	p,d
34. Rova	Niemelä & Tynni, 1979	m, redep.	p,d

Table 6.1. *(cont.)*

Localities	References	Deposits	Biostratigraphical evidence
35. Norinkylä, exposure cores	Niemelä & Tynni, 1979; Donner, 1988	m/f, displ.	p,d
	Grönlund, 1991	f/m	d
36. Viitala, Peräseinäjoki	Grönlund, 1991	f/m	p,d
37. Harrinkangas	Gibbard et al., 1989	f	p,d
38. Vimpeli II	Aalto et al., 1989	f, peat	p
39. Ollala, Haapavesi	Forsström et al., 1987, 1988	m/f	p,d
40. Vesiperä, Haapavesi	Hirvas & Nenonen, 1987	f, humus	p
41. Tepsankumpu, Kittilä	Hirvas, 1983	f	p
42. Paloseljänoja, Sodankylä	Hirvas, 1983	f	p
43. Kurujoki, Sodankylä	Hirvas et al., 1977	f, org. mat. in sand & silt	p
44. Härkätunturi, Savukoski	Hirvas, 1991	f	p
45. Loukoslampi, Pelkosenniemi	Hirvas, 1991	f	p
46. Sivakkapalo 1 & 2, Kolari	Hirvas, 1991	f	p

Deposits: b = brackish   Biostratigr. evidence: d = diatoms
               f = freshwater   fa = foraminifera
               m = marine   mK = mollusks
                                              p = pollen

zonation of Jessen and Milthers (1928), during the time of zones E 3b–E 4b in the zonation used by Zagwijn, and the time of zone 4 in the zonation used in Denmark (S. T. Andersen, 1965). The Eemian transgression reached its peak in The Netherlands during the *Carpinus* maximum in the pollen diagrams, in Jessen and Milthers' zonation zone g, and in Andersen's zone 5. In describing the subsequent zone f, Jessen and Milthers noted that the Eemian sea at the time submerged large coastal areas of northwestern Europe. The Eemian coastline during its highest position in the North Sea region was in Fig. 6.1 drawn on the basis of the map by Oele, Schüttenhelm, and Wiggers (1979), on the Baltic side near Lübeck on the basis of the map by Duphorn used by Königsson (1980), and for Denmark mainly on the map by Houmark-Nielsen (1989). The highest position of the Eemian coastline elsewhere in the Baltic Sea area can only roughly be indicated on the basis of the few sites with marine Eemian sediments. Before discussing the further evidence of the Eemian in Denmark, Norway, Sweden, and Finland, some marginal sites with marine Eemian deposits, which are important in the reconstruction of the Eemian palaeogeography, will be noted. A problematic site that has been difficult to place is Kap Arkona on the island of Rügen. Here a marine clay was found between tills in a borehole down to a depth of over 22 m. A combined pollen diagram of the sediments from two boreholes (Erd, 1970) cannot with certainty be placed in a particular warm stage. Erd considered that it represented a separate Rügenian warm stage between the Dömnitzian and the Eemian (see Table 5.1), a division also put forward by Frenzel (1973), but later Erd (see T. Nilsson, 1983) pointed out that it was difficult to differentiate

between the Rügenian and Dömnitzian on the basis of the pollen diagrams. The absence of *Azolla filiculoides* in the Kap Arkona profile may, according to Erd, be a result of the marine environment. Even if the stratigraphical position of the Kap Arkona profile is still uncertain, it was included in the list of Eemian sites as this age has also been suggested for it (oral communication by Dieter Lange, 1983) and does not nesessitate the introduction of an extra warm stage. In the pollen diagram, however, it is difficult to trace the features characteristic for the Eemian. But as the diagram only covers the lower part of a warm stage and the pollen in the marine sediment may not give a true picture of the vegetational history, the typically Eemian pollen assemblages may not be found.

In the Gulf of Finland, marine sediments occur outside the coast of Estonia 60–75 m below sea level. A pollen diagram was obtained from sediments cored from the island of Prangli near Tallinn where the marine Eemian beds at a depth of 68–75 m overlie fresh water sediments and a brown till, and are covered by a gray Weichselian till. The analyzed samples of the marine sediments have a thermophilous diatom flora and the pollen diagram is very similar to the earlier mentioned Eemian diagram from Röngu, with pronounced maxima of *Corylus* and *Carpinus* (Liivrand, 1984, 1991). The marine sediments at the Prangli site and the surrounding area show that the Eemian sea level at the north coast of Estonia was above the level of −60 to −75 m during most of the warm stage. In the lowermost part of the Eemian, however, in the early *Betula* maximum corresponding to zone c in the zonation of Jessen and Milthers (1928), the diatoms show that there was a transgression similar to that found

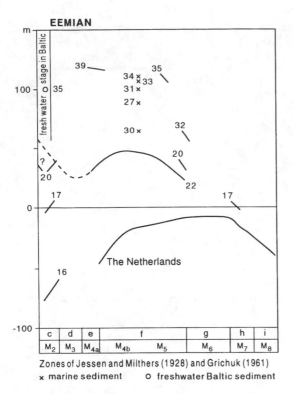

Figure 6.2. Sea-level changes during the Eemian, as interpreted on the basis of sites listed in Table 6.1. Curve for The Netherlands after Zagwijn (1983).

in The Netherlands in the North Sea region (Fig. 6.2), following a short period with freshwater (Liivrand, 1991).

Further east, in the St. Petersburg area and on the Karelian Isthmus, there are more sites with marine Eemian sediments. At Kyyrölä in the central part of the Karelian Isthmus a core from 37.7 m above present sea level was studied by Sokolova et al. (1972). The pollen diagram analyzed covers the silty sediments from a depth of 18 m to 41 m, of which the Eemian represents the lower part from 25 m downwards. The upper part of the diagram represents the early Weichselian (Valdaian). The Eemian pollen diagram shows all the main features that characterize this warm stage and includes the lowermost *Betula* zone. The diatoms analyzed show that there was a strong marine influence until a change in zone M 7 in the zonation used for the Mikulian, corresponding to zone h in the zonation of Jessen and Milthers, when the basin became a separate lake or a lagoon of the Baltic. Sokolova et al. (1972) interpreted the marine sediments as having been deposited during a transgression of the Eemian sea, the marine diatoms being absent in the lowermost marine silt. Even if the evidence for a rise in sea level is not very strong, the Kyyrölä profile clearly registers the change from a marine environment to a lake or lagoon. The marine sediments lie between +2.7 m and −2.2 m in relation to present sea level and the Eemian sediments above between +2.7 m and +12.3 m. This record was included in Fig. 6.2, which shows the sea-level

changes during the Eemian. In addition to the diatoms in the marine sediments at Kyyrölä, shells of the marine mollusks *Macoma calcarea, Mytilus edulis,* and *Portlandia arctica* were recorded. The last mentioned lives at present in High Arctic waters, whereas *Mytilus* has a wider distribution, from Lusitanian to Low Arctic waters, according to the compilation by Feyling-Hanssen (in O. Holtedahl, 1960). There is thus a mixed, death assemblage of marine shells in Kyyrölä Eemian silts. What is striking, however, is the strong marine influence in the eastern parts of the Gulf of Finland, which is also reflected at the other Eemian sites. It is interpreted as showing that there was a connection between the Baltic Sea and the Arctic Ocean, over Lake Ladoga, Lake Onega, and The White Sea. This explains the contrast between a relatively warm Eemian climate, reflected in the pollen diagrams, and the presence of cold arctic water in the Baltic, reflected in the marine diatoms and shells, a contrast which must be taken into account in the interpretation of Eemian sites in the more central part of the formerly glaciated area where only parts of the Eemian sediments have been preserved. The presence of arctic shells in the marine sediments led to the separation of the Eemian Sea from the Portlandia Sea, named after *Portlandia (Yoldia) arctica,* but as *Portlandia* shells were found in clearly Eemian sediments of Mga near St. Petersburg the assumption of the separation of the Portlandia Sea from the Eemian is not tenable. The cold water penetrated mostly into the Baltic as a bottom current (see T. Nilsson, 1983) and the influence of arctic cold water was particularly pronounced in the beginning of the Eemian, but also already in the late Saalian in the areas submerged at that time (Gross, 1967). *Portlandia* thus reflects the first penetration of cold artic water into the Baltic but it seems to have persisted for some time into the Eemian, depending on the sediment facies, and cannot therefore be used in delimiting a separate Portlandia Sea period in the late Saalian–Eemian history of the Baltic. At Mga the marine Eemian (Mikulian) clay, reaching an altitude of 10 m above present sea level, is covered by 11.4 m thick Weichselian (Valdaian) sediments, including till. Underneath the marine clay, 8 m thick, there is 2 m of early Eemian varved clay, from the level of the present sea level to +2 m, on top of 2 m of sand and 10 m of late Saalian varved clay which overlies till. The pollen diagram, as reproduced by Gross (1967), shows that the marine sediments represent the entire Eemian, in addition to the late Saalian, with a strong *Artemisia* component. The Eemian part of the pollen diagram has a pronounced *Corylus* maximum followed by a peak in the *Carpinus* curve. The marine Eemian sediments are at about the same altitude as the sediments at Kyyrölä on the Karelian Isthmus, but at Mga there are no Eemian freshwater sediments in the upper part of the profile. This difference must be due to differences in the land/sea-level changes. At Kyyrölä the isostatic uplift after the Saalian glaciation resulted in a relative regression of sea level at the end of the Eemian, whereas at Mga, which is situated more marginally to the area of uplift, just above the level of the present sea level, there was not a similar change. The whole of the Eemian is here represented in marine sediments, just as at Prangli in Estonia (Fig. 6.2). In the St. Petersburg area Eemian marine sediments other than those found at Mga have also been studied pollen analytically, as for instance at Jukki north of St. Petersburg, where an Eemian diagram has been obtained from silts, also having mollusk shells (Maljasova, personal communication). The pollen diagram is very similar to the two diagrams from the same area. If the marine sediments in Estonia are compared with those in St. Petersburg area, from west to east, it can be seen that the shallow-water Eemian sediments in Estonia are at a greater depth than the deep-water sediments in the St. Petersburg area, with an irregularly high position of the Mga sediments (Liivrand, 1991). This irregularity may, according to Liivrand, be caused by later displacements of some of the sediments from their original positions.

At Petrozavodsk on the western shore of Lake Onega there is a section with Eemian sediments covered by Weichselian sediments, including at least one till (Lukashov, 1982). The Eemian sediments, reaching an altitude of 32–40 m, overlie late glacial marine sands and varved clays of the Saalian Moscovian glaciation. The Eemian pollen diagram has even at this eastern locality a pronounced *Corylus* maximum, followed by a *Carpinus* maximum as well as other features that make a correlation with the pollen diagrams possible. In addition to local zones, the divisions by Grichuk (1961) and Jessen and Milthers (1928) were used by Lukashov (1982). According to Lukashov there was a regression of the sea in the beginning of zone c of the last-mentioned zonation, followed by a transgression at the end of zone c, called the Boreal transgression. The strong marine influence continued until the end of zone f, marking the beginning of the regression recorded in the sediments. But it is not quite clear from the description of the site whether the uppermost Eemian sediments should be described as marine or freshwater sediments. There is, however, a distinct difference between the sediments described as representing the transgression and those formed after the regression. The former contain marine foraminifera and marine shells of *Mytilus edulis, Macoma (Tellina) baltica, M. calcarea,* and *Leda pernula,* whereas shells of *Portlandia arctica* had earlier been recorded. The sediments formed after the regression contain plant remains. The section thus records a marine phase of the Eemian with a relative regression of the sea level beginning somewhat earlier than at Kyyrölä that is closer to the present sea level (Fig. 6.2). The few sites with Eemian sediments described above reflect the Eemian regression toward the end of the warm stage in the marginal parts of the isostatic uplift. The presence of *Portlandia arctica* in the Eemian sediments at Petrozavodsk is in keeping with the occurrences in the St. Petersburg area and does not here necessitate a separation of the Eemian Sea from the Portlandia Sea. The indications of an early Eemian transgression both at Kyyrölä and at Petrozavodsk, as well as at Prangli, most probably reflects a strong eustatic rise, which was later overtaken by the isostatic uplift when the sea level had reached its Eemian peak.

These sites give examples of the evidence for a connection between the Baltic Sea and the Arctic Ocean during the Eemian, a connection outlined on maps for the Eemian Sea (see Königsson, 1980). Lavrova (1961) constructed a map showing the extent of the sea during the Eemian (Mikulian) in the northern parts of Russia, where it has been called the Boreal Sea. The Boreal transgression at Petrozavodsk therefore marks, as already mentioned, the relative rise of the Eemian sea-level culminating in the highest position of the Boreal Sea along the arctic coast of Russia. The Eemian coastline for the areas east and southeast of Finland in Fig. 6.1 is based on the map by Lavrova, whereas the coastline in the other parts of the Baltic is based on the map by Grichuk (as given by Königsson, 1980) with later additions (Donner, 1991), using sites with marine Eemian sediments in the area. Considering that the Eemian sea level during the climatic optimum was higher than the Flandrian sea level and that the earth's crust in the central parts of the Saalian glaciation was more downwarped than during the Weichselian glaciation, it is likely that larger areas were submerged during the Eemian, particularly in Finland but also in Sweden, than is indicated in Fig. 6.1. Southern and central Finland were probably almost entirely submerged in early Eemian time, as they were during the Flandrian. Furthermore, there was an oceanic influence throughout the Baltic Sea, in contrast to the Baltic after the Weichselian, when the connection with the ocean in the west, narrow at best and sometimes even broken, turned the Baltic into a brackish bay and at times a lake. However, in the Baltic the Eemian marine transgression was preceded by a short fresh water stage.

The different distribution of land and sea during the Eemian, as compared with the time after the Weichselian glaciation, had an obvious effect on the climate in the area of the Baltic Sea. Not only did the cold waters of the Arctic Ocean penetrate into the Baltic from the northeast, as earlier concluded, but the oceanic influence with the west was probably also stronger. The connection between the Baltic Sea and the Arctic Ocean still existed, however, as mentioned, during the Eemian climatic optimum. The temperate forest with *Carpinus betulus* and *Quercus robur,* as well as with *Corylus avellana,* which prefers an oceanic climate (see Godwin, 1975), was common beyond its present northern boundary during the climatic optimum, not only in the western parts, but also in the east. The Eemian pollen diagrams in the St. Petersburg area and the diagram from Petrozavodsk all have pronounced maxima of *Corylus,* with high percentages, followed by comparatively high percentages of *Carpinus.* There was a greater uniformity in the Eemian vegetation throughout the area than during the Flandrian climatic optimum. As the deciduous forests with their strong representation of *Corylus* and the presence of *Carpinus* were widespread during the Eemian, it has been possible to pollen analytically identify remnants of Eemian organic freshwater sediments or marine silts and clays in the more central parts of Scandinavia.

In addition to the evidence obtained about the extent and

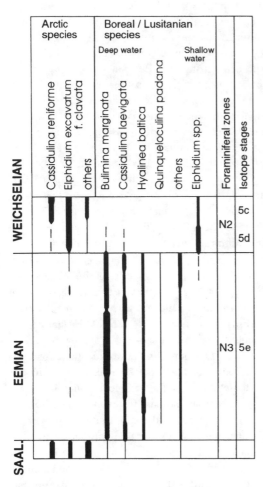

Figure 6.3. Foraminiferal stratigraphy with main species of late Saalian, Eemian, and Early Weichselian marine sediments at Apholm in northern Jutland (after Knudsen, 1986b).

nature of the Eemian Sea from the pollen-analytically studied marine sediments, combined with studies of their shells and diatoms, and at the Petrozavodsk site of foraminifera as well, foraminifera alone or together with mollusks have been used at several sites bordering the North Sea basin and in the study of cores from the basin itself. In the Kattegat area there are marine Eemian sediments, especially in northern Jutland, as in Skaerumhede. Marine sediments in situ have also been reported from Ejby in the northern part of Sjaelland, but these have not been studied in detail (Krog, 1979). In several cores taken from boreholes along the east coast of northernmost Jutland a detailed foraminiferal stratigraphy has been established, which continues into the marine Weichselian sediments. At Apholm, just north of Fredrikshavn, the lower part (140–190 m) of a 200 m core represents the Eemian, lying on top of nonmarine, glacial Saalian sediments (Fig. 6.3). The Eemian is here characterized by such boreal–lusitanian foraminifera as *Bulmina marginata, Cassidulina laevigata, Hyalinea baltica,* and *Quinqueloculina padana,* in contrast to the arctic species dominating the overlying

Weichselian sediments, the changeover from one assemblage to another being very distinct (Knudsen, 1984, 1986b). The significance of the change in faunal assemblages is clear when viewed against the present distribution of the faunas along the European coasts. The same stratigraphy as at Apholm has been found in a number of boreholes in the Fredrikshavn area, the Eemian sediments representing the foraminiferal biozone N 3 in the division used for a profile called Skaerumhede I, and the Weichselian sediment zones N I and N 2, counting from above (Knudsen & Lykke-Andersen, 1982; Knudsen, 1985c). The fauna indicates that the water depth during the Eemian was about 100 m. Of particular importance in the recognition of the Eemian in these marine sediments is that it can be linked to the stages of the deep-sea sediments. A correlation of the foraminiferal zone N 3 in Jutland with deep-sea stage 5e, as suggested by Knudsen (1984), seems well founded.

In addition to the preceding sites with Eemian sediments in northern Jutland, there are several sites along the west coast of southern Jutland, from Esbjerg southwards, with marine sediments which on the basis of their foraminiferal faunas are Eemian. The coastline for the Eemian transgression, as drawn on the map in Fig. 6.1, is based on these localities (Knudsen, 1986b) and, as mentioned earlier, for the whole of Denmark on the map drawn by Houmark-Nielsen (1989). The foraminifera have also been studied in Eemian sediments in Germany further south. Near Oldenbüttel by the Kiel Canal in Schleswig-Holstein, for instance, there are mainly shallow-water, brackish-marine sediments representing part of the Eemian (Knudsen, 1985b). The foraminifera in the Eemian sediments in north Germany generally show that the marine transgression was restricted to the warm part of the Eemian (Knudsen, 1986b), which is in keeping with the earlier mentioned results from The Netherlands and also with the results from the sites at Kyyrölä and Petrozavodsk, where the beginning of the Eemian transgression was recorded (Fig. 6.2).

The sediments in the cores from boreholes in the Danish sector of the central parts of the North Sea, at Roar, Skjold and Dan, all have an Eemian horizon at a depth of about 70–80 m, with a boreal-lusitanian foraminiferal fauna (Knudsen, 1985a, 1986b). It is a shallow-water fauna, in contrast to the fauna of deeper water further north in the North Sea (Knudsen, 1986a). In cores from outside the coast of Norway, in the Statfjord Field, a warm stage, named the Statfjord interglacial, was identified on the basis of its foraminiferal assemblage and correlated with the Eemian and the deep-sea stage 5e (Feyling-Hanssen, 1981, 1982), similar to what was done in the cores from northern Jutland.

The almost complete sequences of the Eemian in these areas, of both freshwater and marine sediments, form the background for the interpretation of the few sites in Norway, Sweden, and Finland correlated with the Eemian, an area which, as pointed out earlier, was an island during that time. In the southeastern part of Hardangervidda in Norway there are two sites, Førnes and Hovden, near Lake Møsvatn at about 900 m a.s.l., with till-covered sediments. The combined stratigraphy from these sites consists of an upper till separated from a lower till by sand and gravel and lowermost by a lacustrine clay (T. O. Vorren & Roaldset, 1977). The pollen spectra of the clay, in an area now above the coniferous forest, have in addition to *Betula, Pinus,* and *Picea,* 2.7–11.3 percent of *Corylus,* which indicates that hazel grew close to the lake. As the pollen composition reflects a climate at least as warm as that of today, it was concluded that the clay represents a warm stage, in the local stratigraphical division called the Hovden thermomer, and that is likely to represent the Eemian (T. O. Vorren & Roaldset, 1977). If this correlation is correct it shows that there are two Weichselian tills in this part of Hardangervidda. A more complete stratigraphical sequence than that near Møsvatn has been described in detail from Fjøsanger at the coast, a section on a steep slope at the fjord of Nordåsvatn at the edge of the city of Bergen (Mangerud et al., 1979a, 1976b). The marine sediments correlated with the Eemian, reaching an altitude of 15 m a.s.l., consist of silt, sand, and gravel, the fine material deposited through suspension and the coarse originating from the shore above. The Eemian sediments overlie a till, the Paradis till, and an older till, the Straume till, was found in two excavations. Both were correlated with the Saalian and overlie the bedrock. On top of the Eemian there is silt and gravel, and another till, the Bønes till, representing the first Weichselian in the area. According to the biostratigraphy of the marine minerogenic sediments, for which assemblage zones were separated on the basis of foraminifera, mollusks, and pollen, it was concluded that the sediments represent a complete cycle of a warm stage, which was named the Fjøsangerian (Mangerud et al., 1981a). If the boundaries of the warm stage were drawn in agreement with the definitions for a warm stage as earlier stated, the lowermost assemblage zone in the pollen diagram, the *Artemisia–Poaceae* zone, would belong to the Saalian cold stage and the *Betula* zone would represent the first warm-stage assemblage zone, the change in vegetation being similar to that at the Weichselian-Holocene boundary. The Fjøsangerian pollen diagram has above the *Betula* zone a *Pinus–Juniperus–Quercus* zone, with a slowly rising *Quercus* curve, followed by a *Corylus–Quercus–Alnus* zone, with a relatively late spread of *Alnus* and *Corylus,* and uppermost a *Picea* zone. *Carpinus* is not present. The characteristic features for Eemian diagrams further south are missing, but as the Fjøsangerian pollen diagram resembles Eemian pollen diagrams more than any other warm-stage diagrams the correlation was considered certain (Mangerud et al., 1981a). The differences in the Eemian vegetational history between the Bergen area and Denmark is then the result of the northern position of Fjøsanger where, for instance, *Quercus* did not become as widespread as further south. *Tilia* did not reach Norway; its Eemian northern limit lay in southern Denmark. It was therefore not present at Hollerup in northern Jutland, but in the Eemian diagrams from Estonia and areas further east *Tilia* is regularly present. Even if the pollen diagram from Fjøsanger does not closely resemble the cited Eemian diagrams,

the Eemian age of the marine sediments is supported by the assemblages of mollusks and benthic foraminifera. They show that the coastal waters were as warm or warmer than during the Holocene and that the sediments in which these assemblages are found therefore represent the oxygen-isotope stage 5e and the Eemian (Mangerud et al., 1979b). The Eemian transgression reached up to 30–45 m a.s.l. at Fjøsanger, higher than the Holocene Tapes transgression that reached an altitude of 15 m. This relatively high sea level also supports the conclusion that the marine sediments at Fjøsanger were formed during the Eemian.

The sea-level curve constructed for the Fjøsanger site (Mangerud et al., 1981a) shows a transgression that reached its peak at a time broadly corresponding to the zone boundary e–f if Jessen and Milthers' (1928) zonation is used for the pollen diagram and compared with the beds studied. The marine Eemian sediments at Fjøsanger are at a higher altitude than the previously mentioned Eemian sites, but the general pattern of the transgression and subsequent regression is at Fjøsanger similar to that earlier recorded (Fig. 6.2). The transgression, however, reached its highest point earlier at Fjøsanger than in The Netherlands. Further, there is an indication of an early Eemian regression before the transgression at Fjøsanger, where an early isostatic uplift can be expected to have affected the land/sea-level changes before the eustatic rise of sea level overtook the land uplift, as was the case in the same area after the Weichselian glaciation.

Marine sediments were also found in Norway south of Bergen, in a clay pit at Bø on the eastern side of Karmøy south of Haugesund. Here the pollen composition and the mollusk assemblages in the Avaldsnes Sand showed it to represent a warm stage, correlated with the Eemian (B. Andersen, Sejrup, & Kirkhus, 1983). During the deposition of the 3 m thick marine sand the sea level was estimated to have been 15–20 m higher than today. On the basis of the pollen assemblages the sand was correlated with the upper Eemian sediments at Fjøsanger. The pollen diagram from Bø, as that from Fjøsanger, lacks the typical Eemian features, which may partly be due to the nature of the sediments. The sedimentation of pollen was probably disturbed and the assemblages distorted in the marine minerogenous sediments and they are therefore not as representative as organic freshwater sediments. The mollusks, among them *Ostrea edulis,* show that the sea-water temperature was as warm or warmer than today. The Eemian sand was covered by marine sand, a gravelly till, the Karmøy Diamicton, silts, and sands, and an upper clayey till, the Haugesund Diamicton (B. Andersen et al., 1983). At Bø there are thus two post-Eemian tills, whereas there is only one at Fjøsanger. The molluskan and foraminiferal stratigraphy of the marine sediments at Bø was also studied by Sejrup (1987), who demonstrated that there is a clear difference in the composition of the benthic foraminifera between the sediments correlated with the Eemian and those correlated with the Weichselian. The boreal–lusitanian foraminifera, *Bulimina marginata* and *Cassidulina laevigata* among others, were also recorded at Bø, two species which, as mentioned earlier, were also recorded from the Eemian marine sediments at Apholm near Fredrikshavn in Jutland. The detailed study of the mollusks revealed a difference as clear as that of the foraminifera between the Eemian and Weichselian, the Eemian sediments having a number of lusitanian species in addition to *Ostrea edulis* (Sejrup, 1987).

The importance of the two coastal sites at Fjøsanger and Bø lies in the possibility here of linking the terrestrial Eemian sequence, as evidenced by pollen diagrams, with the marine sequence, both with the coastal molluskan fauna and with the foraminiferal stratigraphy, which in turn can be linked with that of the North Sea sediments and those of the deeper Norwegian Sea. The correlation of the warm-stage sediments at the two sites with the Eemian is supported particularly by the combined biostratigraphical evidence, whereas the applications of some relative dating methods, such as amino acid racemization dating, have given results not fully in agreement with this correlation. The dating results, however, are not considered conclusive and do not necessitate a different correlation of the sediments (Sejrup, 1987).

When discussing the Eemian development of Sweden and Finland the correlations are primarily based on terrestrial sediments. At Stenberget in southern Scania in Sweden, till-covered sediments represent a warm stage called the Romele interglacial and correlated with the Eemian (Berglund & Lagerlund, 1981). The approximately 3 m thick Eemian sediments, encountered in cores, consist of silty clay and clay-mud at the bottom, covered by peat, mud, and a sandy solifluction sediment with organic material on top. The site is at an altitude of about 160 m, well above the level of the Eemian Sea. The pollen diagram, where samples were combined from three cores, is not complete because a major part of the climatic optimum is missing as a result of a hiatus in the sequence consisting of an eroded peat surface. In the lower part, there is a *Betula* maximum followed by a *Pinus* maximum, and in the upper part, above the hiatus, high percentages of *Alnus* and *Picea,* and some *Corylus* and *Carpinus*. In the solifluction sediment the pollen assemblages were influenced by rebedded pollen, somewhat distorting the percentages. Even if the diagram is incomplete its correlation with the Eemian seems well founded. The site, as with many of the previously mentioned sites, is also important because the sediments above the Eemian can stratigraphically be placed in the Weichselian. The Stenberget site could be used, as stated earlier, as evidence that Saalian tills were preserved in Scania. In the Alnarp Valley with sediments correlated with the Holsteinian (U. Miller, 1977) the Eemian is only represented by a horizon with rebedded pollen, but even here the Saalian could be separated from the Weichselian, even if the site gave no additional information about the Eemian in Scania. As the two sites – Stenberget and the Alnarp Valley – as well as the sites in Norway, often have incomplete sequences correlated with the Eemian, the correlation can at some sites still be questioned. The use of local or regional stratigraphical divisions is particularly necessary at these sites but the names used for the warm stages,

called thermomers or interglacials, were not included in the list of sites in Table 6.1.

At Margreteberg in southwestern Sweden a complex sedimentary sequence was investigated biostratigraphically (Påsse et al., 1988). Underneath a "till-like" sediment a 1 to 10 cm thick, disturbed layer of laminated peat in sand was interpreted, on the basis of the pollen spectra, as representing the upper part of the Eemian, with high percentages of *Pinus* and *Picea,* but with *Corylus* and *Carpinus* still present. Clayey sediments underneath the Eemian represent the late Saalian or late Elsterian, according to Påsse et al. (1988). The 200 m section at Margreteberg is an example of how sediments predating the last Weichselian glaciation have in places partly been preserved even if they have been disturbed. The sediments can with some certainty be linked to the general regional stratigraphical scheme, in this case with that of Denmark, but give little additional evidence of, for instance, the Eemian environmental changes.

In Bollnäs, on the east coast of central Sweden, till-covered brackish-water silts and sands containing organic material were described by Erikson already in the beginning of the century (Erikson, 1912) and later reinvestigated by himself, under a name change to Halden (1948), and recently by Garcia Ambrosiani (1990). The sediments underneath a several m thick till were disturbed and partly displaced by the ice that deposited the till. The old pollen diagram (Halden, 1948) probably shows the end of the climatic optimum with a strong representation of *Betula* and *Alnus,* and with some *Corylus,* and a deterioration of climate with an increase of *Betula* and a corresponding decrease of other deciduous trees. *Pinus* and *Picea* are present throughout the diagram. Even if the pollen diagram can be considered to represent a warm stage it lacks the features typical for the Eemian diagrams further south. The macrofossils earlier reported by Halden and later by Garcia Ambrosiani, summarized by the latter, however, show that the climate was warmer than today, and the presence of a brackish-marine diatom flora at the Bollnäs site, at 88 m a.s.l., shows that the sea level was comparatively high at the time of the climatic optimum. Thus, taking into account the warm character of the warm stage and the high position of sea level, the till-covered sediments at Bollnäs were interpreted as probably representing the Eemian (Garcia Ambrosiani, 1990). This agrees with the results from the west coast of Finland, with similar sites (Fig. 6.1). Viewed against the distribution of remains of Eemian marine sediments it is also likely that the lump of clay found in till-covered glaciofluvial silty sand near Nyköping in Sweden at 25–30 m a.s.l. is also Eemian (U. Miller & Persson, 1973). The corroded diatoms showed it to be a marine clay and the pollen, with a strong representation of *Tilia,* that the clay represents a warm stage. The predominance of *Tilia* was interpreted as a result of weathering, which would have destroyed other pollen more than the robust *Tilia* grains. But it may be noted that *Tilia* is strongly represented in Eemian pollen diagrams from Estonia (Liivrand, 1984), across the Baltic from Nyköping.

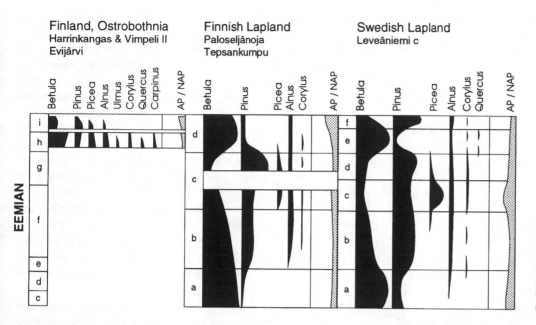

Figure 6.4. Schematic Eemian pollen diagrams from Ostrobothnia in Finland (after Donner, 1988), Finnish Lapland (after Hirvas, 1983), and Swedish Lapland (after Lundqvist, 1971). (Further details of sites are given in Table 6.1.)

In addition to the coastal sites in Sweden containing remnants of marine sediments there are two sites in Swedish Lapland with Eemian freshwater sediments, of which those at Leveäniemi 40 km southeast of Kiruna cover the whole warm stage. The stratigraphical position of the organic sediments of this stage, consisting of drift peat, peat, and mud on top of sand and weathered bedrock, was established by combining the evidence from different sections (Lundqvist, 1971). The thickness of the organic sediments varies between 70 cm and 160 cm in the three sections under pollen analysis. Some small disturbances in these sediments were caused by the pressure of the overriding ice. There were two ice advances over the site, the first depositing a lower till on top of the organic sediments by an ice movement from the northwest, the second depositing an upper till by a movement mainly from the southeast. The two tills are separated by glaciofluvial or glaciolacustrine minerogenous sediments laid down during a nonglacial interval, with no sediments that could represent a warm stage. The upper till is covered by Flandrian organic sediments, mainly fen peat.

The most representative pollen diagram of the till-covered organic sediments is that from section C (Lundqvist, 1971), on which the schematical diagram in Fig. 6.4 is based. The diagram is essentially similar to Flandrian diagrams from the same area. After an initial Betula-dominated assemblage zone (a) recording a vegetation with open birch forest, follow a Pinus–Betula–Alnus zone (b), with some Corylus, and a Picea–Pinus(–Betula–Alnus) zone (c) with a pronounced increase of Picea. The upper part of the diagram was divided into a Pinus–Betula(–Alnus) zone (d), a Betula(–Pinus) zone (e), and a Pinus–Betula(–Alnus) zone (f?). The last-mentioned assemblage may have been influenced by rebedded pollen (Robertssen in Lundqvist, 1971) and cannot therefore be used as showing a climatic amelioration before the subsequent ice advance. There is thus from Leveäniemi, as

already mentioned, a pollen diagram covering a whole warm stage, with a possible presence of Corylus and of mixed oak forest south of Leveäniemi during the climatic optimum. This, and the evidence from beetle remains and macroscopic plant remains representing species that now have a more southerly distribution, such as Carex pseudocyperus, led to the conclusion that the till-covered sediments at Leveäniemi represent the Eemian (Lundqvist, 1971). As it is the only more-or-less complete Eemian freshwater sequence from northern Scandinavia, it is a very important locality for comparisons with less complete sequences elsewhere in this area, as well as for correlations of the till beds near the former ice divide in Swedish and Finnish Lapland.

At Seitevare, about 130 km southwest of Leveäniemi in Swedish Lapland, an approximately 10 cm thick peat layer was found at nearly 4 m depth, covered by two till beds (Robertsson & Rodhe, 1988). On the basis of the till-fabric analyses and the petrographic composition of the tills it was shown that the lower till was deposited by an ice advance from the northwest and the upper till by an advance from west–southwest, thus giving the same sequence as that found at Leveäniemi. At Seitevare, however, a till was also found underlying the peat. The pollen diagram shows that the vegetation during the formation of the peat was a Pinus-dominated forest, with some Betula, Alnus, Corylus, and Picea. Then followed a more open vegetation with an increase of Betula and nonarboreal pollen, but with some pollen of deciduous trees interpreted as having been redeposited. The sediment formed during the deterioration of climate was a silty sediment with organic material. By comparison with the Leveäniemi site, the correlation of the pollen diagram, Seitevare A, with the end of the Eemian (Robertsson & Rodhe, 1988) is well founded, even if features typical for Eemian pollen diagrams further south cannot be traced in diagrams from as far north as Swedish Lapland.

In contrast to Norway and Sweden a number of sites have been described from Finland with either redeposited or replaced Eemian sediments or with preserved till-covered Eemian sediments, but which mostly represent only a part of the warm stage. It has been assumed that large parts of southern and central Finland were submerged during the time of the Eemian climatic optimum. Compared with the Flandrian history of uplift, it is likely that even larger areas than those indicated on the map in Fig. 6.1 were submerged during the Eemian. It is in any case clear that the Eemian sediments in the coastal areas of Finland are principally marine; the freshwater sediments could not be formed until the end of the Eemian when these parts had ermeged from the Baltic Sea as a result of the isostatic uplift after the Saalian. The first site with evidence of the Eemian submergence was the find of lumps of Eemian clay in the glaciofluvial material in an esker in Rouhiala, now in Russia (Brander, 1937, 1943). The site, about 13 km southeast of Imatra in Finland, at an altitude of 60–65 m, is close to the Eemian sites on the Karelian Isthmus described earlier. The diatoms in the lumps showed them to be marine clay. The pollen assemblage in one lump had high percentages of *Alnus* and *Corylus*, in addition to *Betula* and *Pinus*, and with some *Ulmus*, *Quercus*, and *Carpinus*. This led to the conclusion that this redeposited lump at least partly represents the Eemian zone f in the division of Jessen and Milthers (1928). Another lump had a different pollen composition with high percentages of *Picea* and therefore it represents another, later part of the Eemian (Brander, 1943). Brander assumed, by taking into account the transport of material generally, that the lumps of marine Eemian clay had been carried by ice and meltwaters southeastwards from the southern part of the Lake Saimaa area about 20 km away. As the level of Lake Saimaa is at present 76 m a.s.l., this hypothesis implies that the altitude to which the sea reached during the Eemian climatic optimum was 40–50 m above that during the corresponding time of the Flandrian, that is, the level of the Litorina Sea. This difference is in agreement with the results from the earlier mentioned site at Bollnäs in Sweden and also with the results from Somero in southwestern Finland (Donner & Gardemeister, 1971). At Somero the early Flandrian clay and silt in a 33.3 m deep borehole (borehole B) had a pollen assemblage with high percentages of *Alnus* and *Corylus*, and with some *Carpinus*, which led to the conclusion that the clay and silt was redeposited Eemian material, including sediments from zones f and g of the Eemian climatic optimum, as defined in Denmark. In addition, the clay and silt contained marine diatoms typical for the Eemian. The water was, according to the diatoms, more saline than during the early Flandrian deglaciation and the connection of the Baltic Sea with the White Sea was supported by the presence of the arctic species *Grammotophora arcuata* (Tynni in Donner & Gardemeister, 1971). Radiocarbon measurements of the organic material in the Somero clay and silt gave an age which confirmed that it was mostly redeposited old material (Donner & Jungner, 1973). It was assumed by Donner and Gardemeister that the redeposited clay and silt with its

microfossils had first been part of the till carried by the ice and later, during the deglaciation nearly 10 ka ago, had been washed out from the till by meltwaters and deposited outside the retreating ice margin. It was further assumed that it came from an area not far from Somero, from the direction of the main movement of the ice, from the northwest. The level reached by the sea at Somero during the Flandrian climatic optimum at the time of the Litorina Sea was about 55 m a.s.l., whereas the clay and silt in Somero from the time of the Eemian climatic optimum is from an altitude of about 90–100 m. Thus the difference in altitude of 35–45 m is similar to that estimated on the basis of the Rouhiala find.

In addition to the evidence of the Eemian submergence obtained from Rouhiala and Somero there are several sites in Ostrobothnia (Pohjanmaa), on the west coast of Finland, with remnants of marine, mostly minerogenic, sediments as well as with organic freshwater sediments, either in situ or disturbed by the Weichselian ice sheet that later overrode them and deposited a till on top of them. The coast of Ostrobothnia belongs, together with Finnish Lapland, to the area in which till-covered eskers have been preserved (Niemelä, 1979). Most of the organic sediments found so far have been exposed in sections in these eskers. Some of the sediments have been correlated with the Eemian (Niemelä & Tynni, 1979) and others considered as representing an Early Weichselian interstadial (Donner, 1983). This is because relatively warm Early Weichselian interstadials, among them Brørup, are known in areas further south, whereas no similar interstadials immediately preceded the Eemian. The underlying eskers were interpreted as having been formed during the Saalian deglaciation. Further, there is only one till on top of the organic sediments in Ostrobothnia (Hirvas & Nenonen, 1987). The separation of the Eemian sediments from the interstadial sediments has been based on pollen analysis; nowhere have they been found in the same section. Additional evidence has, however, been obtained from the Eemian marine deposits. As has been observed, the Eemian coastline during the climatic optimum was above that of the Flandrian climatic optimum. Thus, if organic limnic sediments or peats that represent the beginning of a nonglacial interval occur at or below altitudes at which Eemian marine sediments are found, they cannot also be Eemian. Because of this, and because the vegetational history reflected in the pollen diagrams indicates that they were deposited during an interstadial, they have been dated as Early Weichselian (Donner, 1983, 1988), and will be discussed later. This dating is based on the assumption that the level of the Baltic Sea, as well as the ocean level, was comparatively low during the Early Weichselian. The possibility of an independent freshwater body during some time of the Eemian, as during the Flandrian, has, however, also to be taken into account, as shown by the evidence of the diatoms.

By using the patchy evidence from Ostrobothnia, the vegetational history during part of the Eemian was reconstructed on the basis of pollen diagrams from organic sediments and put together into a schematic pollen diagram (Fig. 6.4). At Evijärvi a 90 cm thick, muddy silt and an about 50 cm thick, shallow, water mud, separated from each other and probably slightly dislocated, were found in a well at over 6 m depth underneath till. Their pollen assemblages include *Ulmus, Corylus, Quercus, Carpinus,* and over 10 percent *Corylus,* which show that the sediments were formed at the end of the Eemian (B. Eriksson, Grönlund, & Kujansuu, 1980). As it is a well documented site in spite of its disturbed sediments it was used as a type locality for Ostrobothnia and named the Evijärvi warm stage, and correlated with the Eemian (Donner, Korpela, & Tynni, 1986). The diatoms in the muddy silt were nearly all marine forms including, for instance, *Grammatophora oceanica,* which was common in the Rouhiala sediments, and also arctic diatoms interpreted as confirming the Eemian connection between the Baltic Sea and the White Sea (B. Eriksson et al., 1980). The diatoms in the mud have also freshwater diatoms in addition to saltwater forms, which shows that the sediment was formed in a shallow bay in which the sea water was mixed with fresh water. The sediments as well as the diatoms thus show that at this site, about 60 m a.s.l., the water depth decreased towards the end of the Eemian. At Evijärvi the late Eemian relative regression of the sea, estimated to correspond to the zone boundary f/g (Donner, 1988), is thus recorded in the area close to the center of the isostatic uplift, assuming the regression to have been similar to that after the Weichselian glaciation, whereas the marine sediments at Rouhiala and Bollnäs were from the Eemian climatic optimum from a time of greater submergence, as seen from Fig. 6.2. The pattern of Eemian land/sea-level changes is, however, similar to that in more marginal areas with less uplift.

The type locality at Evijärvi was a section exposed in a well; the stratigraphy could therefore not be documented in detail. The stratigraphy of the other localities with till-covered sediments on top of eskers is therefore better known. At some sites the till-covered sediments consist of fine-grained minerogenous material in which the microfossil content is not as representative of the conditions surrounding the sites as is the case when dealing with organic sediments. But by making comparisons of the pollen diagrams from the minerogenous sediments with those from organic sediments, the former can often be placed in their right stratigraphical position. But if the pollen diagrams show no vegetational succession there is the possibility that the minerogenous sediments, including their microfossils, were reworked in a way similar as to the silt and clay at Somero. As examples of sites with till-covered minerogenous sediments interpreted as Eemian, two studied by Niemelä and Tynni (1979) are included here. At Hietakangas in Alajärvi (Fig. 6.1) a disturbed silt bed about 50 cm thick, lying underneath till, at a depth of about 3.5 m, had a pollen assemblage with predominantly *Betula* and *Alnus,* but also with high percentages of *Corylus* and with *Carpinus,* apart from low percentages of *Pinus* and *Picea.* It is thus an assemblage that can be considered as representing the Eemian. The diatoms in the silt showed it to be clearly marine, similar to the lower sediment at Evijärvi. Even if the silt at Alajärvi was disturbed it shows that this site, at 106 m a.s.l., was still submerged at the time of the Eemian climatic optimum.

There is a similar site further north along the coast of Ostrobothnia, at Rova (Fig. 6.1). There a silt bed about 1.5 m thick was in places preserved on top of a till-covered esker; the thin till, however, did not cover the silt. The pollen assemblage is at this site somewhat problematic because in addition to the taxa used as evidence for an Eemian age, such as *Corylus* and *Carpinus,* the silt has older redeposited pollen, such as *Tsuga, Podocarpus,* and *Abies,* partly representing a Tertiary flora (Niemelä & Tynni, 1979). The diatoms show that the silt is marine. It is at an altitude of about 110 m a.s.l. and is, when compared with the site at Alajärvi, possibly Eemian, but the biostratigraphical evidence from this site alone cannot be considered conclusive. Nor is its original stratigraphical position in relation to the till covering the esker quite clear.

Whereas the sediments interpreted as Eemian are minerogenous at the two sites, Alajärvi and Rova, the sediments at Norinkylä, also correlated with the Eemian, include an up to 0.5 m thick freshwater mud, first described by Niemelä and Tynni (1979) and later reinvestigated (Donner, 1988). The section at Norinkylä is at the side of an esker and shows that the till-covered sediments have been displaced and folded while being sandwiched into the basal part of the till deposited on top of the esker (Fig. 6.5). On top of the glaciofluvial material of the underlying esker there is about 0.5 m of till with sandy lenses, separated from the about 2 m thick till on top by the less than 1 m thick disturbed sediments representing the Eemian (Fig. 6.6). In the basal part of these sediments of organic material in sandy silt, a disturbed 12 cm thick layer of clay with marine diatoms was reported by Niemelä and Tynni (1979), which was not found later when the section was reinvestigated. But it showed that the basin from which the sediments were displaced included marine sediments which, as will be seen, were later found in the cores from the depression nearby. On top of the sandy silt there is a folded mud, in places 0.5 cm thick, covered by sandy silt with organic layers and an up to 30 cm thick layer of sand, partly merged into the mud. The boundary to the overlying till is sharp, similar to the unconformity at the base of the till on the top of the esker material. A schematic pollen diagram for Norinkylä was produced by reconstructing the presumably original sequence of the layers, taking into account the folds of the sediments (Donner, 1988). It was concluded that the sediments represent the later part of the Eemian, with high percentages of *Alnus* in addition to *Betula* and with a strong representation of *Corylus* in the beginning and an increase of *Picea* toward the top. *Carpinus* was also represented, in addition to *Ulmus* and *Quercus.* The *Pinus* percentages are relatively low in the diagram. When compared with Danish diagrams the Norinkylä sediments were correlated with zones f, g, and h in the zonation of Jessen and Milthers (1928). As there are remnants of marine sediments at Norinkylä the site must have emerged from the Baltic before the end of zone f. The marine sediments are now at 107.5 m but were presumably originally at a lower altitude before being displaced. The time estimated for the emergence is slightly earlier than that estimated for Evijärvi at a lower altitude (Donner, 1988).

Figure 6.5. Displaced and folded organic Eemian sediments in sands underneath Weichselian till at Norinkylä in Ostrobothnia, Finland. Profile of section in Fig. 6.6. (Photo J. Donner)

Undisturbed Eemian sediments were later found northwest of the Norinkylä sections in boreholes through the bog now bordering the esker. According to the diatoms the basal clay on top of the till was deposited in freshwater, whereas the overlying silty mud is marine (Grönlund, 1991). Thus there is a succession similar to that at Prangli in Estonia, described earlier. At another site, Viitala, Peräseinäjoki, at a lower altitude (89 m) than Norinkylä, there is similarly a till-covered 40 cm thick Eemian clay in which the lowermost part has freshwater diatoms and the overlying clay is marine (Grönlund, 1991). The pollen diagram shows that the freshwater period corresponds to the initial *Betula* zone, as in the Prangli sequence.

At Harrinkangas, south of Norinkylä, there is a kettle hole in an esker with about 4 m of till-covered sediments mainly clayey silt but with some organic bands, on the top of the esker gravel (Gibbard et al., 1989). The sediments were formed in a freshwater basin, as shown by the diatom assemblage. The pollen diagram is dominated by *Pinus* and *Betula,* with some *Picea,* and shows a deterioration of climate, a change from forest toward a treeless vegetation with a marked increase of nonarboreal pollen. The climatic deterioration is further reflected by the finds of periglacial cryoturbation and thermal contraction cracks in the gravel and sand covering the silt but underlying the till. The till-covered sediments in the kettle hole were interpreted as representing the end of a warm stage, possibly the Eemian, the esker being Saalian, in agreement with results from elsewhere is Ostrobothnia. The site, which is about 140 m a.s.l., is thus likely to represent a part of the Eemian, as well as the earliest Weichselian, with freshwater sediments formed after the regression of the sea, in an area that presumably was submerged during the Eemian (Fig. 6.1). It is, however, at a higher altitude than any of the previously discussed sites in Ostrobothnia. But the presence of reworked marine diatoms in the till, 97 percent of the total flora (Gibbard et al., 1989), is an indication that marine Eemian sediments were present at this altitude but were later eroded.

At Risåsen, in the same area as Norinkylä and Harrinkangas but closer to the present coast, there is a till-covered esker reaching an altitude of about 60 m a.s.l. In the sand underlying the till there are lenses with charcoal and pieces of wood, and in one section a preserved soil horizon (Niemelä & Tynni, 1979; Donner, 1988). Because of its stratigraphical position, similar to the organic deposits at the two sites already described, it is likely that it was formed at the end of the Eemian, after the regression of the sea below the altitude of the esker.

In addition to the exposures in eskers a section about 300 m long, in places about 15 m deep, was exposed in the limestone quarry at Ryytimaa in Vimpeli (Fig. 6.1). A till-covered layer of compressed and partly disturbed drift peat 20–100 cm thick was first investigated (Aalto et al., 1983). Later a new section, in which a 10–20 cm thick layer of compressed peat with pieces of wood and cones of pine and spruce in a 40 cm thick layer of sand, was exposed underneath a till bed of about 10 m and overlying about 1 m of gravel and a lower till on top of the bedrock (Aalto et al., 1989). As the two sections, named Vimpeli I and Vimpeli II, have only remnants of organic sediments of nonglacial intervals their correlation with other sites in Ostrobothnia has been problematic. The surface of the exposure at Vimpeli is at about 125 m a.s.l. and the peat in the Vimpeli II section at about 115 m. The short pollen diagram is dominated by *Pinus,* but with about 20 percent of *Betula,* about 10 percent of *Picea,* and a few percentages of *Alnus* and *Corylus,* the latter possibly represent-

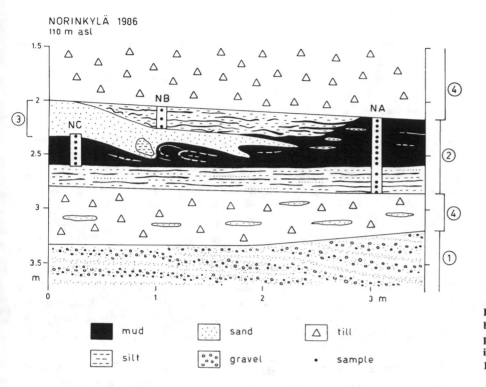

NORINKYLÄ 1986
110 m asl

Figure 6.6. Section at Norinkylä in Ostro-bothnia, Finland, showing displacement and partly folded Eemian sediments sandwiched into the basal part of the overlying till (Donner, 1988, Fig. 2).

ing *Myrica* – of which there were drupelets among the macro-fossils. The pollen assemblage is similar to that at Harrinkangas and was correlated with the end of the Evijärvi warm stage in Ostrobothnia, and thus with the Eemian. The macrofossils in the peat included some southern species growing in areas with a higher summer temperature than that at Vimpeli today, such as the water plants *Oenanthe aquatica* and *Najas flexilis,* and the forest shrub *Sambucus nigra.* The till-covered peat in the Vimpeli II section, if Eemian, would, as at Harrinkangas, have been formed after the regression of the sea. Among the numerous fragments of wood in the section there were several sticks cut by beaver, with clear tooth marks. They were interpreted as having come from a beaver dam or lodge in the close vicinity of the section studied. The organic sediments, including the pieces of wood and the cones, were all to some extent disturbed and flattened by the ice moving over the site and by the weight of the deposited till. The finer minerogenic sediments were similarly disturbed, resulting in small faults.

The Vimpeli I section differs from the one described above. The compressed drift peat shows the same sequence of tree pollen assemblages in three profiles, even in the one represent-ing only 20 cm. A *Betula* maximum is replaced by a *Pinus* maximum, with only sporadic occurrences of other tree pollen. The peat was clearly formed at the beginning of an interval, which is also shown by the decrease in the relative amount of nontree pollen toward the top of the diagrams, as well as of shrubs, among the *Juniperus* (Aalto et al., 1983). When, as seen earlier, remnants of Eemian marine sediments have been recorded at altitudes close to that at Vimpeli, the peat in the

Vimpeli I section can hardly represent the beginning of the Eemian, unless there was a low position of water level in Ostro-bothnia before the Eemian submergence. The Vimpeli I and Vimpeli II sections are nearly 200 m apart and their till-covered sediments could not stratigraphically be connected with each other. It was therefore not possible to examine the strati-graphical relationship between the organic layers of the two sections. The age of the till-covered Vimpeli I peat could not on the basis of the Vimpeli site alone be placed in any particular warm stage or interstadial (Aalto et al., 1989), but by comparing it with other sites depicting a similar vegetational history its likely stratigraphical position can be assessed, as will be seen later in the account of the Weichselian interstadials.

At another site in Ostrobothnia, at Ollala in Haapavesi northeast of Evijärvi and Vimpeli (Fig. 6.1), sand, marine silt, and mud merging into freshwater mud, in places covered by undefined organic material, were found between tills, the upper till nearly 8 m thick, in an excavation and in six boreholes (Forsström, Eronen, & Grönlund, 1987; Forsström et al., 1988). The till-covered sediments have been displaced, eroded, de-formed, and mixed. It is therefore understandably difficult to trace the natural forest history in the two pollen diagrams from the site, the first diagram being a combined diagram of samples from the excavation and from a borehole, and the second diagram being from another borehole. The diagrams, which cover about 2 m of sediment each, including the silt, differ from each other but have some features that can be used in placing them in a certain part of the Eemian. The topmost parts show higher percentages of *Corylus* and *Picea,* and an increase of nonarboreal pollen, whereas the silt has mainly *Betula, Pinus,* and *Alnus,* with some *Quercus, Ulmus, Corylus,* and *Picea.* The presence of macrofossils of *Najas tenuissima, N. flexilis, Carex pseudocyperus,* and *Lycopus europaeus,* with their present northern limits of distribution south of Ollala, confirm the correlation of the sediments with the Eemian. The diatoms, with a brackish-marine littoral flora in the silt changing into a freshwater flora in the mud can be used to determine the horizon in the sediments corresponding to the emergence of the site from the Eemian Sea. The top of the mud is at about 116–117 m a.s.l. The horizon of the emergence in the two pollen diagrams most likely corresponds to the beginning of the Eemian (Forsström et al., 1988), probably to the time corresponding to the top of zone e (Donner, 1988). This would be somewhat prior to the emergence of the sites at Norinkylä and Evijärvi at lower altitudes (Fig. 6.2). The reconstruction of the emergence of the Ostrobothnian coast during the Eemian is, however, only tentative and only gives an idea of the general pattern of land/ sea-level changes that were essentially similar to those after the Weichselian glaciation.

At Vesiperä, at another site in Haapavesi not far from Ollala, there is silt, gravel, sand, and a humus layer about 15 cm thick between two tills, the upper till being about 3 cm thick (Hirvas & Nenonen, 1987). The pollen diagram of the silt and the upper part of the sand, and the humus layer, including a hiatus, cannot be correlated with any particular Eemian zone of diagrams further south (Donner, 1988), but comparatively high percentages of *Alnus* and *Corylus,* and the presence of *Ulmus* and *Quercus* make a correlation with the Eemian likely, as for many of the previously mentioned sites. The site does not, however, give any additional evidence about land/sea-level changes but it does show, as mentioned earlier, that there is also here a till that may be Saalian.

Even before all these till-covered sediments in Ostrobothnia were known and investigated, Heinonen (1957) had studied the pollen content of tills particularly in Ostrobothnia, but also in tills from other parts of Finland as well as from sites in Sweden. By assuming that the microfossil content, including pollen, of a particular till originated mainly from sediments of a nonglacial interval preceding the glaciation represented by the till, Heinonen could draw conclusions about the nature of this interval. In Ostrobothnia there was a clear difference in the pollen frequency between a till described as a loose surface layer and the undisturbed basal till and also between their pollen spectra. Similar differences were found in other parts of Finland, differences that could be confirmed at sites with two tills on top of each other. Thus, in many parts of Finland there is an upper part of the till with a *Pinus*-rich pollen flora and a lower part with a *Betula*-rich pollen flora, often with relatively high percentages of pollen of deciduous trees, including *Corylus* and *Carpinus.* The tills also contained diatoms, including marine forms. By making comparisons of the pollen composition of the tills with the assemblages at known sites outside Finland, Heinonen concluded that the *Betula*-rich pollen flora came from Eemian

deposits and that the *Pinus*-rich pollen flora most likely represented a subsequent Weichselian interstadial, a conclusion supported by the results from the studied sample from sites in Sweden.

In this study of tills Heinonen was able to indicate the nature of the Eemian vegetation in Finland before any organic sediments representing this warm stage had been found, as well as the nature of Weichselian interstadial. Further south in Estonia, where complete Holsteinian and Eemian sedimentary sequences are preserved, their pollen assemblages have been compared with those of tills (Liivrand, 1991). Even if the tills also contain microfossils rebedded from older deposits, even from pre-Quaternary sediments, it seems possible to place the studied tills in particular cold stages on the basis of their pollen assemblages, a stratigraphical method earlier used, for instance, in the study of the already mentioned sediments in the Alnarp Valley in southern Sweden (U. Miller, 1977).

The sites with sediments considered to represent a warm stage equivalent to the Eemian are in northern Finland, including the regions described as Peräpohjola and Finnish Lapland in the north, in many respects different from those in Ostrobothnia. Most areas in northern Finland were not submerged during the Eemian; marine sediments are therefore lacking. In addition, the vegetational history, as during the Holocene, differs from that in areas further south, as already noted in the discussion of the Eemian site at Leveäniemi in Swedish Lapland. The influence of the mixed oak forest did not reach these northern areas during the Eemian climatic optimum. On the other hand, the warm-stage sediments are bounded by two till beds, above by till bed III and below by till bed IV, the younger interstadial sediments lying above till bed III. Thus the sediments described as Eemian are separated from younger interstadial sediments by a till, in contrast to areas further south where this till bed is lacking (Hirvas & Nenonen, 1987; Hirvas, 1991).

By the beginning of the 1980s about 80 sites with till-covered organic sediments had been found in Finnish Lapland, of which 39 out of 49 pollen-analysis sites can be described as representing a warm stage (Hirvas, 1983, 1991). Of these sites only a few have organic sediments thick enough to show some vegetational succession in their pollen diagrams. The organic sediments consist either of compressed peat, often sandy, or of mud. The pollen assemblages are dominated by *Betula* and *Pinus,* but often also have *Alnus* and *Picea.* In addition, low percentages of *Corylus* and occasional occurrences of *Quercus* show the influence of long-distance transport of pollen from areas further south. The warm stage these assemblages represent was named the Lapponian and correlated with the Eemian (Hirvas, 1983). In contrast to the previous Leveäniemi diagram from Swedish Lapland the diagrams from Finnish Lapland also often have pollen of *Larix, Abies,* and *Tsuga.* The origin of these occasional occurrences in the partly minerogenous sediments is not quite clear. They may have been reworked from older sediments. The pollen diagrams referred to as representing the Lapponian show some differences in the proportion of the main tree pollen and six types of pollen assemblages were therefore separated in Finnish Lapland (Hirvas, 1983, 1991). But in addition four pollen zones were defined by combining the results from several sites, as seen in the schematic diagram in Fig. 6.4. Zone a is characterized by a *Betula* maximum and low percentages of *Pinus*. Zone b starts with the spread of *Alnus* and zone c with the rise of the *Picea* curve. The beginning of zone d is marked by an increase of *Betula* and a decrease of *Pinus*. There is still some *Alnus* but *Picea* is practically absent. Zone d reflects the deterioration of climate after the climate optimum, a development also reflected by an increase in the relative amount of nonarboreal pollen. The broad outlines in the forest history in these diagrams are similar to those in the Eemian diagram from Leveäniemi, which covers a longer period of the warm stage than any of the diagrams from Finnish Lapland.

As examples of sites representing the Lapponian warm stage three sites may be mentioned. The pollen diagram from Tepsankumpu in Kittulä (Fig. 6.4) of the over 2 m thick till-covered mud represents zones a, b, and the beginning of zone c, with a dominance of *Betula* throughout the diagram (Hirvas, 1983, 1991; Hirvas & Nenonen, 1987). The diagram from Paloseljänoja in Sodankylä, also with about 2 m thick till-covered mud, shows zones c and d, with c dominated by *Pinus*, as shown in Fig. 6.4, where the diagram is shown above that from Tepsankumpu. As the latter is a comparatively complete diagram it was suggested as a stratotype section for the Tepsankumpu warm stage in Finnish Lapland (Donner, Korpela, & Tynni, 1986), also later adopted by Hirvas (1991), instead of using the general term Lapponian, without a type site. They have both, however, been correlated with the Eemian. Of sites studied earlier, Kurujoki in Sodankylä has thin till-covered organic sediments in sands and silts from the end of the warm stage (Hirvas et al., 1977), with pollen assemblages similar to those at Paloseljänoja and thus representing zones c and d in the division used later by Hirvas (1983). Other sites referred to as representing the Tepsankumpu warm stage, with pollen diagrams similar to those described above (Table 6.1), are Härkätunturi, Loukoslampi, and Sivakkapalo (Hirvas, 1991).

Many of the till-covered sediments in Finnish Lapland were stratigraphically placed in the warm stage correlated with the Eemian on the basis of the till stratigraphy. Using the same criteria, till-covered sediments in Finnmarksvidda in North Norway, northwest of the sites in Finnish Lapland, were correlated. Thus sands covered by 3 tills in Volgamasjåkka were preliminarily separated as representing a warm stage, the Vuolgamasjåkka, possibly corresponding to the Eemian (Olsen, 1988; Garcia Ambrosiani, 1990). One pollen spectrum with 21 percent of *Betula* and 79 percent of nonarboreal pollen (Olsen, 1988) neither confirms nor contradicts this conclusion.

All the sites in Finnish Lapland and the site in North Norway were above the level reached by the Eemian Sea, as far as can be judged from the sediments studied. The evidence from Eemian sites further south summarized in Fig. 6.2, however, allows some conclusions to be drawn of the general pattern of land/sea-level

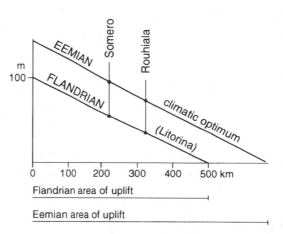

Figure 6.7. Schematic relationship between the Eemian and Flandrian areas of uplift and the altitudes of the shorelines of the Eemian climatic optimum and that of the Flandrian in Finland and areas bordering it in the southeast.

changes, particularly in the Baltic Sea area. The marine Eemian sediments at Bollnäs, Rouhiala, Somero, and Evijärvi, as well as at Fjøsanger on the Atlantic coast, all show that the sea level was higher during the Eemian climatic optimum than during the Flandrian climatic optimum, corresponding to the Litorina time in the Baltic and Tapes in Norway. This relationship was clearly demonstrated in Ostrobothnia (Donner, 1983; Forsström et al., 1988). In addition, it seems that the difference between the level reached by the sea during the Flandrian climatic optimum and that during the Eemian was particularly great in southwestern Finland. This led to the conclusion that the center of the isostatic uplift was situated in this area, assuming that the center of the Saalian ice sheet was displaced to this area during the extensive Saalian glaciation (Forsström & Eronen, 1985). As the pattern of uplift after the Saalian glaciation was similar in Ostrobothnia to that after the Weichselian glaciation, it is likely that this area also had the greatest uplift during the Eemian. The relatively high level reached by the Eemian Sea during its climatic optimum – at Somero an estimated 35–45 m above the corresponding Flandrian level and at Rouhiala 40–50 m – would then be the result of a different interplay between the eustatic sea-level changes and the isostatic uplift than was the case after the Weichselian. The Eemian land/sea-level changes in these areas can perhaps at present best be explained by taking into account that a much larger area of the earth's crust was depressed during the Saalian glaciation than during the Weichselian, resulting in a larger area of uplift after deglaciation. The uplift during the Eemian as compared with that since the Flandrian climatic optimum is schematically shown in Fig. 6.7. The results from Somero and Rouhiala indicate that the 100 m isobase for the Eemian climatic optimum was just northwest of Somero, whereas the Litorina isobase for the Flandrian was close to the Ostrobothnian coast, at least 200 km further to the northwest. In addition, the Eemian transgression of sea level reached higher in relation to present sea level than the Flandrian transgression. The Eemian transgression was, as seen, demonstrated at Fjøsanger in Norway in the west and at Petrozavodsk at the shore of Lake Onega in the east. In the Baltic basin, in Estonia and in Ostrobothnia in Finland, there is, as mentioned earlier, evidence of an initial freshwater period before the Eemian salt water penetrated into this area. There was thus in the Baltic basin an early Eemian freshwater lake before the marine stage resulting from the Eemian transgression.

If the vegetational history, particularly the forest history of the whole area treated in the present account, is considered, some general features typical for the Eemian can be detected. The Eemian had a strong oceanic influence with a rather uniform vegetation in a large area (Zagwijn, 1990). Further, the climatic optimum had a warmer climate than that of the Flandrian climatic optimum, seen in the northward spread of the deciduous trees and also in the distribution of certain species. Thus, for instance, macrofossils at Leveäniemi in Swedish Lapland showed the Eemian to have been relatively warm even at the northern latitudes. Compared with the present vegetational zones (see Fig. 3.3) the Eemian forest vegetation within the zones was different. In Swedish and Finnish Lapland the forest was either dominated by *Betula* or *Pinus,* but with *Alnus* and *Picea* also present. The influence of the mixed oak forest was stronger in areas such as Ostrobothnia in Finland during the Eemian than during the Flandrian, but the composition of this forest was not that of the Flandrian mixed oak forest. *Tilia* occurs in Eemian diagrams from Estonia and Petrozavodsk in the east but is generally absent in diagrams from Finland, as in diagrams from Denmark. *Quercus* and particularly *Corylus,* on the other hand, are strongly represented in Eemian diagrams from Ostrobothnia, more so than in Flandrian diagrams. *Carpinus* spread during the Eemian into the southern parts of Sweden, presumably also further north, and occurs regularly in diagrams from Finland as far north as Ostrobothnia. *Carpinus* pollen are seldom found in Flandrian diagrams from these areas. These examples show that analogies with the present forest zones and present northern limits of individual trees are of limited use when dealing with the Eemian vegetation as well as with older warm stages. In a broadly outlined reconstruction of the Eemian vegetation, Central Europe, Denmark, southern Sweden, and southernmost Finland, the last mostly submerged, had, during the *Carpinus* zone time, a *Carpinus–Quercus* forest, and the coastal areas of southern Norway, central Sweden, and the coastal areas of northern Sweden, as well as most of Finland south of Lapland, including Ostrobothnia, a *Picea–Quercus* forest with *Betula* (Zagwijn, 1990). The schematic pollen diagrams of the Eemian (Fig. 6.4) show the pollen assemblages representing these forests.

7 Division of the Weichselian stage and the application of radiocarbon dating

Whereas the Eemian stage, like the older warm stages, was characterized by a uniform climatic cycle with a climatic optimum, the relatively cold climate of the much longer Weichselian stage was interrupted by several interstadials of comparatively short duration, but long enough for organic sediments to have had time to form. The fluctuations in temperature, which can be identified as interstadials and stadials, were not unique features for the Weichselian. Similar conditions presumably prevailed during earlier cold stages, but the stratigraphical evidence from those is incomplete and the details therefore less known than those from the Weichselian.

In the division of the Weichselian into substages the boundary between the Early Weichselian and the Middle Weichselian has been placed above the initial comparatively warm interstadials, which were followed by a pronounced cooling of the climate (Mangerud et al., 1974), in The Netherlands corresponding to the boundary between Early Glacial and the Pleniglacial (van der Hammen, Wijmstra, & Zagwijn, 1971). These interstadials were first defined in Denmark, Schleswig-Holstein in Germany, and in The Netherlands. The most pronounced interstadial is Brørup in Denmark, defined on the basis of the stratigraphy and pollen diagram of the Brørup Hotel Bog in Jutland (S. T. Andersen, 1961, 1965). Here the Eemian peat is overlain by interstadial and stadial peats, muds, clay-muds, and clays, all covered by sand several meters thick. During the interval defined as the Brørup Interstadial, forests again invaded the area after a period of treeless vegetation. The pollen diagram has first a maximum of *Betula* which is then replaced by a *Pinus* maximum. This part of the diagram also has *Picea,* partly representing *Picea omoricoides,* a species considered to be extinct but related to the present southern European series *P. omorica* now found in the mountains of Serbia, and *Larix* (S. T. Andersen, 1961). The deterioration of climate after the Brørup Interstadial is shown by an increase of nonarboreal pollen, a change to an open vegetation resulting in the sedimentation of sand on top of the organic sediments. The pollen diagram from Brørup Hotel Bog, which starts in the upper part of the Eemian, shows between the Eemian and Brørup a thin horizon that was taken to represent a short interstadial called the Rodebaek in

Denmark and correlated with a similar short interstadial, Amersfoort, in The Netherlands (S. T. Andersen, de Vries, & Zagwijn, 1960; S. T. Andersen, 1961; Zagwijn, 1961). In the Brørup Hotel Bog diagram there is only an increase of pollen of water plants, the pollen assemblage otherwise depicting a treeless vegetation. The Rodebaek Interstadial was named after the site at Rodebaek in Jutland, where a thin layer of clay-mud is separated from the Eemian wood peat by sand (S. T. Andersen, 1961). The pollen composition is also here dominated by nonarboreal pollen, with some *Betula* and with some rebedded tree pollen, which are generally present in Early Weichselian sediments and are related to the amount of mineral matter in these sediments.

With increased knowledge of the Early Weichselian the earlier conclusions have later been modified. During the Amersfoort Interstadial there was forest in The Netherlands and therefore the Dutch Amerfoort–Brørup complex has been correlated with Brørup in Denmark, a conclusion supported by sites in northern Germany (Behre, 1989). The first well-documented Early Weichselian interstadial, during which a forest vegetation spread into Denmark, is thus Brørup (Table 7.1), which leaves out the treeless Rodebaek Interstadial from the generally recognized scheme. There is, however, some evidence of a climatic deterioration within Brørup that can be seen as an expression of the difference between Amersfoort and Brørup as originally defined in The Netherlands (Behre, 1989). A similar difference can be seen in pollen diagrams from other areas in Central Europe, where a short interstadial referred to as Amersfoort preceded the warmer and longer interstadial named Brørup (Welten, 1981). Thus, together, these two are named the Brørup Interstadial.

The second distinct Early Weichselian interstadial was demonstrated on the basis of the stratigraphy and pollen diagram from Odderade in Schleswig-Holstein (Averdieck, 1967). On top of the Eemian muds and peats, recovered in a boring that penetrated more than 15 m of sediments, the peats representing the Brørup Interstadial were covered by minerogenous sediments and peats of the second interstadial, named the Odderade. A more complete Eemian pollen diagram was later presented for

Table 7.1. *Subdivision of the Weichselian*

Substages	Interstadials		Ages for boundaries
			10 ka BP (radiocarbon years)
		11.0 ka	
	Allerød		
		11.8 ka	
Late		12.0 ka	
	Bølling		
Weichselian		13.0 ka	
			25 ka BP (radiocarbon years)
	Denekamp	c. 28–32 ka	
Middle	Hengelo	c. 36–39 ka	
Weichselian	Moershoofd	c. 46–50 ka	
	Glinde		
	Oerel		
			74 ka (= stage boundary 4/5 in oxygen isotope curve, Martinson et al., 1987)
Early	Odderade		
Weichselian	Brørup, incl. Amersfoort		
			115 ka (peak of stage 5e in oxygen isotope curve = c. 125 ka, Martinson et al., 1987)
	Eemian		

For Late Weichselian division, see Table 11.2.

this site (Averdieck et al., 1976), as earlier mentioned (see Table 6.1). The pollen diagram of the Odderade Interstadial shows a forest history similar to that of Brørup, with an initial *Betula* maximum followed by an increase of *Pinus* and with *Picea* in the upper part, but with less *Alnus* than during Brørup. There has been some erosion of the upper parts of both interstadial sediments, but the older of the two, Brørup, with *Picea omoricoides* as in Brørup in Denmark, had presumably a somewhat warmer climate than the younger Odderade, and was also longer than the latter (Welten, 1981). *Larix*, which was recorded from the Brørup Interstadial in Denmark, was also present in the Brørup Interstadial at Odderade.

The two Early Weichselian interstadials Brørup and Odderade followed the Eemian warm stage, which ended about 115 ka ago. After the introduction of the radiocarbon method in dating organic sediments, attempts were made to extend its range as far back in time as possible and to date the Early Weichselian interstadials that had at the same time been identified, in the beginning of the 1960s. As at that time there was not a generally accepted age for the Eemian, the radiocarbon dates, interpreted as finite, resulted in a chronological frame for the Early Weichselian (van der Hammen, Wijmstra, & Zagwijn, 1971) that has later been abandoned, both because of the too-young ages recorded and because of doubts as to the possibility of dating sediments as old as those of the Early Weichselian with the radiocarbon method. The effect of contamination of younger material for samples older than 20 ka is already considerable (Mook & Waterbolk, 1985) and

the limit of the method can generally be held to be 40 ka but may be extended to 60 ka (Délibrias, 1985). Some laboratories using special proportional counters have, however, reported finite ages up to 75 ka (Stuiver, Robinson, & Yang, 1979). The use of accelerator mass spectrometry (AMS), which measures the ^{14}C concentration instead of the ^{14}C decay, and which uses very small samples, has not extended the range of radiocarbon dating. It is thus clear that this method can neither be used in dating the organic sediments of the Early Weichselian nor the upper boundary of this substage.

It was earlier noted that the Eemian can be correlated with stage 5e in the oxygen-isotope curve of the deep-sea stratigraphy. The terrestrial stratigraphy shows a pronounced climatic deterioration at the time chosen to mark the end of the Early Weichselian and similarly there is a marked change in the deep-sea record at the boundary between stages 5 and 4. This change has therefore been correlated with the boundary between the Early and Middle Weichselian (West, 1988; Behre, 1989), and its age determined to 74 ka (Martinson et al,. 1987), as seen in Table 7.1. Further, the oscillations in the oxygen-isotope curve during stage 5, after the Eemian stage 5e, have been correlated with the interstadials and stadials of the Early Weichselian. In this correlation stages 5c and 5a, marking the rises in the ^{18}O values, have been linked with the interstadials Brørup and Odderade, and stages 5d and 5b with two stadials (Behre, 1989). But it must be remembered that such a detailed correlation between the oxygen-isotope curve reflecting global changes of ice volume with the terrestrial evidence from northern Europe is not based on direct evidence, even if such a correlation seems likely, as the fluctuations during stage 5 correspond to the number of climatic fluctuations in the terrestrial sequence.

In the proposed subdivision of the Weichselian by Mangerud et al. (1974) the boundary between the Middle Weichselian and the Late Weichselian was placed at 13 ka, in radiocarbon years BP, corresponding to the age of the lower boundary of the Bølling Interstadial, which is the older of the two interstadials in the period called the Late Glacial in the vegetational history. Using the available radiocarbon ages for the two interstadials their chronostratigraphical boundaries were defined by conventional radiocarbon years B.P. (Mangerud et al., 1974). In this division the lower and upper boundaries of the Bølling Interstadial were defined as 13 ka and 12 ka, and of the Allerød Interstadial as 11.8 ka and 11 ka (Table 7.1). The Older Dryas Stadial is between the interstadials and the Younger Dryas Stadial above Allerød. These interstadials and stadials will, however, be treated in more detail in Chapter 11, dealing with the Late Weichselian and Early Flandrian deglaciation.

In a later discussion of the subdivision of the Weichselian by Mangerud and Berglund (1978) it was proposed that the Middle–Late Weichselian boundary be placed at 25 ka, which would then include the last glacial advance with it maximum at around 18 to 22 ka B.P. As the date of 25 ka is more in agreement with divisions used elsewhere, as in central Europe (Welten, 1981) and in western Europe (Shotton, 1973; West, 1988), and as it has already been used in the Scandinavian area (Lundqvist, 1986b), it was also adopted here (Table 7.1), even if arguments for the later date of 13 ka have been put forward (S. T. Andersen, 1979).

The Middle Weichselian substage between 74 ka and 25 ka was generally cooler than the Early Weichselian and therefore the vegetation during the interstadials was open in northern and western Europe, as in The Netherlands, for instance, and with no forest. Thus, already on the basis of their pollen diagrams these interglacials were more difficult to identify than the two Early Weichselian interstadials. Further, as seen from the preceding discussion, the Middle Weichselian interstadials represent a time interval in which radiocarbon dating has the limit of its range. Even if finite dates from organic sediments fall within the boundaries of the Middle Weichselian, erroneous dates are likely, caused mainly by contamination of the samples both by younger material and reworked older material. The possibilities of successfully correlating Middle Weichselian interstadials are therefore limited and should be supported by stratigraphical evidence. An organic sediment does not, however, by definition represent an interstadial, as noted earlier; it may merely represent a sediment formed in a favorable place at any time during the Middle Weichselian (Behre, 1989). But in areas where extremely cold conditions prevailed during the stadials, the likelihood of finding organic sediments not representing interstadials is small.

Three Middle Weichselian interstadials were originally identified in The Netherlands, all representing sediments with organic material and with pollen compositions of an open vegetation. The two younger, Hengelo and Denekamp (van der Hammen et al., 1967), have been dated at about 37 ka and 30 ka B.P. respectively (Menke & Tynni, 1984), Hengelo covering the period 39 ka to 36 ka and Denekamp 32 ka to 28 ka, according to Behre (1989), as given in Table 7.1. The first and oldest interstadial, Moershoofd (Zagwijn & Paepe, 1968), which is less distinctly an interstadial (Behre, 1989), has been dated at about 46–50 ka B.P. (Menke & Tynni, 1984). Later two older interstadials were identified at Oerel in Lower Saxony in northwest Germany in a basin with organic layers intercalated by sand, also formed in an environment with an open vegetation (Behre & Lade, 1986; Behre, 1989). The sediments of both interstadials overlie sediments of the Eemian Stage and the Odderade and Brørup Interstadials. The first Middle Weichselian interstadial was named Oerel, with high percentages of nonarboreal pollen and pollen of *Betula nana,* but with no tree birches. The vegetation was a treeless shrub tundra, as also during the following interstadial found at Oerel and named Glinde. No ages have been given for these two interstadials.

In Scandinavia these Weichselian interstadials are separated either by cold stage nonglacial sediments or by glacial sediments, and in some areas by tills, representing periods of various lengths. In many areas these stadials have been given local names. These will only be mentioned in the following chapters when describing some particular areas and were therefore not included in Table 7.1. The importance of the interstadial localities of a cold stage apart from giving information of the vegetation of that period and thus of the climatic conditions, is in showing which areas were ice-free at the time of each interstadial, provided that they have been correctly identified. As the Early Weichselian and older Middle Weichselian interstadials cannot be directly dated this is often difficult and their stratigraphical position has to be assessed on the basis of comparisons with other areas. In the more central parts of the Scandinavian glaciations the till beds can be grouped and separated with the help of the interstadial sediments, provided that these sediments have been correctly identified.

8 Early Weichselian substage

8.1. Biostratigraphy of freshwater deposits

During the two Early Weichselian interstadials Brørup and Odderade, as identified in southern Jutland and northern Germany, there was a forest vegetation with *Betula* being replaced by a *Pinus* dominance and with *Alnus, Picea*, partly *Picea omoricoides*, and *Larix* also present (Fig. 8.1). The difference between the two interstadials is not great even if Brørup, as noted earlier, has been considered to have been somewhat warmer and longer than Odderade, the length of the former estimated to 5.8–10.5 ka (Behre, 1989). Some differences have also been detected in the vegetational succession. The expansion of *Picea* and *Latrix* took place in the birch period in the Brørup but in the pine period in Odderade, and the sequence was normally *Larix–Picea* in the former and *Picea–Larix* in the latter (Behre, 1989). *Larix,* however has not been identified in all diagrams of the Odderade Interstadial (the absence of *Larix* in some older pollen diagrams can also be due to the fact that it was not identified).

Although the identification and correlation of Weichselian interstadial sediments is seldom complete, in Norway, Sweden, and Finland they can often, because of their relatively warm character, be recognized as representing the Early Weichselian and not the Middle Weichselian, but the separation of Brørup from Odderade is more difficult, unless they are found at the same site. The relative sea-level was low during the Early Weichselian, as mentioned in connection with the Eemian sites, and therefore the sediments at most sites are freshwater sediments. Early Weichselian marine sediments have only been recorded from some areas, with northernmost Jutland in Denmark the best known.

The freshwater sediments placed in the Early Weichselian are listed in Table 8.1. The sequence of the vegetational history for each site is also mentioned. If the interstadial sediments overlie Eemian sediments this is also noted. When the pollen diagram starts with a *Betula* maximum it is marked with a B, and if it is followed by an assemblage zone with more trees they are listed in the order of their relative frequences, with those in brackets so poorly represented that they may not have grown in the

neighborhood of the sites. Further, assemblages that are dominated by nonarboreal pollen are marked with NAP. If local names have been used for the interstadials they are mentioned, and alternative correlations suggested for the sites are also included. All sites are shown on the map in Fig. 8.2. As seen from the table there is a site with Brørup and Odderade overlying Eemian at Rederstall in Holstein in North Germany (Menke & Tynni, 1984) in addition to the mentioned sites of Odderade and Oerel. At two previously studied sites in Schleswig-Holstein, at Geesthacht and Loopstedt, Brørup was found on top of the Eemian (Schütrumpf, 1967). All diagrams for Brørup show the succession described earlier, and the three diagrams of Odderade are also similar. The sites further north have generally been placed in the Brørup. Only in cases where the vegetation reflects a cooler climate than in other Early Weichselian diagrams from the same area, has a correlation with Odderade been considered as more likely. At Stenberget in Sweden the Eemian peats are overlain by silt and sand, and a mud formed at a time with a forest of *Pinus, Betula,* and *Picea*, described as a subarctic woodland (Berglund & Lagerlund, 1981). The mud, the Slätteröd Mud, was correlated with the Brørup. In the detailed study of the lithostratigraphy of the two drumlins at Dösebacka and Ellesbo near Gothenburg, Hillefors (1969, 1974) distinguished horizons showing that they represented nonglacial intervals. Thus a deflation surface, called the Older Dösebacka–Ellesbo Interstadial, has been held to represent the Brørup Interstadial. The Dösebacka and Ellesbo sites have yielded a number of bones, mainly of mammoth, which have been used in dating the younger Weichselian intervals. In Brumunddal in Norway there is a till-covered 0.6–1 m thick compressed peat at 395 m a.s.l. that shows the succession from an initial *Betula*-dominated composition to one with *Pinus,* with *Picea* and *Larix* in addition, and toward the top an increase of nonarboreal pollen (Helle et al., 1981). This rather complete nonglacial sequence, representing locally the Brumunddal Interstadial, was correlated with Brørup. So were the lacustrine minerogenic sediments covered by two tills, 900 m a.s.l., at Førnes near Møsvatn in Hardangervidda in South Norway, but not quite on equally good grounds. The pollen assemblage is

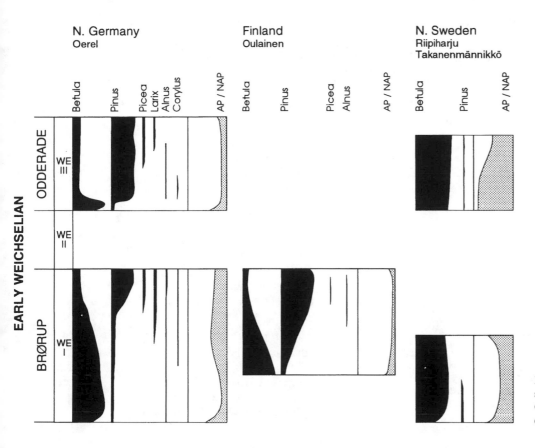

Figure 8.1. Schematic Early Weichselian pollen diagrams from north Germany, Finland, and north Sweden (sites listed in Table 8.1).

dominated by nontree pollen, with some *Betula,* which may represent *B. nana* (T. O. Vorren & Roaldset, 1977). As the site is at a relatively high altitude it shows that most of Scandinavia was deglaciated during the nonglacial interval represented by the lake sediments. The interval was named the Førnes Thermomer and, through comparisons with sites in Sweden, correlated with the Brørup Interstadial. As may be seen, the reason for this correlation is based not only on biostratigraphical evidence. In Gudbrandsdalen, also in South Norway, northwest of the Brumunddal site, there are several localities with till-covered glaciofluvial sediments with permafrost features and with mammoth remains (Bergersen & Garnes, 1971). The sediments represent a nonglacial interval named the Gudbarndsdalen Interstadial that has also been correlated with the Brørup Interstadial (U. Miller, 1986; Lundqvist, 1986a). The evidence for this correlation is, indeed, as purely circumstantial as for the site at Førnes. There are, however, several sites in northern Sweden for which the evidence for an Early Weichselian age is stronger. At Pilgrimstad a till-covered lake sediment with alternating layers of silty mud, sandy mud, silt, and sand, about 1.5 m thick, was studied by pollen analysis. It had in the beginning a *Betula* maximum with some *Prinus* and with an increasing amount of nonarboreal pollen followed, after a gap in a silt layer poor in pollen, by an assemblage with a strong representation of nonarboreal pollen, but also with *Pinus* and some *Picea* (Robertsson, 1988). A broadly similar pollen diagram, also

analyzed by Robertsson, was earlier published by Lundqvist (1967), who held it to represent the locally identified Jämtland Interstadial, which was correlated with the Brørup Interstadial as defined in Denmark. Another possible correlation is, according to Robertsson (1988; Garcia Ambrosiani, 1990), that the lower part of the Pilgrimstad diagram represents Brørup and the upper Odderade, being separated by sediments formed during cooler conditions, perhaps being partly aeolian. This possibility is also given in Table 8.1. The coastal site at Härnösand has a bed of till-covered sandy mud, pollen-analytically studied by Robertsson (Garcia Ambrosiani, 1990; Garcia Ambrosiani & Robertsson, 1992). The pollen assemblage is dominated by non-arboreal pollen, the tree pollen being mainly *Betula,* but with some *Pinus* and *Picea,* and with a few pollen of *Larix.* The site was correlated with Brørup. At Långsele, inland and northwest of Härnösand, an old pollen diagram (Sundius & Sandegren, 1948) from an over 1 m thick till-covered mud has a *Betula*-dominated tree pollen composition throughout, with *Pinus* and some *Picea.* The site was placed in the Jämtland Interstadial and then correlated with the Brørup Interstadial (Lundqvist, 1967). The site at Tåsjö is even further inland, at about 300 m a.s.l., and has an about 10 cm thick bed of peaty silt and sand covered by sand and gravel, and a till (Fig. 8.3). The pollen diagram is dominated by non-arboreal pollen, but with some *Betula* and *Pinus,* representing an arctic environment (Lundqvist, 1978). The site was also first correlated with the Jämtland Interstadial but

Table 8.1. *Early Weichselian freshwater sites, with suggested correlations and main pollen components and sequences, or other evidence*

Sites	Presence of Eemian	Brørup	Odderade
Denmark			
1. Brørup (S. T. Andersen, 1961)	Eemian	B – P B Pc L (A)	
2. Rodebaek (S. T. Andersen, 1961)		pre-Brørup, treeless	
Northern Germany			
3. Odderade (Averdieck, 1967)	Eemian	B – P B Pc L A	B – P B Pc (A)
4. Oerel (Behre & Lade, 1986)	Eemian	B – P B Pc L A	B – P B (Pc L A)
5. Geesthacht (Schütrumpf, 1967)	Eemian	B – P B Pc A	
6. Loopstedt (Schütrumpf, 1967)	Eemian	B – P B Pc A (C?)	
7. Rederstall (Menke & Tynni, 1984)	Eemian	B – P B Pc L A	B – P B Pc (L A)
Southern Sweden			
8. Stenberget, Slätteryd (Berglund & La-gerlund, 1981)	Eemian	P B Pc (A C)	
9. Dösebacka-Ellesbo (Hillesfors, 1969, 1974)		deflation surface (Older Dösebacka-Ellesbo Ist.)	
Southern Norway			
10. Brumunddal (Helle et al., 1981)		B – B P Pc L (A C) (Brumunddal Ist.)	
11. Førnes (T. O. Vorren & Roaldset, 1977)		NAP (Førnes thermomer)	
12. Gudbrandsdalen (Bergersen & Garnes, 1971)		till-covered glaciofluvial sediments (Gudbrundsdalen Ist.)	
Northern Sweden			
13. Pilgrimstad (Lundqvist, 1967)		B – NAP B (P Pc) (Jämtland Ist.)	
new diagram (Robertsson, 1988; Garcia Am-brosiani, 1990)		B (P Pc)	NAP B (P Pc)
14. Härnösand (Garcia Ambrosiani & Rob-ertsson, 1992)		NAP B P (Pc L)	
15. Långsele (Sundius & Sandegren, 1948)		B P (Pc)	
16. Tåsjö (Lundqvist, 1978) (Garcia Ambrosiani, 1990)		NAP (B P) ————————→	
17. Vålbacken (Lundqvist, 1967)		till-covered clay with microscopic remains of arctic–subarctic vegetation	
(Garcia Ambrosiani, 1990)		————————————→	
18. Boliden (Robertsson & Garcia Ambrosiani, 1988)	Eemian?	NAP (B P) ——————→ ?	
19. Gallejaure (Lundqvist, 1967)		B – NAP B (P)	
20. Seitevare (Robertsson, 1988)	Eemian	NAP B	
21. Leveäniemi (Garcia Ambrosiani, 1991)	Eemian	NAP (incl. Betula nana)	
22. Takanenmännikkö (Lagerbäck & Rob-ertsson, 1988)		NAP B (P)	
23. Riipiharju (Lagerbäck & Robertsson, 1988)			NAP (B) (Tärendö Ist.)
Northern Norway			
24. Vuoddasjavri (Olsen, 1988)	Eemian (Vuolgamasjåkka Igl.)	till-covered glaciofluvial sediments (Eiravarri Ist.)	
25. Sargejåk (Olsen, 1988)			(Sargejåk Ist.) possibly younger

Table 8.1. *(cont.)*

Sites	Presence of Eemian	Brørup	Odderade
Finland			
26. Vimpeli I (Aalto et al., 1983)		B – P B (Pc)	
27. Oulainen (Forsström, 1982; Donner, 1983; Donner et al., 1986)		B – P B (Pc A) (Oulainen Ist.)	
28. Marjamurto (Peltoniemi et al., 1989)		B – P B (A)	
29. Permantokoski (Korpela, 1969)		B (P) (Peräpohjola Ist.)	
30. Ossauskoski (Korpela, 1969)		B (P) (Peräpohjola Ist.)	
31. Maaselkä (Hirvas, 1991)		NAP B (Maaselkä Ist.)	
Estonia		(Harimäe Substage)	
32. Otepää (Liivrand, 1991)		B P Pc (reworked pollen excluded)	
33. Peedu (Liivrand, 1991)		_" _	
34. Tõravere (Liivrand, 1991)		_" _	
35. Harimäe (Liivrand, 1991)		_" _	
Sites with marine sediments			
Northern Denmark			
36. Skaerumhede I (Knudsen & Lykke-Andersen, 1982)			
37. Skaerumhede II (Bahnson et al., 1974)	Eemian	Early Weichselian	
38. Apholm (Knudsen, 1984)			
Southwestern Norway			
39. Fjøsanger (Mangerud et al., 1981a)		(Fana Ist.)	
40. Bø, Karmøy (B. Andersen et al., 1983) (Larsen & Sejrup, 1990)		(Torvastad Ist.)	(Torvastad Ist.)

Note: B, *Betula;* P, *Pinus;* Pc, *Picea;* A, *Alnus;* L, *Larix;* C, *Corylus;* NAP, nonarboreal pollen, in parentheses when poorly represented. Presence of Eemian underlying interstadial sediments is also indicated.

because of its cool character later interpreted as representing the Odderade Interstadial (Garcia Ambrosiani, 1990; Lundqvist & Miller, 1992). The site at Vålbacken, not far from Pilgrimstad (Fig. 8.2), has till-covered clay with macrofossil plant remains of an arctic-subarctic vegetation. This site was also considered to represent the Jämtland Interstadial but was reinterpreted as being a sediment from the Odderade Interstadial. At Boliden further north and near the coast there is a layer of silt and sand about 30 cm thick containing organic material and covered by a several m thick till bed. Boliden was the first site in Sweden described as interglacial but later by Lundqvist (1967) placed in the Jämtland Interstadial and reinvestigated by Robertsson and Garcia Ambrosiani (1988). The pollen diagram of the till-covered sediments with organic material has high percentages of nonarboreal pollen with only small amounts of *Betula* and *Pinus*, the former at least partly representing *B. nana*. The sediments could, according to Robertsson and Garcia Ambrosiani be interglacial, Eemian, or represent either of the two Early Weichselian interstadials. However, Garcia Ambrosiani (1990) placed

it in the Brørup Interstadial, but with a mention of the alternative correlations mentioned above. Robertsson and Garcia Ambrosiani pointed out that the climatic conditions during the formation of the till-covered sediments at Boliden were as severe as in Denmark and The Netherlands during the Middle Weichselian interstadials. A pollen diagram from Gallejaure northwest of Boliden from a till-covered muddy silt showed a vegetational succession similar to the pollen diagram from Pilgrimstad, that is, the end of an interstadial (Magnusson, 1962; Lundqvist, 1967). A *Betula*-dominated basal part changes into an upper part dominated by nonarboreal pollen. Here, as at other sites, *Betula* is partly represented by *B. nana;* the few percentages of *Pinus* are due to long-distance transport. The interval with its sparse birch forest vegetation was placed by Lundqvist (1967) in the Jämtland Interstadial, thus in Brørup, a correlation also accepted by Garcia Ambrosiani (1990).

In addition to these sites described as Early Weichselian there are a few sites in northernmost Sweden placed in the same substage. At two sites, Seitevare and Leveäniemi, the inter-

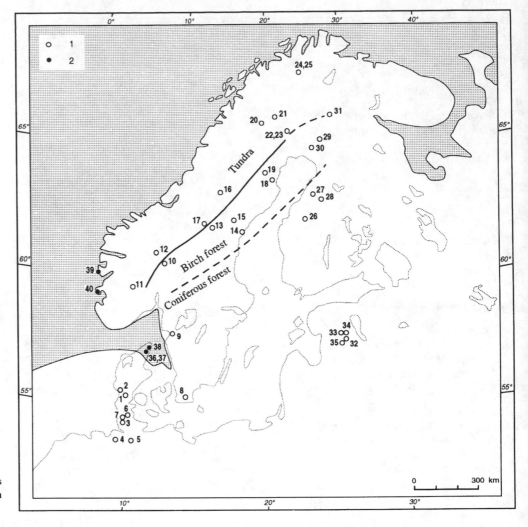

Figure 8.2. Early Weichselian sites (listed in Table 8.1) with vegetation zones. 1, freshwater; 2, marine.

stadial sediments overlie a till and sediments dated as Eemian, described earlier. At Seitevare the compressed Eemian peat is covered by a diamicton, a till-like sediment, which may be a solifluction sediment and a silty sandy sediment containing organic matter, all covered by till (Robertsson, 1988; Robertsson & Rodhe, 1988). The pollen diagram from the sediment with organic material has high percentages of nonarboreal pollen and the arboreal pollen are mainly *Betula,* with some *Pinus.* This interval with an open vegetation and with some birch was correlated by Robertsson with the Early Weichselian Jämtland Interstadial, and would thus also represent Brørup. At the nearby site of Leveäniemi, the Eemian peat and mud is overlain by two tills, in between which there are about 1 m thick minerogenic sediments which yielded a pollen diagram. The sediments consist of silt with organic matter, partly laminated (Garcia Ambrosiani, 1991). The pollen assemblage is dominated by shrubs of which *Betula nana* is the most common, but with a strong nonarboreal component. The interval that this open vegetation represents was correlated with the Brørup Interstadial by Garcia Ambrosiani (1990, 1991).

Detailed studies of the Quaternary stratigraphy in northeastern Norrbotten in northern Sweden enabled Lagerbäck and Robertsson (1988) to identify two Early Weichselian interstadials. They correlate the older with the Jämtland Interstadial, as defined in Central Sweden, and with Brørup, including Rodebaek, further south. They correlated the younger interstadial, locally named the Tärendö Interstadial, with Odderade. In the stratigraphical scheme there were ice advances over the area between the Eemian and Brørup and Odderade, and after the last mentioned interstadial. The first of the three ice advances left esker material from an Early Weichselian deglaciation, the second ice advance a thin till, and the third advance mainly deformed older sediments, without in this area depositing a till of any significant thickness (Lagerbäck & Robertsson, 1988). The stratigraphical evidence of the two interstadials was obtained from several sites, of which two are mentioned in Table 8.1 and shown on the map in Fig. 8.2. At Takanenmännikkö the pollen diagram of a 40 cm thick peat between silt and sand, covered by a sandy till, has a pollen assemblage in the middle part with a strong representation of nonarboreal pollen, but also with pollen of both tree birches and dwarf birch, *Betula nana,* which shows that the site was close to the northern limit of birch forests. The few pollen of *Pinus* are, however, from areas further south. Stratigraphically above these sediments that are correlated with Brørup are those of the Odderade Interstadial, best documented at Riipiharju. Here the interstadial sediments consist of sands, nearly 7 m thick, containing organic matter, overlying a sandy till, and covered by a diamicton. The pollen assemblage of this interstadial is more strongly represented by nonarboreal pollen than the older interstadial, with high percentages of *Artemisia.* This was interpreted as reflecting a continental climate. The evidence from Norrbotten, as described by Lagerbäck and Robertsson (1988), thus shows that two rather similar Early Weichselian interstadials can be identified in northern Sweden, of which the younger, correlated with Odderade, is slightly cooler than the older one correlated with Brørup. This is very much in keeping with the evidence from the areas further south where these interstadials were first identified.

A stratigraphical frame for the Early Weichselian similar to that used for Norrbotten in northern Sweden was used for the glacial and nonglacial sediments in Finnmarksvidda in North Norway (Olsen & Hamborg, 1983, 1984; Olsen, 1988). Sands between tills were correlated with ice-free periods further south, also taking into account the till stratigraphy in surrounding areas, particularly in Finnish Lapland. Thus, a sand at Vuoddasjavri covered by two tills was referred to the Eiravarri Interstadial, and a sand at Sargejåk between the upper two tills to the Sargejåk, correlated with the Brørup and Odderade Interstadials respectively. The Sargejåk Interstadial, if defined as such, is, however, likely to be younger than Odderade when the till stratigraphy and its correlation with Finnish Lapland is taken into account (Olsen, 1988). A sand below three tills was referred to as the Vuolgamasjåkka Thermomer and correlated with the Eemian. A few pollen counts, though of limited use as they were of minerogenous sediments, showed pollen assemblages dominated by nonarboreal pollen for all intervals, which is in agreement with the results from Norrbotten in northern Sweden. The conclusions about Weichselian interstadials in these northern areas have been based almost entirely on lithostratigraphical evidence, primarily on the number of ice advances depositing tills, and on the assumption that they correspond to Early Weichselian stadials separating the Eemian and the two subsequent interstadials from one another. Even if the correlations are correct these sites cannot be used in a biostratigraphical comparison of Early Weichselian sites.

Of the sites in Finland Vimpeli in Ostrobothnia was already discussed in connection with the Eemian sites. All three pollen diagrams from the till-covered compressed peat in the Vimpeli I section show a *Betula* maximum being replaced by a *Pinus* maximum, at the same time as there is a decrease in the relative amount of nonarboreal pollen, even in a profile only 20 cm thick; the thickest peat analyzed was 1 m. There are also low percentages of *Picea* and of *Corylus,* which here, however, most likely represent *Myrica* as shown by the macrofossil plant

remains (Aalto et al., 1983). As it seems likely, on the basis of other sites in Finland, that the Vimpeli site would have been submerged during the early part of the Eemian, it was concluded that the peat in the Vimpeli I section, with its succession typical for the early part of an interval, should be correlated with the Brørup Interstadial (Donner, 1988). Vimpeli may thus be a site where both Eemian and Early Weichselian freshwater organic sediments are present, although in two sections nearly 200 m apart, both covered by a thick till bed. Vimpeli I has a pollen diagram similar to that from Oulainen further north in Ostrobothnia, used as a type locality for the Early Weichselian interstadial, the Oulainen Interstadial, correlated with Brørup (Donner, Korpela, & Tynni, 1986; Donner, 1988). The Oulainen section in an esker has an about 50 cm thick layer of undisturbed mud and drift peat with some sand layers, on top of muddy sand and mud, and overlain by muddy sand and sand, with a sharp contact to a till on top. The site, which was studied in great detail by Forsström (1982), was originally correlated with the Eemian but later with Brørup (Donner, 1983). The analyzed pollen diagrams have lowermost a *Betula* assemblage zone, followed by a *Pinus* zone in the mud, with a relative decrease of the nonarboreal pollen, as in the Vimpeli I diagrams. In this zone there are a few percentages of *Picea* and *Alnus.* There is an increase of *Betula* and nonarboreal pollen in the upper sandy mud, in which some pollen of *Carpinus, Corylus, Quercus,* and *Ulmus* were also encountered. As the sediment changes into sand and there is an increase of nonarboreal pollen toward the top, the presence of pollen of the last-mentioned deciduous trees was interpreted as an increase in the amount of reworked pollen (Donner, 1988). The pine-dominated forest vegetation was thus interpreted as representing the climatic optimum of the interstadial cycle. As the Oulainen site has the most complete sequence of sediments interpreted as Early Weichselian in Ostrobothnia it was, as mentioned, used as a type site (Fig. 8.1). In Haapavesi southeast of Oulainen, and also southeast of the Eemian site at Ollala, there is another site, Marjamurto, on a till-covered esker with sediments similar to those at Oulainen and also with a similar pollen diagram (Peltoniemi et al., 1989). In the section from which the most complete pollen diagram was obtained there is an over 2 m thick bed of silty mud with pieces of wood on top of sand and also covered by sand, underneath a till bed up to 1.5 m thick. In the pollen diagram the lowermost *Betula* maximum is replaced by a pollen assemblage in which *Pinus* dominates, but still with high percentages of *Betula,* and with a few percentages of *Alnus.* The relative amount of nonarboreal pollen is low throughout the diagram, in contrast to the Oulainen diagram with its higher percentages both at the base of the diagram and at the top. The Oulainen diagram also has an increase of *Betula* towards the top, with a corresponding decrease of *Pinus.* Otherwise the Marjamurto diagram shows a similar forest history, a development towards a pine-dominated forest with birch.

In Peräpohjola, northern Finland, several sites with till-covered organic sediments were investigated by Korpela (1969),

who referred to them as representing the Peräpohjola Interstadial and similar to the sediments of the Jämtland Interstadial in
Sweden. At the time of Korpela's investigation a Middle Weichselian age for the Peräpohjola Interstadial was thought likely, but
later studies of the stratigraphy of Sweden and northern Finland
made a correlation with the Early Weichselian Brørup Interstadial more probable. At Permantokoski (Korpela, 1969), later
used as a type site for the Peräpohjola Interstadial (Donner,
Korpela, & Tynni, 1986), a till-covered about 1 ın thick peat
yielded a pollen diagram dominated by *Betula* throughout,
mostly reaching over 90 percent, with a few percentages of
Pinus, and occasional grains of *Picea* and *Alnus* pollen. There is
no indication of a change toward the top of the diagram which
would reflect a development toward warmer conditions with a
spread of *Pinus.* On the other hand, the organic sediment does
not represent the whole ice-free period; the initial and final parts
with high percentages of nonarboreal pollen are missing.
Another pollen diagram from an approximately 1 m thick till-
covered peat at Ossauskoski (Korpela, 1969) also has a tree
pollen assemblage dominated by *Betula.* Both diagrams have 20–
30 percent of pollen of shrubs and herbs, which is an indication
that an open birch forest prevailed in the surroundings of the
sites during the time of the Peräpohjola Interstadial.

In the study of the Pleistocene stratigraphy in Finnish Lapland,
Hirvas (1991) identified interstadial organic sediments stratigraphically separated from the underlying Eemian sediments by
a till, III, and overlain by two tills, I and II. He named it the
Maaselkä Interstadial after a site with an about 50 cm thick till-
covered mud, with a pollen diagram in which the arboreal pollen
is almost exclusively represented by *Betula,* but with about 50
percent nonarboreal pollen of the total pollen sum, thus
reflecting an open birch forest vegetation near the northern
forest line. The diagram is essentially similar to the diagrams
from Permantokoski and Ossauskoski, representing the Peräpohjola Interstadial defined by Korpela (1969) and with which
Hirvas (1991) correlated his Maaselkä Interstadial.

In addition to the evidence of the Early Weichselian development in Finland there are sites further south, in Estonia, which
show that the area was ice free at that time (Liivrand, 1991).
Thus, at Peedu Eemian sediments are directly overlain by Early
Weichselian minerogenous sediments that have a pollen assemblage dominated by *Betula* and are covered by tills. At Otepää
there are similar sediments covered by two tills and with a pollen
composition dominated by *Betula,* but with *Pinus* and some
Picea as well, in a pollen diagram from which the pollen
interpreted as being reworked has been excluded. Furthermore,
these sediments, described as periglacial deposits by Liivrand
(1991), have further been found at other sites, such as Tôravere
and Harimäe, both with pollen assemblages similar to those at
Otepää. In the chronostratigraphical scheme for Estonia, Liivrand (1991) called the Early Weichselian interstadial the Harimäe Interstadial, which included both Brørup and Odderade;
no interstadials were separated within the Harimäe Substage.

As may be seen from this discussion, most interstadial sites in

Norway, Sweden, and Finland placed in the Early Weichselian
have been correlated with the Brørup Interstadial. Some sites,
however, from which the biostratigraphical evidence points to
climatically more severe conditions than those which are
believed to have prevailed during Brørup, have been placed in
the Odderade. It is only in the north, in Norrbotten in northern
Sweden and in northern Norway, that there is evidence of two
subsequent nonglacial intervals, but their correlation with
Brørup and Odderade is primarily based on lithostratigraphical
evidence. In Norrbotten, however, there is strong evidence of
two interstadials which probably can be placed in the Early
Weichselian. It can then generally be concluded about the fresh
water interstadials that many of the comparatively warm Early
Weichselian interstadials probably correspond to the Brørup Interstadial, but that some may also be referred to Odderade. The
relatively cool character of an interstadial is not, however,
evidence for its being Odderade; the site may merely have only a
part of the interstadial sediments preserved, with the climatic
optimum missing. The sites with evidence of an arctic or
subarctic vegetation, presumed to represent Odderade, generally have fine-grained minerogenous till-covered sediments
formed in a cold treeless periglacial environment. These sediments show that they were formed during nonglacial intervals
but are difficult to place in a particular Early Weichselian interstadial. They may in some areas even represent parts of ice-free
stadials. There is thus at many sites a degree of uncertainty as to
which Early Weichselian interstadial they represent.

If all the Early Weichselian sites in Scandinavia are viewed
together, most sites reflect a general pattern of vegetational
zones (Fig. 8.2) in a southwest–northeasterly direction, zones
first referred to as being interglacial but later referred to as
representing the Jämtland Interstadial (Lundqvist, 1967). The
map showing these zones was later completed and redrawn
(Lundqvist, 1978), of which a simplified version was used by U.
Miller (1986). The map in Fig. 8.2 is slightly modified from the
previous ones, showing the northern limit of the coniferous
forest, consisting mainly of *Pinus* but with some *Picea,* in
addition to *Betula,* as well as the boundary between the birch
forest and the open tundra vegetation. The map depicts the
conditions during the warmest time of the Early Weichselian and
therefore most likely of the Brørup Interstadial. But some
individual sites may represent the Odderade Interstadial, such
sites as Boliden, which is within the birch forest zone but has
only sediments from a time with an open vegetation. The
evidence of the nature of this younger Early Weichselian interstadial from southern Jutland and northern Germany clearly shows
that this possibility cannot be ruled out.

In Sweden and Finland, as seen from Table 8.1, the zone of
coniferous forest had as main components, after an initial birch
period, *Pinus* and *Betula,* with a relatively small representation
of *Picea,* in pollen diagrams probably often as a result of long-
distance transport, and with *Larix* barely recorded. *Alnus,* which
was present further south, did not reach these areas, nor can the
few records of *Corylus* pollen be taken to show that it was

present in southern Norway or Sweden. The evidence from Estonia, in spite of being incomplete, fits into the general pattern described above, a pattern that is also reflected in the schematical Early Weichselian pollen diagrams in Fig. 8.1. It was assumed in the comparison that the sequences in Sweden and Finland are incomplete and represent shorter time intervals than those at Oerel in Germany. This is clearly reflected by the diagram from Oulainen.

8.2. Marine sequences

The position of the sea-level during the Early Weichselian was, as earlier noted, relatively low when compared with the level of the Eemian Sea. In northernmost Jutland in Denmark, however, the Eemian marine sediments are overlain by Weichselian marine sediments, which shows that the submergence of this area continued after the Eemian. In addition, there is some evidence from southwestern Norway on an Early Weichselian marine influence. The Early Weichselian position of the shoreline shown in Fig. 8.2 is according to Houmark-Nielsen (1989).

The marine Eemian sediments encountered in borings at several localities in northern Jutland were already mentioned in Chapter 6. At Skaerumhede, where the Skaerumhede I core reached a depth of 200 m (Knudsen & Lykke-Andersen, 1982), the foraminiferal zone N3 with a boreal-lusitanian fauna was, as mentioned earlier, correlated with the Eemian (Knudsen, 1986b). The overlying sediments of zone N2, with a boreo-arctic fauna of more shallow water, were correlated with the Early Weichselian and the sediments of zone N1 with an arctic fauna, with *Elphidium excavatum* f. *clavata* and *Cassidulina reniforme*, with the Middle Weichselian. Some of the foraminiferal and mollusk zones separated in another boring, Skaerumhede II, were first correlated with Early Weichselian interstadials and stadials (Bahnson et al., 1974; Feyling-Hanssen & Knudsen, 1979), but this correlation was later modified, with no detailed subdivision of the Early Weichselian foraminiferal zone N2 (Knudsen & Lykke-Andersen, 1982; Knudsen, 1986b). There is thus no evidence of the interstadial-stadial fluctuations in these marine sequences (Houmark-Nielsen, 1989). The same foraminiferal zones as were identified in the Skaerumhede cores were also found at the earlier mentioned site at Apholm near Fredrikshavn (Knudsen, 1984, 1986b) and in cores from other sites in North Jutland (Knudsen, 1985c). In the boreholes of sediments of Roar, Skjold, and Dan in the central North Sea (Fig. 8.2) the marine Eemian sediments are overlain by Weichselian glacio-lacustrine sediments (Knudsen, 1985a, 1986b). The marine Eemian to Early and Middle Weichselian sequences are thus more complete in the deep boreholes of North Jutland than in the comparatively shallow waters of the North Sea, with the top of the Eemian sediments being at a depth of about 70 m, in an area where the water depth is about 40–50 m.

At the Eemian site of Fjøsanger in southwestern Norway the marine warm-stage sediments are covered by silty gravel, silt, gravel and silt, all overlain by a till, the Bønes till (Mangerud et al., 1981a). The gravel bed (F) was interpreted on the basis of

the marine foraminifera and mollusks representing a predominantly cool arctic fauna, but with *Chlamys islandica* included, as having been formed during a first local Weichselian interstadial, the Fana Interstadial, corresponding to the Rodebaek Interstadial in The Netherlands and thus to the beginning of the Brørup Interstadial (Mangerud et al., 1981a). As the Early Weichselian marine beds reach an altitude of 15 m in the Bergen area surrounding Fjøsanger, it was further concluded that the coastal area was glacio-isostatically depressed at this time, an indication of the presence of an already large Early Weichselian ice sheet in parts of the Scandinavian mountains soon after the Eemian warm stage.

The second Eemian site with Early Weichselian sediments in southwestern Norway is Bø on Karmøy (B. Andersen, Sejrup, & Kirkhus, 1983). Immediately overlying the Avaldsnes Sand correlated with the Eemian is a bed of stratified sand about 40 cm thick called the Torvastad Sand, which is overlain by two diamictons separated by silts and sands referred to as the Bø Sand Formation. The pollen diagram of the minerogenous sediments shows a change from the Avaldsnes Sand to the Torevastad Sand, the latter having a *Betula* assemblage zone with a rise of the nonarboreal pollen curve and representing a cool period. Similarly, molluscs of relatively warm water found in the Avaldsnes Sand are in the Torvastad Sand replaced by cold-water species, but with *Mytilus edulis*, suggesting that the coastal areas were icefree at that time. In a general correlation of the stratigraphy at Bø the Torvastad Sand was placed in the Early Weichselian and the Bø Sand in the Middle Weichselian, in the local division called the Torvastad Interstadial and Bø Interstadial respectively. The Middle Weichselian age of the latter was mainly based on radiocarbon datings of marine shells and will be discussed later. The two Early Weichselian interstadials in southwestern Norway, Fana at Fjøsanger and Torvastad at Bø, are however beyond the reach of radiocarbon dating. During the Torvastad Interstadial the sea level was estimated to have been between the level of the present sea level and about 20 m above it (Sejrup, 1987), which is about the same as that estimated for the level during the Fana Interstadial and showing at both sites that the sea level had fallen from its Eemian level.

On the basis of the biostratigraphical criteria used so far in separating the interstadials, combined with lithostratigraphical evidence, the two interstadials in Norway described as Early Weichselian cannot be placed with any certainty in either of the two interstadials Brørup or Odderade, even if Fana Interstadial was correlated with the beginning of the former. A later detailed study of the foraminifera and mollusks at Bø gave further details about the seawater temperatures during the two interstadials found at this site (Sejrup, 1987), but no conclusive evidence as to the age of the Torvastad Interstadial in relation to the Fana Interstadial. The evidence used in dating these two interstadials is that from amino acid ratios in shells, a method which will be discussed later. The amino acid analysis results were interpreted as showing that the Torvastad Interstadial is younger than the Fana Interstadial (B. Andersen, Sejrup, &

Figure 8.3. Early Weichselian peat layer in outwash material north of Lake Tåsjö, central Sweden (Site 16, Table 8.1), correlated with the Odderade Interstadial. (Photo J. Lundqvist)

Kirkhus, 1983), and they were therefore correlated with the Odderade and Brørup Interstadials respectively (B. Andersen & Mangerud, 1990). This correlation is exclusively based on using aminostratigraphy as a tool for comparisons, and the correlation thus differs from the methods used for the previously mentioned Early Weichselian sites. In another scheme the Fana Interstadial is correlated with the beginning of the Brørup Interstadial, in accordance with the suggestion by Mangerud et al. (1981a), and the Torvastad Interstadial with the Brørup Interstadial, that is, its upper part, the Bø Interstadial also being included in the Early Weichselian and correlated with the Odderade Interstadial (Larsen & Sejrup, 1990).

It can be seen from the preceding discussion that whereas the marine sediments in northern Denmark cannot be correlated with any particular Early Weichselian interstadials or stadials, the sediments at Fjøsanger and Bø in southwestern Norway have beds likely to represent interstadials. The dating of these sediments, however, must still be considered tentative and open to revision. The few sites with Early Weichselian sediments are listed in Table 8.1 with suggested correlations with interstadials.

8.3. Glaciation

At some sites the Early Weichselian interstadial beds are separated from underlying Eemian sediments by a till, and at a few sites a till occurs between sediments correlated with the two interstadials Brørup and Odderade. Thus, assuming that the preceding correlations are correct, there was a glaciation, in places consisting of two advances, during the Early Weichselian

shortly after the Eemian. By studying the lateral extent of particular tills related to this glaciation its extent can be traced, but as later Weichselian ice advances covered a larger area the limit of the Early Weichselian glaciation is difficult to determine. It is, for instance, not limited by a well-defined end moraine.

The largest extent of the Early Weichselian glaciation as shown in Fig. 8.3 is mainly after B. Andersen and Mangerud (1990), but with minor changes and some additions. If the deflation surface near Gothenburg interpreted as the Older Dösebacka-Ellesbo Interstadial represents Brørup, the underlying till, on top of the rock basement with warm-stage weathering from the Eemian, represents an Early Weichselian ice advance (Hillefors, 1974, 1983), generally, as will be seen later, referred to in Sweden as Stadial I of the Weichselian glaciation (Lundqvist, 1986a). The possible early ice advances as far as Gothenburgh is included in the map in Fig. 8.4. Equally uncertain is the interpretation of the proximity of the ice margin at the coast of southwestern Norway. If the Fana Interstadial at Fjøsanger (Mangerud et al., 1981a) and the Torvastad Interstadial at Bø (Larsen & Sejrup, 1990) are both correlated with the Brørup Interstadial, as earlier concluded, then the presence of the glaciomarine silt at Fjøsanger shows that the ice margin was close to the site during the Gulstein Stadial between the Fana–Brørup Interstadial and the Fjøsanger–Eemian Stage. At Bø there is a hiatus between the Eemian sediments and those of Torvastad Interstadial. An Early Weichselian age for the Bønes Till overlying the Fana Interstadial at Fjøsanger has also been suggested, but in a correlation where the Fana and Torvastad Interstadials represent Brørup and Odderade respectively (B. Andersen & Mangerud, 1990).

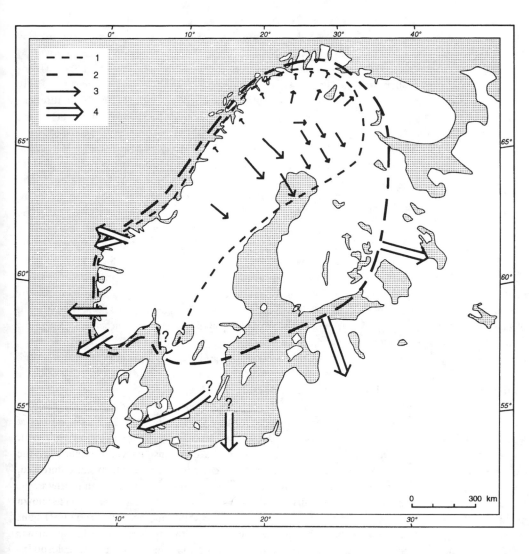

Figure 8.4. Extent of Early Weichselian glaciation, 1, and Middle
Weichselian glaciation, 2 (after B.
Andersen & Mangerud, 1990). 3,
direction of Early Weichselian ice
movement; 4, early Middle Weichselian ice advance.

The difficulties encountered in the interpretation and dating of the coastal sites were already discussed earlier, but it may be added that the response of the ice margin to climatic fluctuations of stadials and interstadials was more pronounced at the west coast of Scandinavia than on the lee side of the mountains. Therefore several local ice advances down the fjords have been recorded, with tills alternating with marine sediments at the coastal sites such as Fjøsanger and Bø (Larsen & Sejrup, 1990). This can already be seen in a comparison of the glaciation curves of the western maritime side of the Weichselian ice sheet and the eastern continental side (B. Andersen & Mangerud, 1990).

These conclusions about the extent of the Early Weichselian glaciation agree with those made by Lundqvist (1967, 1969a, 1974) on the basis of results from Jämtland in Sweden. The Jämtland Interstadial was, as earlier noted, correlated with Brørup and the preceding glaciation considered to be Early Weich-selian, representing the first glacial, W I. This scheme agreed with the division of the Weichselian glaciations proposed by Ljungner (1949), which was based on a study of the striae in the mountains of northern Sweden and on morphological features of glacial

abrasion in Scandinavia generally. Ljungner's "Prime Glacia-tion," succeeded by an "Interval" correlated with the Jämtland In-terstadial, had its ice-divide west of the highest Scandinavian mountains, along the Norwegian coast (Lundqvist, 1974).

These conclusions were based on the assumptions that the Jämtland Interstadial corresponds to the Brørup Interstadial and that the tills underlying the interstadial sediments are Early Weichselian. As this is by no means certain, later conclusions about their age are more cautious. Thus Lundqvist (1981, 1986a) mentioned that there is no conclusive evidence for an Early Weichselian glaciation in most parts of Sweden, except in northern Sweden. Subsequent studies of interstadials that have already been described, and their relationship to Eemian sediments, have brought back the concept of a limited Early Weichselian glaciation in and close to the Scandinavian moun-tains and in the northern parts of Sweden and Finland, as shown on the map in Fig. 8.4. This agrees with the conclusions about the glaciations of Hardangervidda in southern Norway, where an early ice movement, Phase I, was correlated with the first glacial, W I, as suggested above (T. O. Vorren, 1979), because the till laid down during this phase is stratigraphically between the Hovden warm stage correlated with the Eemian and the Førnes Interstadial correlated with the Jämtland Interstadial and the Brørup (T. O. Vorren & Roaldset, 1977; T. O. Vorren, 1979).

The evidence from these areas is still weak and open to revision. Some of the uncertainties stem from use of radiocarbon dates for correlation, which led to the short chronology for the Weichselian, with the upper part of the Eemian at about 70 ka, as shown in a comparison of some correlation tables (T. O. Vorren, 1979). The stratigraphical evidence from southern Norway, however, has not been changed. According to studies of several sites and areas, the presence of Early Weichselian ice advances in the northern parts of Norway, Sweden, and Finland is well established.

Both at Seitevare and at Leveäniemi in northernmost Sweden the sediments described as representing Brørup are separated from the underlying Eemian sediments by a till bed, which – if these conclusions are correct – means that these parts had an Early Weichselian glaciation. This is substantiated by the results from Norrbotten, where a thick till layer was formed during the stadial between the Eemian and the first Early Weichselian interstadial correlated with Brørup, a till in a position similar to that at Leveäniemi (Lagerbäck & Robertsson, 1988). A till bed was also identified further south in the Bothnian Bay and at the Swedish coast, close to the assumed outer limit of the Early Weichselian glaciation (Andrén, 1990). In addition to this till, there is evidence in the form of a thin till of a second Early Weichselian ice-advance between the Brørup Interstadial and Tärendö Interstadial correlated with Odderade. The extent of this second ice advance in northern Sweden is not known because there are no similar observations from neighboring areas, as mentioned by Lagerbäck and Robertsson (1988). But the glacial sequence in Finnmarksvidda in northern Norway, as interpreted by Olsen (1988), has also two Early Weichselian till beds of an age similar to those in Norrbotten, connected with sands formed during interstadials. Alternatively, however, the second till may be of Middle Weichselian age if the Sargejåk Interstadial does not correspond to the Odderade Interstadial but is younger (Olsen, 1988). In Finnmarksvidda, as in Norrbotten, the main Early Weichselian ice advance occurred during the stadial between the Eemian Stage and the Brørup Interstadial, and it is evidently this glaciation that was clearly recorded at Seitevare and Leveäniemi by their till beds.

In northern Finland the sequence of till beds is well known because of the extensive excavations made to establish a regional glacial stratigraphy. In the lateral tracing of the different tills, lithological and fabric studies were combined, but their dating was based on their relationship to the organic deposits referred to earlier. In this way, till bed III in northern Finland was placed in the Early Weichselian, because it is between sediments correlated with the Eemian and sediments of the Maaselkä (Peräpohjola) Interstadial correlated with Brørup (Hirvas &

Nenonen, 1987; Hirvas, 1991). In the stratigraphy for northern Finland the warm stage correlated with the Eemian has, as already mentioned, been named the Lapponian (Hirvas, 1983) or – after the stratotype section – the Tepsankumpu Stage (Donner, Korpela, & Tynni, 1986; Hirvas, 1991), whereas the Early Weichselian interstadial was first called the Peräpohjola Interstadial (Korpela, 1969) and later the Maaselkä Interstadial (Hirvas, 1991).

The outer limit of the Early Weichselian glaciation in Fig. 8.4 was, as mentioned, drawn on the basis of the map published by B. Andersen and Mangerud (1990), but with minor changes. The directions of ice movements during this glaciation were added to the map according to some regional studies, such as that of the Bothnian Bay by Andrén (1990). Assuming that Phase I of ice movements in Hardangervidda in South Norway represents the Early Weichselian time (T. O. Vorren, 1979) the ice-divide there was near the coast. The movements in northern Sweden, Norway, and Finland were mainly drawn according to a map by Olsen (1988, 1990) of the "Till 3" glaciation. The name refers to till bed III in northern Finland, which had directions of ice flow shown on the map (Hirvas et al., 1977; Hirvas & Nenonen, 1987; Hirvas, 1991). Here the ice-divide was in an east–west direction, away from the coast, which resulted in ice advances toward the north and northeast in northernmost Norway and Finland. In northern Sweden southeast of the mountains the direction of flow was from the northwest, as shown by till fabric studies in Leveäniemi (Lundqvist, 1971) and Seitevare (Robertsson & Rodhe, 1988), where the Late Weichselian till was related to the pollen-analytically studied Eemian and Early Weichselian sediments, as mentioned above. It can be assumed that during the comparatively restricted Early Weichselian glaciation the main movement of the ice further south in Sweden was similarly from the northwest, as indicated on the map, and that it was only in the coastal areas of Norway that there was a flow toward the northwest. But as the initial movements of ice during earlier and subsequent glaciations were presumably also away from the ice-divide along or close to the highest Scandinavian mountains, a movement from the northwest in the northern parts of Sweden may not represent an Early Weichselian glaciation. The age of the "Prime Glaciation" (Ljungner, 1949), or W I (Lundqvist, 1974), recorded as an old movement from the northwest in Jämtland (Lundqvist, 1967, 1969a), is therefore not necessarily Early Weichselian but either older or much younger, probably Middle Weichselian, a possibility already mentioned and taken into account in later studies (Lundqvist, 1981, 1986a).

If all uncertainties of correlation are considered it is possible that the Early Weichselian glaciation was more restricted than indicated on the map in Fig. 8.4. It may have been limited to a smaller area in northernmost Sweden, Norway, and Finland, with perhaps a smaller ice sheet in the mountains of Norway further south.

9 Middle Weichselian substage

9.1. Biostratigraphy of freshwater deposits

The beginning of the Middle Weichselian Substage at 74 ka, as defined on the basis of the deep-sea record (see Table 7.1), marks the beginning of a general global cooling. In northern Europe the climate in the ice-free areas was cool throughout the substage until the extremely cold beginning of the Late Weichselian at 25 ka BP that resulted in an extensive Late Weichselian glaciation. The vegetation was open, without forest, even during the many Middle Weichselian interstadials in northern Germany and The Netherlands. This is the background against which the interpretations of nonglacial intervals in Scandinavia must be viewed. Furthermore, the ages of some of the intervals are, as earlier noted, outside the limit of the radiocarbon-dating method. The main problem in dealing with the glaciated area of Scandinavia is therefore whether there are any nonglacial sediments that can with certainty be placed in the Middle Weichselian. If they do occur their distribution shows how large an area was deglaciated at that time. But the evidence from this comparatively long period of nearly 50 ka is meager compared with what is known about the Early Weichselian, which in itself reflects the cooling of the climate. The likelihood of organic sediments being preserved from this time is small. And those that do remain have a microfossil content difficult to interpret, as an admixture of minerogenous material into the sediments also means that rebedded pollen, among others, were introduced into them.

The few sites in Scandinavia that can with some certainty be interpreted as having Middle Weichselian fresh water sediments are all in southern Scandinavia (Fig. 9.1, Table 9.1). In a cliff at Kongshøj on Sejerø, on the southeast coast of the island, there is a 3 m thick bed with thin layers of fine sand, silt, and clay with plant fragments underneath three till beds with intercalated sands (Houmark-Nielsen & Kolstrup, 1981), in an area of Denmark with a complicated sequence of Weichselian ice advances, as will be seen later. The pollen diagram of the fine sediments underlying the lowermost till is dominated by non-arboreal pollen, mostly about 90 percent, represented mainly by *Cyperaceae* and *Gramineae,* but with a strong influence of reworked pollen including pre-Quaternary taxa. Because of this, no detailed information of the surrounding vegetation could be obtained, but it is clear that the sediments were formed in a treeless environment. The radiocarbon ages at about 36–37 ka BP, as shown in Table 9.1, were used as evidence for placing these sediments in the Middle Weichselian and interpreted by Houmark-Nielsen and Kolstrup as having been formed during the Hengelo Interstadial. This correlation is thus based on the radiocarbon dates, here accepted as giving the true ages of the sediments.

A Middle Weichselian age has also been given to the about 5 m thick clayey silts underlying even thicker sands, the Gärdslöv beds, below 80 m in the 100 m deep borehole in Gärdslöv in the previously mentioned Alnarp Valley in Scania in southernmost Sweden (U. Miller, 1977; Berglund & Lagerlund, 1981). The pollen diagram studied by Miller shows an interstadial flora with an arctic–subarctic vegetation dominated by herbs, but with a high proportion of reworked older pollen and spores, common in fine-grained minerogenous sediments with a low content of organic matter. The radiocarbon dating gave five finite and two infinite dates, listed in Table 9.1, which led to the conclusion that the Gärdslöv beds represent a Middle Weichselian interstadial. It was named the Gärdslöv I Interstadial, as higher up in the core another, younger, Late Weichselian interstadial was identified (U. Miller, 1977). The Gärdslöv beds are covered by several tills, here as at Sejerø formed duing ice-advances from different directions. The true age of the Gärdslöv I Interstadial is, however, difficult to determine because it is clear from the radiocarbon ages, of which two are infinite, that at least some of them have been affected by reworked organic matter and are therefore too old, as mentioned by U. Miller (1977). Whether the younger dates are correct cannot with certainty be ascertained.

In the complex stratigraphy of the previously noted drumlins at Dösebacka and Ellesbo near Gothenburg, a Younger Dösebacka–Ellesbo Interstadial was identified and placed in the Middle Weichselian because of the radiocarbon age (Hillefors, 1969, 1974). At Dösebacka the organic material obtained from 2 kg of clay gave an age of $24,020 +450 -1800$ (Lu-104). At Ellesbo the organic material from sand, containing some pollen, gave an age of 30,300

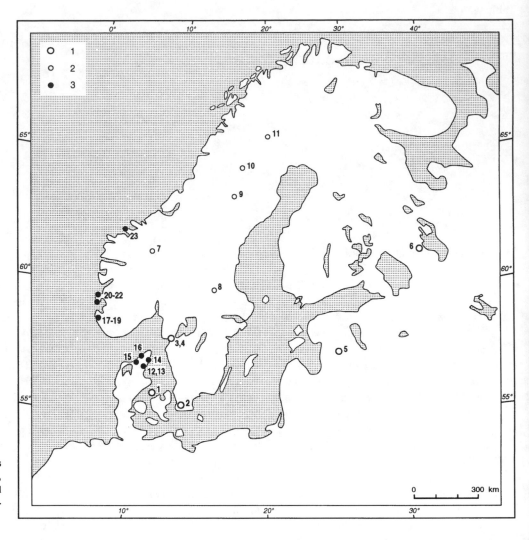

Figure 9.1. Middle Weichselian sites (listed in Table 9.1). 1, freshwater; 2, freshwater sites earlier interpreted as Middle Weichselian but later rejected; 3, marine.

+950 −850 (Lu-280), with the margins of error later estimated at +7600 and −1800 (Hillefors, 1974). It is therefore difficult to judge if there is a real difference in age between the two sites. The interstadial was considered by Hillefors to be approximately contemporaneous with the Hengelo and Denekamp Interstadials in The Netherlands, the latter being closer in age to that of the younger Dösebacka–Ellesbo Interstadial than the former. An exact correlation is, however, not possible. At Ellesbo there is a deflation surface on top of the interstadial sand, which is covered by a till, as is the interstadial clay at Dösebacka.

These three sites have generally been accepted as having Middle Weichselian interstadial freshwater sediments, thus showing that southernmost Sweden and Denmark were ice-free at some time during this substage (Lundqvist, 1986a; B. Andersen & Mangerud, 1990; Garcia Ambrosiani, 1990). The evidence of a deglaciation of central and northern Norway and Sweden, and of Finland, is, on the other hand, more ambiguous. The radiocarbon ages, earlier used as evidence for a Middle Weichselian interstadial, are often for redeposited material and, furthermore, the dated material consists in some

cases of wood of taxa, which – from the evidence from areas further south – could hardly have grown in these areas at that time; the evidence for a Middle Weichselian deglaciation is therefore highly uncertain (Lundqvist, 1981; Garcia Ambrosiani, 1990). One site in Norway and four sites in Sweden earlier interpreted as having Middle Weichselian interstadial sediments, however, are included in Table 9.1 and marked on the map in Fig. 9.1, to show how they are seemingly Middle Weichselian if the radiocarbon ages alone are used.

At Gråmobekken and Djupdalsbekken in Hedmark in south-western Norway there are till-covered glaciofluvial and lacustrine sediments described as representing the Gråmobekken Interstadial, the pollen spectra in three samples reflecting an open vegetation (Thoresen & Bergersen, 1983). The radiocarbon ages of between 30 ka and 40 ka for the soluble and insoluble organic fractions at Gråmobekken led the authors to conclude that the sediments were formed during the Middle Weichselian. The sediments are stratigraphically in a position similar to the till-covered glaciofluvial sediments, described as preglacial, in the Gudbrandsdal Valley somewhat south of Hedmark, with features

Table 9.1. *Middle Weichselian sites*

Location	References	Radiocarbon ages B.P.		
		Freshwater sites		
Denmark				
1. Sejerø	Houmark-Nielsen & Kolstrup, 1981	GrN-9456	35710	± 460
Sweden				
2. Gärdslöv, Alnarp valley (Gärdslöv I Interstadial)	U. Miller, 1977	St-4273	21305	± 3000
		St-4938	22835	± 1680
		St-4271	27535	± 5000
		St-3158	32730	± 2040
		St-4946	32880	± 1770
		St-4274	> 34000	
		St-4272	> 35000	
3. Dösebacka	Hillefors, 1969, 1974	Lu-104	24020	+ 450 − 1800
4. Ellesbo (Younger Dösebacka-Ellesbo Interstadial)	Hillefors, 1969, 1974	Lu-280	30300	+ 950 (+ 7600) − 850 (− 1800)
Estonia				
5. Volguta	Liivrand, 1991			
Russia				
6. Petrozavodsk	Lukashov, 1982			
		Sites earlier interpreted as Middle Weichselian but later rejected		
Norway				
7. Hedmark, Gråmobekken (Gråmobekken Interst.)	Thoresen & Bergersen, 1983	T-3556A(sol)	37330	+ 640 − 590
		T-3556B (insol)	32520	+ 650 − 590
Sweden				
8. Borlänge	Lundqvist, 1978	U-4000	32200	+ 1700 − 1400
		U-2667	41800	+ 4100 − 2800
9. Vojmå	Lundqvist, 1978	St-5178	32685	± 4200
10. Juktan	Lundqvist, 1978	St-4814	32060	± 5135
11. Gällivare	Lundqvist, 1978	St-5405	41655	± 4350
		Sites with marine sediments		
Denmark				
12. Skaerumhede I	Knudsen & Lykke-Andersen, 1982			
13. Skaerumhede II	Bahnson et al., 1974			
14. Apholm	Knudsen, 1984			
15. Nørre Lyngby	Lykke-Andersen, 1987	shells		
		K-3611	34030	+ 1310 − 1130
		GrN-10024	40550	± 900
		GrN-10023	40700	± 900
		plant material		
		GrN-9707	47300	+ 1500 − 1200

Table 9.1. *(cont.)*

Location	References	Radiocarbon ages B.P.			
16. Hirtshals	Knudsen & Lykke-Andersen, 1982				
Norway					
17. Oppstad, Jaeren (Sandnes Ist.)	B. Andersen et al., 1981	shells			
		T-3422A	39200	$+1800$ / -1600	outer
		T-3422B	38600	$+1600$ / -1300	inner
		T-922	41300	$+6200$ / -3500	
18. Foss, Eigeland (Sandnes Ist.)	B. Andersen et al., 1981	T-3423A	27900	$+850$ / -760	outer
		T-3423B	31330	$+700$ / -640	inner
19. Hove, Sandnes (Sandnes Ist.)	B. Andersen et al., 1981	Y-2377	38100	±1000	
20. Bø, Karmøy (Nygaard Ist.)	B. Andersen et al., 1981	T-2953A+B	38500	$+1700$ / -1400	outer
		T-2953C+D	42700	$+2600$ / -1800	inner
21. Nygaard, Karmøy (Nygaard Ist.)	B. Andersen et al., 1981	T-796	39200	±2000	
		T-797	42000	±2000	
22. Bø, Karmøy (Bø Ist.)	B. Andersen et al., 1983		37000	$+800$ / -700	
			39600	$+1100$ / -1000	
			40600	$+1200$ / -1000	
			41300	$+1200$ / -1000	
23. Sunnmøre (Ålesund Ist.)	Mangerud et al., 1981b	from T-2306	28400	±800	
		to T-2658B	37900	±1600	inner

interpreted as permafrost structures and with numerous remnants of mammoth (Bergersen & Garnes, 1971). These sediments, representing the nonglacial interval named the Gudbransdal Interstadial, is, as earlier concluded, most likely Early Weichselian, of the same age as the Brumunddal Interstadial in the same area, correlated with Brørup (see Table 8.1). An Early Weichselian age for the till-covered sediments at Hedmark is therefore also possible, but their true age is still not established. If the sediments of the Gråmobekken, Gudbrandsdal, and Brumunddal Interstadials were Middle Weichselian, it would mean that most parts of the Scandinavian mountains were deglaciated at that time (B. Andersen & Mangerud, 1990), an interpretation of the Weichselian glacial history that has gained less support from recent investigations than one of a continuous ice cover in the Scandinavian mountains throughout the Middle Weichselian.

At Borlänge in Sweden redeposited wood fragments were found in a 45 m thick sand covered by 21 m of silt (Gustafson, 1976). The radiocarbon age for fragments of *Picea*, *Juniperus*, and *Salix* gave an age about 32 ka BP and for *Juniperus* wood – from another boring – an age of about 42 ka BP (Lundqvist, 1978). At Vojmå, further north, the organic material in a 2 m thick silt bed, covered by till, gave an age of just over 32 ka BP and at Juktan a similar till-covered silt with organic material an age of about 32 ka BP (Lundqvist, 1978). Near Gällivare in Swedish Lapland organic material in till was dated at nearly 42 ka BP (Lundqvist, 1978). These four sites in Sweden were used for separating an interstadial complex, 30 ka to 40 ka old, from the Jämtland Interstadial, a separation later questioned (Lundqvist, 1981, 1986a), but in some general summaries considered as a possible alternative (U. Miller, 1986). Taking into account the arguments mentioned before, that is, that the material dated at the four sites in Sweden is redeposited and that the wood found cannot be Middle Weichselian, and that there is no clear biostratigraphical evidence for any interstadial of that age, the sites cannot be used as evidence for a Middle Weichselian deglaciation of large parts of central and northern Sweden.

When comparing the radiocarbon ages from Hedmark in Norway and the four sites in Sweden with those from Dösebakka-Ellesbo and Gärdslöv in southern Sweden and with those from Sejerø in Denmark (Table 9.1), it can be seen that most of them fall into the age bracket between 30 ka and 40 ka BP, some being younger than 30 ka. But, as shown, the ages alone are not enough for accepting that the sediments dated are Middle Weichselian; the conclusion must be supported by biostratigraphical evidence. In this respect the sequences at Sejerø and Gärdslöv, from which there are pollen diagrams, are more convincing than at Dösebakka–Ellesbo, even if the stratigraphy at the latter site is well documented.

The confusion that may arise from a strict use of radiocarbon dates in placing sediments in the stratigraphical scheme can be demonstrated with two examples. At Göta Älv near Gothenburg radiocarbon ages of between about 27 ka and 29 ka BP of shells in a marine clay were used to describe the sediment as having been formed during a Middle Weichselian interstadial, the Göta Älv Interstadial (Brotzen, 1961). Two, later, accelerator-mass spectrometry (AMS) dates of *Portlandia arctica*, 12,350 ±265 (Ua-701) and 12,620 ±225 (Ua-702), after a sea correction of 400 years, however, showed that the marine clay is Late Weichselian from the time of the deglaciation of the Swedish west coast (Klingberg, 1989). At the previously noted sites at Somero in southern Finland the organic material in the clay, a mixture of Eemian material and material about 10 ka old, gave an age of 32,600 +1500 −1240 (Hel-209), with a difference of about 5 ka when the insoluble and soluble fractions were dated separately (Donner & Jungner, 1973). Both examples show how extremely sensitive radiocarbon ages are to contamination of organic material in the samples. Even if a comparison of the finite ages from Norway and central and northern Sweden, listed in Table 9.1, with the ages quoted for the Middle Weichselian interstadials in The Netherlands in Table 7.1 makes correlation with the Hengelo and Denekamp Interstadials tempting, such correlations cannot be accepted when possible errors in dating are taken into account. The same problem arises in the interpretation of the radiocarbon ages of mammoth remains, which will be discussed later.

Some old finite radiocarbon ages, in addition to a greater number of infinite ages, have also been obtained from sites in northern Finland, but they can be rejected on the same grounds as those from Norway and central and northern Sweden, and cannot be used as evidence for Middle Weichselian nonglacial intervals. Among the ages reported is the age of 42,000 ±2000 (Su-153) for pieces of charcoal from a till-covered contorted soil profile at Vuotso (Kujansuu, 1972). At another site, Marrasjärvi, organic material and charcoal in the lower part of a till bed

was first dated at 42,300 +6000 −1700 (Su-236) and 48,000 +4700 −2300 (Su-237) respectively (Heikkinen, 1975), but the dates were later changed to the infinite ages >42,000 and >48,000 (Heikkenen & Äikää, 1977). At Kauvonkangas, Tervola, a till-covered peat, interpreted as representing the Peräpohjola Interstadial and thus Brørup, was dated at 48,000 +4100 −2400 (Su-688) and 48,100 +7200 −3800 (Su-689) (Mäkinen, 1979), dates which must be considered to be infinite, as is that of 63,200 +5500 −3200 (GrN-7982) for the till-covered organic deposits of Oulainen further south (Aario & Forsström, 1979), also correlated with Brørup. In addition, from Finnish Lapland, Hirvas (1991) listed 7 finite dates of the insoluble fraction from Early Weichselian deposits (Maaselkä Interstadial) and 13 from Eemian deposits (Tepsakumpu Interglacial) which, as he pointed out, can all be rejected.

As a result of a reinterpretation of the radiocarbon ages of the organic fraction in minerogenous sediments or pieces of wood or charcoal in southern Norway, central and northern Sweden, and in northern Finland, earlier interpreted as being Middle Weichselian, the stratigraphical scheme for these central parts of Scandinavia, with several glacial advances interrupted by interstadials (Lundqvist, 1986a; U. Miller, 1986), has to be modified. It seems likely that most parts of Norway, Sweden, and Finland were glaciated throughout the entire Middle Weichselian Substage, with only southernmost Sweden and Denmark ice-free, and occasionally some parts of the coast of Norway, as will be seen from the evidence from marine deposits (B. Andersen & Mangerud, 1990). The only evidence of the Middle Weichselian vegetation was obtained from the minerogenous sediments at Sejerø and Gärdslöv, where the pollen diagrams showed that the vegetation was treeless and was described as arctic–subarctic. The two sites cannot, however, be correlated with any of the interstadials listed in Table 7.1, nor can the Younger Dösebacka–Ellesbo Interstadial near Gothenburg be placed in this scheme.

The conclusion that southernmost Sweden and Denmark were ice-free in Middle Weichselian time agrees with the results from the three Baltic states Estonia, Latvia, and Lithuania, and from Russia, as summarized by Liivrand (1991). There are several sites in Estonia with minerogenous sediments between a lower violet-brown till and in places up to three tills formed during a second Weichselian ice advance. At Volguta east of Vortsjärv (Fig. 9.1), for instance, a pollen diagram of the till-covered silts and clays has, if the rebedded Eemian pollen is excluded, an assemblage dominated by *Artemisia* and other nonarboreal pollen and with some *Betula*, thus a composition typical for a periglacial environment (Liivrand, 1991). This nonglacial interval was named Toravere after a similar site at Volguta, with silty clays between two till beds, the upper till here only consisting of one single bed. The interval was placed in the Middle Weichselian between the first Weichselian ice advance, named Mägiste, and the second, named Vargjärve, both stratigraphically represented by tills.

Further east, at the Petrozavodsk site on the western shore of

Lake Onega (Fig. 9.1), borings showed that the Eemian (Mikulian) marine clays are overlain by a till on which there are freshwater sediments that are partly covered by a second Weichselian (Valdaian) till (Lukashov, 1982), with the freshwater sediments thus in a stratigraphical position similar to those in Estonia. The freshwater sediments at Petrozavodsk consist of many meters of varved clay, 2 m of clay with organic material and sandy silty clay on top. The pollen diagram of the thick freshwater sediments has high percentages of *Betula* in the beginning and at the end, but is interrupted by a *Pinus* period, with *Betula, Picea,* and *Alnus* and with low percentages of *Corylus, Quercus,* and *Ulmus,* but they are clearly less frequent than in the Eemian marine sediments (Lukashov, 1982). It is difficult to assess to what extent the pollen diagram of the freshwater sediments has been influenced by reworked pollen from older sediments, as in similar sediments in Estonia, but the diagram shows that throughout most of the nonglacial interval the vegetation was dominated by *Betula,* but with a strong influence of herbs, such as *Artemsia.* This is in agreement with what is known from the Middle Weichselian periglacial area elsewhere, as already described. The freshwater sediments at Petrozavodsk were placed in the Middle Weichselian (Valdaian), in the Mologo–Sheksninian horizon between the Lower and Upper Valdaian horizons, each represented by a till (Lukashov, 1982), in agreement with the stratigraphical scheme of Eastern Europe as summarized by Liivrand (1991).

Even if there were probably some fluctuations of the ice margin in these areas, it seems likely that most parts of Scandinavia were covered by an ice sheet during the whole 50 ka-long Middle Weichselian Substage. The most extensive glaciation, reaching far outside Scandinavia, took place during the Late Weichselian, during the globally recorded short cold period before the Holocene. The Middle and Late Weichselian glaciation is treated as a whole in Chapter 10, as a boundary between the two substages cannot be established in the glacial deposits in Scandinavia.

9.2. Marine sediments

The detailed studies of foraminifera and mollusks of the previously described thick marine sediments in northern Jutland in Denmark, encountered mostly in deep boreholes, show that there is a complete marine succession from Eemian, in places even Saalian sediments, to sediments formed during the Early and Middle Weichselian. The marine sediments on top of those of the relatively warm Eemian Sea have been referred to as representing the cooler Skaerumhede Sea (L. A. Rasmussen, 1982), but in the detailed studies the foraminiferal assemblage zones have been correlated with the Weichselian substages, with some attempts at making correlations with interstadials found in the terrestrial sequences. It is especially the Middle Weichselian marine sediments that are well represented in northern Jutland (Fig. 9.1). The foraminiferal zones of the sediments of the Skaerumhede I (Knudsen & Lykke-Andersen, 1982) and the Skaerumhede II

(Bahnson et al., 1974) borings and those of the Apholm boring (Knudsen, 1984), all near Fredrikshavn, as well as those of the sediments of the coastal cliff at Hirtshals (Knudsen & Lykke-Andersen, 1982) and of the Nørre Lyngby boring (Lykke-Andersen, 1987) on the west coast of northern Jutland, in Vendsyssel, have been correlated with each other and compared with macrofossil zones based on mollusks (Lykke-Andersen, 1987). The radiocarbon dates from Hirtshals listed in Table 9.1 of mollusks and plant material confirm the Middle Weichselian age of the sediments; the older dates were considered to represent sediments formed during the Moershoofd Interstadial (Knudsen & Lykke-Andersen, 1982; Lykke-Andersen, 1987). Further, the differences between the Middle Weichselian foraminiferal assemblage zones, registered in all marine sequences in Vendsyssel, showing alternating arctic and boreoarctic conditions, have been taken to reflect the stadials and interstadials in the terrestrial sequences. In this stratigraphical scheme (Fig. 9.2) the Moershoofd Interstadial is bounded by two stadials and uppermost there is an interstadial correlated with the Hengelo–Denekamp Interstadials (Lykke-Andersen, 1987; Frederica & Knudsen, 1990; B. Andersen & Mangerud, 1990). There are no similar indications of interstadials and stadials in the older Early Weichselian sediments in the same area, as noted earlier.

Whereas the thick marine sediments in northern Jutland make it possible to follow the development from the Eemian into the Early Weichselian and further to the Middle Weichselian, encountered in several boreholes, the marine sediments of the last substage mentioned are more difficult to define and date at the coastal sites in western Norway (Fig. 9.1). The marine shallow-water sediments are here intercalated with other sediments, often with tills, in very varied sedimentary sequences. The mostly cold to cool foraminiferal and molluskan faunas, in places with a boreal type of fauna, have been placed in the Middle Weichselian Substage and correlated with each other primarily on the basis of radiocarbon dates of molluskan shells. In the study of several sites in Jaeren and on Karmøy, two Middle Weichselian interstadials were identified, an older Nygaard Interstadial and a younger Sandnes Interstadial, by combining the stratigraphical evidence and using radiocarbon dates from a number of exposures (B. Andersen et al., 1981). The dates listed in Table 9.1 are only those that were specifically mentioned as referring to these two interstadials out of a total of 24 radiocarbon dates listed. Three shell dates for the Sandnes Interstadial give the age for a boreoarctic fauna from a section with marine sediments covered by a diamicton in Oppstad in Jaeren. Two dates are for outer and inner fractions of shells from the till-covered clay of the Kaberg clay section of Foss-Eigeland, and one date is shell material from the pit at Håve in Sandnes, from the zone I beds of a section with till-covered glaciomarine sediments. It was on the basis of these sediments, with alternating arctic and boreoarctic to high boreal faunas, that the Sandnes Interstadial was first named by Feyling-Hanssen (1971, 1974), and the site should therefore – by definition – be considered as the type site for this interstadial. The radiocarbon ages

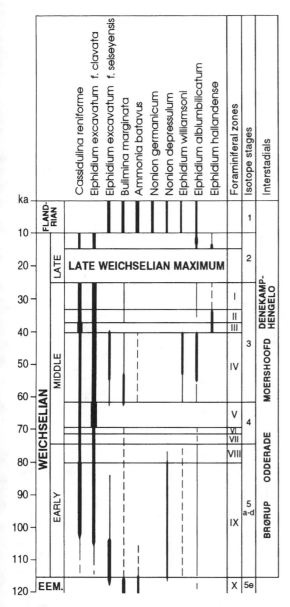

Figure 9.2. Foraminiferal stratigraphy with main species of Weichselian and Flandrian marine sequences in Vendsyssel, Denmark (after Fredericia & Knudsen, 1990).

ent, but a separation of the two is supported by the observation that sea-level was low during the older interstadial and high during the younger interstadial. The altitudes for the marine sediments from which the samples were taken were 5 m and 15 m for those of the Nygaard Interstadial and 65 m and 170 m for those of the Sandnes Interstadial. As it is not likely that this clear difference in altitude was caused by tectonic movements (B. Anderson et al., 1981) the difference has been explained as the result of a glacioisostatic depression caused by an ice advance between the two interstadials (Mangerud, 1983). The main Weichselian ice advance, which covered the whole coastal area, however, took place after the Sandnes Interstadial. Before that the ice margin oscillated near the coast, depositing tills and glaciomarine sediments.

In the detailed study of the site at Bø on Karmøy (B. Andersen, Sejrup, & Kirkhus, 1983), only about 4 km from the site of Nygaard, a diamicton-covered sand formation was identified, consisting of 8 beds of laminated sand, silt sand, and silty clay, with a total thickness of 5 m. The four radiocarbon ages (Table 9.1) of *Mya truncata* from these beds, defined as representing the Bø Interstadial, are about 37 ka to 42 ka old. The pollen assemblages of the marine beds have a strong influence of reworked pollen and are therefore of little help in the identification of the interstadial, but the relatively high values of nonarboreal pollen, including *Artemisia,* suggests that the vegetation was treeless (B. Andersen, Sejrup, & Kirkhus, 1983). The sediments of the Bø Interstadial are younger than those of the Early Weichselian Torvastad Interstadial at Bø, but the relationship of the former to the other interstadials is not quite established (Sejrup, 1987). There is some doubt about the radiocarbon ages being minimum ages (Mangerud, 1981, 1983), but it is possible that in age the Bø Interstadial broadly corresponds to the Nygaard Interstadial. In the identification of the latter, radiocarbon ages were in fact used both from the Bø and the Nygaard sites (B. Andersen et al., 1981), and as seen in Table 9.1 the ages for the Bø Interstadial are similar to those for the Nygaard Interstadial from Bø, all ages deriving from *Mya truncata* shells. The interstadial character of the sediments at Bø were clearly demonstrated in the detailed study of its mollusks and foraminifera, the biostratigraphical sequence starting with the Avaldsnes (Eemian) sediments and including the Early Weichselian Torvastad Interstadial (Sejrup, 1987).

In Sunnmøre, further north along the Norwegian coast, there are several sites with shell-bearing tills from which the shells have been dated. This was used as evidence that there was an ice-free period representing the time when the shells lived in the area, before they were picked up by an advancing ice and interbedded in the till deposited by the ice. The interval was named the Ålesund, first dated with 6 radiocarbon ages (Mangerud et al., 1979). A later more detailed study listed 15 ages (Mangerud et al., 1981b), of which only the youngest and the oldest are given in Table 9.1, showing that the interstadial has, according to these dates, an age of about 28 ka to 38 ka B.P. In general correlations of the Middle Weichselian interstadials

quoted as dating the older Nygaard Interstadial are all from Karmøy, from the profiles at Bø and Nygaard, of which the former was later studied in more detail and will shortly be discussed. At both sites there are marine till-covered sediments, and the outer and inner fractions of thick *Mya truncata* shells from the *Mya* bed at Bø were dated, as well as two samples of *Mya* shells from the sediments at Nygaard, from the *Mya truncata* zone of the Nygaard gravel unit. The Nygaard Interstadial was named after this unit and the profile is therefore the type site for this interstadial (B. Andersen et al., 1981).

Using only the radiocarbon dates listed in Table 9.1 an age difference between the two preceding interstadials is not appar-

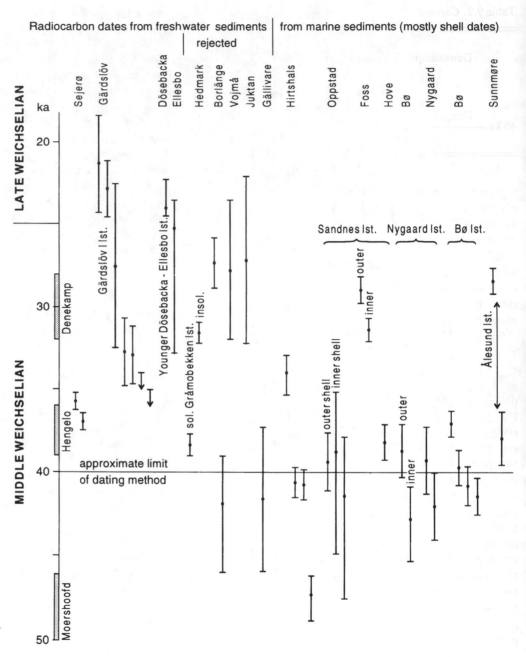

Figure 9.3. Middle Weichselian radiocarbon dates.

identified at the Norwegian coast, the Ålesund Interstadial has been correlated with the Sandnes Interstadial (Mangerud, 1981, 1983). As there is an element of uncertainty in basing the correlations of the interstadials on radiocarbon dates with a large amplitude of ages (Table 9.1), the use of the local names for the interstadials is particularly important here, even if their original definitions vary. On the Norwegian coast there are thus four presumably Middle Weichselian interstadials: Sandnes, Ålesund, Nygaard, and Bø. The correlation of these and the previously described interstadials represented by freshwater sediments will be discussed next, before a discussion of their use in delimiting times of glaciations.

9.3. Correlation of Middle Weichselian interstadials

The sites with sediments described as representing Middle Weichselian interstadials and listed in Table 9.1 have in various summaries been correlated with the interstadials defined further south, shown in Table 7.1. The three uppermost interstadials have been given fairly accurate ages for their lower and upper boundaries, on the basis of radiocarbon dates, but the two older interstadials are clearly beyond the reach for radiocarbon dating, as noted earlier. In comparing the Scandinavian Middle Weichselian interstadials with one another there arises, first, the problem of judging how the ages of such great amplitude – and

Table 9.2. *Correlation of Middle Weichselian interstadials in Scandinavia*

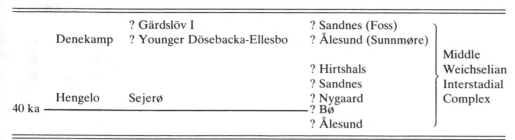

	Denekamp	? Gärdslöv I ? Younger Dösebacka-Ellesbo	? Sandnes (Foss) ? Ålesund (Sunnmøre)	
			? Hirtshals	Middle Weichselian Interstadial Complex
	Hengelo	Sejerø	? Sandnes ? Nygaard	
40 ka			? Bø ? Ålesund	

this applies even to samples from the same site – can be used in placing a local interstadial in the general stratigraphical scheme, as seen in Fig. 9.3, in which the earlier listed radiocarbon ages with their standard errors are shown. Secondly, as the sediments dated and used for correlations were formed in an environment with a cool climate, they do not all necessarily represent conditions that should be defined as interstadial; they merely show that the sites were ice-free and therefore represent nonglacial intervals during the cold Weichselian Stage.

Some of the radiocarbon ages were, as mentioned, rejected because the material on which the dates were based could not be considered as Middle Weichselian. The other dates range from just over 20 ka to nearly 50 ka. Of all the sites shown in Fig. 9.3 only Sejerø in Denmark can be considered to correspond to the Hengelo Interstadial, as is shown by the two dates with a small standard error. All the other dates are more or less difficult to set. The dates for the Gärdslöv I and Younger Dösebacka–Ellesbo Interstadials are, however, mostly in the younger part of the Middle Weichselian and these interstadials, may therefore – at least partly – represent the Denekamp Interstadial. So may some of the sediments described as representing the Sandnes Interstadial, that is, those at Foss, and some of the sediments described as representing the Ålesund Interstadial in Sunnmøre (Table 9.2). The older radiocarbon ages, including those from Hirtshals in Denmark, group themselves close to the age of 40 ka, close to the limit of the radiocarbon method, which, however, may sometimes be extended to 60 ka. It is therefore questionable whether the ages of around 40 ka should be given as finite dates or if they should be considered infinite, as are the two oldest dates from Gärdslöv. In using only radiocarbon dates, the

sediments dated at Hirtshals, Oppstad, Hove, Bø, and Nygaard have to be placed somewhere near 40 ka, perhaps lower in the scheme. The previously mentioned observations of the differences between the water depths during the Nygaard and Sandnes Interstadials have, however, to be taken into account (B. Andersen et al., 1981), even though dates from a particular area taken to represent only one of these interstadials may in fact represent both, as indicated in Table 9.2, or ice-free intervals between them. The conclusions based on dating methods other than the radiocarbon method will be discussed later, to see if the stratigraphical scheme represented here has to be modified or may be made more accurate.

The sites with marine sediments at the Norwegian coast were close to the ice margin, which fluctuated during the Middle Weichselian, and their sediments were therefore sensitive to minor ice advances not necessarily recorded as stadials elsewhere. This is reflected in the difficulties encountered in schematic correlations constructed for this coast (B. Andersen & Mangerud, 1990; Larsen & Sejrup, 1990). Many of the correlations presented in Table 9.2, however, agree broadly with comparisons made for Scandinavia between Norway, southern Sweden, and Denmark (Lundqvist, 1986). But on the basis of the material available it is more realistic to group all interstadials mentioned in Table 9.2 together and take them to represent a Middle Weichselian Interstadial Complex, as presented by Garcia Ambrosiani (1990), with the exception of Sejerø, which is likely to correspond to Hengelo. As already seen, the use of radiocarbon dates has also led to the conclusion that no interstadial sediments in Scandinavia are from the early part of the Middle Weichselian.

10 Middle and Late Weichselian glaciation

10.1. Pattern of glaciation and till stratigraphy

The spatial distribution of interstadial sediments dated as Early or Middle Weichselian and their relationship to till beds have led to the construction of schematic glaciation curves for the whole Weichselian in Scandinavia. The curve presented in Fig. 10.1 summarizes the evidence from the interstadial sites discussed earlier and shows their relationship to the till beds and the glacial advances during which they were formed. The glaciation curve shows the correlation with the Weichselian substages and the Early and Middle Weichselian interstadials as earlier presented in Table 7.1. For a couple of sites there are two alternative correlations. The time scale, which is based on the deep-sea oxygen-isotope curve with its stages (Martinson et al., 1987), is given in the middle of Fig. 10.1, which separates the areas on the western side of the Scandinavian mountains from those on the eastern and southeastern side.

The two restricted Early Weichselian advances from the Scandinavian mountains, of which the older was more extensive, were discussed earlier. Many of the Early Weichselian interstadial sites lie between a lower till (III) and one or two upper tills (I and II), as in Finnish Lapland, or are only covered by one till, as in Ostrobothnia on the Finnish west coast. They can therefore stratigraphically not be referred to either Brørup or Odderade; the previously mentioned, more detailed correlations with mostly Brørup were based on pollen-stratigraphical evidence.

The main Weichselian glaciation began after the Odderade Interstadial, at the beginning of the Middle Weichselian Substage, but did not reach its maximum extent until the Late Weichselian. Formal boundaries between these substages, let alone between interstadials and stadials, cannot, however, be used for the glacial sediments formed during these glacial advances, as these sediments are time-transgressive. A detailed correlation of the till beds identified in the whole area of glaciation is therefore difficult.

As no Middle Weichselian interstadial sites have been found in the central areas of glaciation (Fig. 9.1) it has been assumed that Finland and most parts of Sweden and Norway were glaciated throughout this substage, as well as during the Late Weichselian.

The single glaciation was, however, divided into two glacial advances in the marginal parts where sediments interpreted as representing Middle Weichselian ice-free intervals lie between post–Early Weichselian tills. The extent of the ice sheet during the time it was most restricted in the Middle Weichselian, corresponding to isotope stage 3, is indicated on the map in Fig. 8.4, a map slightly modified from that published by B. Andersen and Mangerud (1990). The extent of the Middle Weichselian glaciation, correlated with isotope stage 4, is difficult to determine on the basis of the present evidence. At Holmstrup in Denmark, Middle Weichselian marine sediments rest on a till bed younger than the Eemian (Petersen, 1984a) and which could thus be correlated with isotope stage 4 (B. Andersen & Mangerud, 1990), a possibility also mentioned by Houmark-Nielsen (1990). This would agree with the interpretation of the stratigraphy in northern Poland, where an early Middle Weichselian ice advance preceded the main Late Weichselian advance, with an ice-free interval in between (Bowen et al., 1986), an interpretation considered uncertain by Sibrava (1986) but which has still been taken into account in later studies (Eissmann, 1990). Further east there is, however, strong evidence of an early Middle Weichselian ice advance, as seen at the sites with periglacial sediments between Weichselian tills postdating the Eemian (Liivrand, 1991). The early Middle Weichselian violet-brown till in Estonia represents the local Mägiste Stadial (Fig. 10.1), which has also been identified in Latvia and Lithuania. The ice sheet did not, however, reach as far as Belarus, to the marginal areas of the later Weichselian glaciation (Liivrand, 1991). As the Late Weichselian glaciation covered a larger area than the early Middle Weichselian glaciation, the position of the ice margin during the latter substage has been difficult to trace. Further north, in Karelia, there is also a till that separates the Eemian (Mikulian) marine sediments from the Middle Weichselian (Valdaian) freshwater sediments, as seen from the site at Petrozavodsk (Lukashov, 1982). Thus, the Weichselian till stratigraphy is here similar to that in Estonia, both areas being outside that covered by ice during the whole of the Middle Weichselian (Fig. 8.4).

As noted, there were probably two ice-advances after the

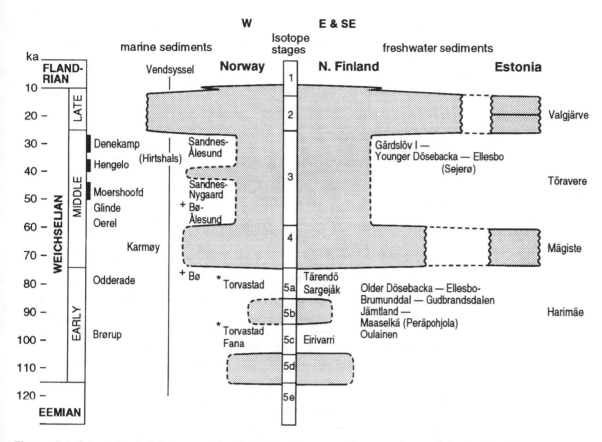

Figure 10.1. Schematic glaciation curves for the Weichselian west and east-southeast of the Scandinavian mountains compared with interstadials and deep-sea isotope stages. (Adapted from B. Anderson & Mangerud, 1990, with additions.)

Early Weichselian on the eastern and southeastern side of the Scandinavian mountains, which can be demonstrated outside the central area that was continuously glaciated over 60 ka until the beginning of the Flandrian (Fig. 10.1). The same main pattern can be seen on the western side of the mountains, at the west coast of Norway, at several marine Middle Weichselian sites already described. The exact correlation of some of the Middle Weichselian interstadials, as those of the Early Weichselian, has, as discussed earlier and as shown in Fig. 10.1, been problematic. The tills or diamictons related to these sites have therefore been difficult to place in the stratigraphical scheme. Further, on the Norwegian coast, close to the Scandinavian mountains, there were several local ice-advances in addition to the main ones. But two Early Weichselian minor ice-advances already described, and two major advances, on the eastern side of the mountains, can be singled out as being stratigraphically well established. An extensive early Middle Weichselian glaciation covering the northern parts of the North Sea has been demonstrated at Karmøy, the till bed representing the Karmøy Stadial, during which tills were also deposited further south in Jaeren (B. Andersen, Wangen, & Østmo, 1987; B. Andersen & Mangerud, 1990). Further north on the west coast of Norway, at Vigra and Godøya in Sunnmøre, there are several tills postdating the Early

Weichselian, and two of these have been placed in the early Middle Weichselian (Landvik & Hamborg, 1987). The till fabric shows that the lower till (lower part of the Synes Lower Till, Vikebugt Till) was deposited by an ice movement toward west-southwest (Fig. 8.4), a movement influenced by the orientation of the fjords, whereas the overlying till (upper part of Synes Lower Till, Till F and Strandkleiv Lower Till) was deposited by an ice movement toward west-northwest and northwest when the ice sheet had grown and was more independent of the underlying morphology (Landvik & Hamborg, 1987). The tills are separated from the Late Weichselian till by inter-till sediments laid down during an ice-free interval at this coast. The general withdrawal of the ice sheet at the time correlated with isotope stage 3 was probably interrupted at the Norwegian coast by an ice-advance at about 40 ka, during which till beds were laid down in a glaciomarine environment (B. Andersen & Mangerud, 1990), corresponding in age with the Skjonghelleren Stadial (Larsen & Sejrup, 1990).

From the map in Fig. 8.4 it can be seen that there is evidence of an ice-advance in southwestern Norway as well as in the southeastern part of the Middle Weichselian glaciation, whereas the possibility of an advance through southern Sweden to Denmark and Poland is uncertain. The stratigraphical sequence in Vendsyssel in northern Jutland, including the Middle Weichselian marine sediments at Hirtshals, however, clearly shows that this area was not glaciated after the Eemian until during the Late Weichselian. Vendsyssel would thus have remained ice-free, and submerged, at a time when the ice sheet spread both northwest of Vendsyssel into the North Sea and possibly east of it into southern Denmark, as shown on maps presented by Petersen (1985) and Houmark-Nielsen (1989). But, as seen later, Denmark was even later an area into which ice-lobes penetrated both from the north and from the Baltic basin in the east.

The schematical glaciation curves in Fig. 10.1 and the map in Fig. 8.4 show, as already mentioned, that most parts of Scandinavia were glaciated from about 74 ka until the deglaciation about 10 ka ago. This means that any Weichselian till within the outer limit of the ice sheet during isotope stage 3 and postdating the Early Weichselian could have been deposited at any time during this period. It has, however, been assumed that the tills at the surface were mainly formed during the extensive Late Weichselian glaciation and its retreat. That they are not older than Middle Weichselian can be confirmed at sites where they overlie Early Weichselian sediments. These sites also show if there are any remnants of tills from the time of the early Middle Weichselian glaciation, tills that could be correlated with those identified outside the outer limit of the ice sheet of isotope stage 3, as could the till from the Karmøy Stadial in Norway and the till from the Mägiste Stadial in Estonia. Some of the early Middle Weichselian tills, however, could have been reworked and redeposited during the Late Weichselian glaciation inside the limit of glaciation.

When dealing with an area as large as Scandinavia the diachronous character of the till beds has to be taken into account, as well as the origin of these beds. By studying the lithology of tills, till beds can be correlated with one another, but in many areas these correlations have chiefly been done with the help of till fabric analyses. The preferred orientation of the clasts reflect the episodes of ice-advances, with often a different flow pattern of ice for each episode. This approach in studies of the till stratigraphy has been particularly successful in the areas east of the Scandinavian mountains, as in Finland. In the mountains themselves and in the Norwegian coastal areas the local topographical variations resulted in more varied flow patterns, more difficult to reconstruct. Each till bed, however, does not represent a glaciation. The best preserved and most common tills are from the last glaciation and it is especially these that lithologically can be divided according to the generally used classification of tills based on the formation and deposition of tills (Dreimanis & Lundqvist, 1984).

There are several examples of how, during a single glacial episode, two distinct till beds were formed, as demonstrated in a study of the Weichselian and pre-Weichselian sediments in southern Finland (Bouchard, Gibbard, & Salonen, 1990). During the last glacial episode a lodgment till, a compact dark gray till with striated clasts, the Siuntio till, was deposited by an advancing ice sheet. After a period of quiescence with subglacial meltwater activity depositing subglacial sands, a reactivated ice

sheet with a lobate flow pattern deposited a basal melt-out sandy till, the Espoo till, with a loose structure and angular clasts (Bouchard, Gibbard, & Salonen, 1990). That last till is widespread in the coastal areas of southern Finland, whereas tills similar to the lodgment till are rare in exposures. The till fabric of the Siuntio lodgment till showed that it was deposited by an ice-advance from the north and of the melt-out till that it was formed by an ice sheet flowing from the northwest, which was the predominant movement during the last event of the glaciation (Fig. 10.2). The lodgment till is similar in origin to the "dark till" described from a number of sites in southern and central Finland (Rainio & Lahermo, 1976, 1984), in southern Finland also called the "old till" (Hirvas & Nenonen, 1987). It is, as mentioned above, separated from the younger till by sand and gravel. East of Hyvinkää, south of the Salpausselkä moraines, a remnant of the lodgment till was found preserved underneath an esker delta, in a position where tills are generally not found, as glaciofluvial sediments normally rest directly on the bedrock (Kurkinen, Niemelä, & Tikkanen, 1989). The till fabric-studies in southern Finland generally show that the ice movement at the time of the deposition of the lodgment till was from the north (Hirvas & Nenonen, 1987), as when the Suintio till was deposited. In addition, more westerly directions were obtained for the "dark till" in southwestern Finland (Rainio & Lahermo, 1976). There are thus differences in the till fabric directions within the same area for the older lodgment till, which show that they were formed at different times during a glaciation. A northern direction of ice-advance (Fig. 10.2) was also demonstrated for the lodgment till in Ostrobothnia (Bouchard, Gibbard, & Salonen, 1990), this till named the Kauhajoki till, with the stratotype at Harrinkangas (Gibbard et al., 1989). In several exposures of this till in Ostrobothnia (Niemelä & Tynni, 1979) there is a facies change near the top into a basal melt-out till (Bouchard, Gibbard, & Salonen, 1990).

A till genetically similar to that in Finland has been found at several localities in central and northern Sweden, called the "old clayey till" (Björnbom, 1979). The dark color of this till, first interpreted as due to a high clay content as a result of incorporated Eemian sediments, has later been explained as having been mainly caused by a ferrous sulphide pigment, precipitated from the ground water (Rainio & Lahermo, 1984). The till fabric studies of the "old clayey till" in Sweden showed that the ice movement was from the west and northwest (Björnbom, 1979), similar to the directions in southwestern Finland (Fig. 10.2).

Even if the deposition of the lodgment till in these areas was time-transgressive, as already seen from the differences in the directions of ice movements, it has been suggested that the till was formed during the initial stage of the last glaciation, during the first ice advance (Rainio & Lahermo, 1976, 1984; Björnbom, 1979; Bouchard, Gibbard, & Salonen, 1990). If, as earlier concluded, the areas in Finland and Sweden in which the dark lodgment till occurs were glaciated throughout the Middle and Late Weichselian, the deposition of the lodgment till started in Early Middle Weichselian time, during the advance correlated

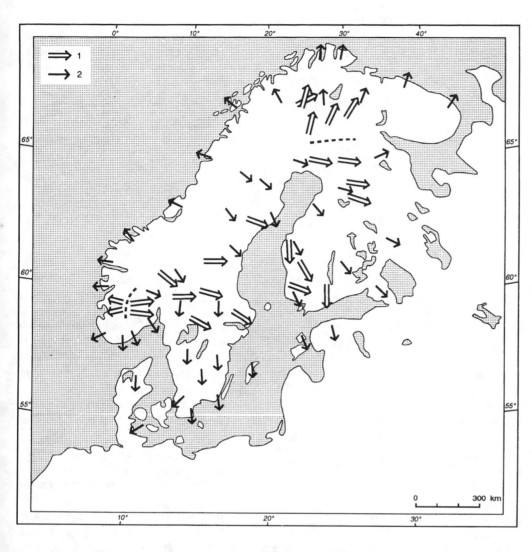

Figure 10.2. Middle and partly Late Weichselian till fabric directions, 1, and main movements of ice during the Late Weichselian as shown by striae, 2.

with isotope stage 4 (Fig. 10.1). The till would thus have been formed during the same ice-advance as that which deposited the early Weichselian till in Estonia, during the Mägiste Stadial. The lodgment till in southern and central Finland, including that in Ostrobothnia, with a till fabric showing an advance from the north, would have been laid down during the initial early Middle Weichselian advance from the north. The lodgment till formed at a time when the ice movement was more from the west would have been deposited at a later stage of the same glacial episode, possibly closer to the time of the Middle Weichselian retreat.

Tills genetically similar to the lodgment tills described above have been found further north in Finland. In Koillismaa, in the Kuusamo and Suomussalmi areas in northeastern Finland, dark gray tills, the Pudasjärvi and Soivio tills, were deposited during an ice-advance from the west-northwest called the Tuoppajärvi flow phase (Aario & Forsström, 1979). Further south, in Kainuu around Suomussalmi and Kuhmo, a similar till (S IV and K III) covered by younger tills (S I–III and K I–II) was also deposited by an ice flow with a similar direction (Saarnisto & Peltoniemi, 1984). Whereas the younger tills in Koillismaa and Kainuu were

deposited during the final deglaciation of the area, ending about 9.5 ka ago, the underlying older tills were formed earlier during the same glaciation, but a more exact dating of them could not be given in this area (Saarnisto & Peltoniemi, 1984). It is, however, clear that they were deposited further away from the ice margin, when the ice sheet was larger and the direction of ice movement was more uniform, in contrast to the varied movements of ice during the deglaciation. But it is also clear that the older tills in these areas are younger than the dark lodgment tills in Ostrobothnia and southern Finland, as is clearly shown by the difference in the directions between the ice-advances that deposited these tills (Fig. 10.2).

In Finnish Lapland there are two tills postdating the Early Weichselian interstadials, till beds I and II (Hirvas, 1991). In addition to the younger till (I), interpreted as a basal till, there is an ablation till or surficial till deposited during the final stages of deglaciation. Till bed II was deposited during a flow stage that covered the whole area, with a direction of flow from west to east in the southern parts and toward the northeast in the northern parts, directions that agree with those for the older

Table 10.1. *Middle and Late Weichselian tills in Sweden, Finland, and northern Norway*

	Sweden	S Finland	Ostrobothnia	Koillismaa & Kainuu	Finnish Lapland	Finnmarksvidda, Norway
Late & Middle Weichselian	W III	Espoo till		S I–III & K I–II	I	Kautokeino till B & C
					sand & gravel	Sargejåk Sand
				Pudasjärvi & Soivio tills (Tuoppajärvi flow phase) S IV & K III	II	Vuoddasjavri till D
	W II	Siuntio till (dark till, old till)	Kauhajoki till			
Early Weichselian						

tills further south in Koillismaa and Kainuu (Fig. 10.2). Till beds I and II are in places separated by sands and gravels, and in the northeastern part, the surface of till bed II has frost wedges and a boulder pavement, which indicate that there was at least an ice-free interval between the deposition of the two till beds (Hirvas, 1991). Whether this interval is only local or has a regional stratigraphical significance cannot be determined, but further south in Kainuu, for instance, sorted sediments were recorded between the older till and the younger tills (Saarnisto & Peltoniemi, 1984). They are different from the subglacial sands between the tills in southern Finland described earlier.

If the main Middle and Late Weichselian tills are schematically and tentatively compared with one another (Table 10.1), it appears that most of the lodgment tills in southern Finland and Ostrobothnia are probably older than till bed II in Lapland and the corresponding tills in Koillismaa and Kainuu in northeastern Finland. Further, there is an indication of an ice-free interval in the northeast and in the far north, or in places of local intervals caused by oscillations of the ice margin during its Late Weichselian retreat. The same pattern of glaciation as in Finnish Lapland has been described from Finnmarksvidda in northern Norway (Olsen & Hamborg, 1983, 1984; Olsen, 1988). The upper formation, the Kautokeino till, including a basal melt-out till, B, and a lower lodgment till, C, are separated from an underlying till, the Vuoddasjavri till, D, by a sand, the Sargejåk Sand (Olsen, 1988). This sand, interpreted as representing an interstadial, was correlated with the Odderade Interstadial (see Table 8.1), but as the underlying till D has been correlated with till II in Finnish Lapland (Olsen, 1988), which is Middle or Late Weichselian (Hirvas, 1991), till D must also postdate the Early Weichselian interstadials and the sand represent a younger ice-free interval. It is thus stratigraphically in the same position as the interval between tills II and I in Finnish Lapland, mentioned above. This correlation is supported by the results from the till

fabric studies. The main direction of ice flow during the deposition of till D in Finnmarksvidda was from the southwest, as in northern Lapland (Fig. 10.2), but the initial movement was more from the west because the ice center at that time was closer to the Norwegian coast in the west (Olsen, 1988, 1990).

Northern Sweden, where both Late Weichselian interstadials were demonstrated and where the younger Tärendö Interstadial corresponds to Odderade (see Table 8.1), was overrun by two ice advances in Middle and Late Weichselian time. There was very little erosion and older landforms from Early Weichselian time were therefore preserved, for example, as irregular moraines (Lagerbäck, 1988a, 1988b).

When the till stratigraphies in these areas are compared with one another (Table 10.1) it seems likely that in the north and northeast there was an ice-free interval interrupting the long Middle and Late Weichselian glaciation. The nature of this interval, however, is not well known, nor can even a tentative age be given for it. The broad outlines of a glaciation with two active ice-advances in Finland generally agrees with the results from Sweden. As noted, there is even no evidence of an ice-free Middle Weichselian interval, except in southern Sweden, where the glaciation after the Early Weichselian Jämtland Interstadial was twofold, being divided into the "Second Weichselian glaciation" (W II), immediately following the Early Weichselian, with the "Third Weichselian glaciation" (W III), dated as Late Weichselian (Lundqvist, 1981, 1986a). The dark, bluish-gray till, similar to the till in southern Finland, would have been deposited during the older of these two ice-advances (Björnbom, 1979; Lundqvist, 1981). According to this till stratigraphy the early Middle Weichselian glaciation would not have reached further south in Sweden than to the latitude of Stockholm, which does not agree with a possible advance at that time to Denmark and northern Poland, shown in Fig. 8.4. There is thus still a discrepancy between the interpretations of the outer limit of the early Middle Weichselian ice-advance and also, to some extent, of the nature of the interval between this advance and the more extensive advance of the Late Weichselian.

The changes in the directions of ice flow during the twofold glaciation in Sweden and Finland, as well as during the older Early Weichselian glaciation, have been linked with the previously mentioned glaciation model put forward by Ljungner (1949), with flow patterns governed by changes in the position of the ice-divide in the Scandinavian mountains (Lundqvist, 1974, 1986b). The Prime Glaciation in Jungner's scheme, correlated with Weichsel I by Lundqvist, was placed between the Eemian and the Jämtland Interstadial and is therefore Early Weichselian. The Posterior Glaciation, which followed the interstadial, was correlated by Lundqvist with the subsequent twofold glaciation, Weichsel II and Weichsel III, of which the former is broadly equivalent to the early Middle Weichselian ice advance described above (Table 10.1). It is not possible to link the inadequately known till stratigraphy with the successive directions of ice movement during the Posterior Glaciation separated by Jungner with the help of striae in and around the Scandinavian mountains. A movement of the ice from west to east over some parts of Sweden and Finland was, however, explained as being the result of the displacement of the center of the ice sheet to an area east of the mountains, whereas at times when the ice-divide was closer to the mountain chain the direction of ice flow was more from the northwest. This influence on the ice flow caused by lateral changes of the ice-divide has, as mentioned earlier, been demonstrated clearly in the study of the till stratigraphy of Finnish Lapland (Hirvas, 1991).

The mountains of southern Norway were continuously glaciated in Middle and Late Weichselian time, but even there the changes of the ice-divide influenced the directions of ice movement. In Hardangervidda east of Bergen three phases were separated in the glaciation, the Førnes kryomer, after the Early Weichselian interstadial, the earlier mentioned Førnes thermomer (T. O. Vorren, 1979). During the oldest phase, II, of glaciation after the interstadial, the ice-divide was near the coast in the west and the main ice-advance had an easterly direction over Hardangervidda; only near the coast was there a movement towards the coast. In Gudbrandsdalen northeast of Hardangervidda several phases of ice movement during the last glaciation were also identified, with several tills of a single glacial cycle after the Late Weichselian. An initial valley glaciation, A, was followed by the main phase, B, during which a till bed up to 30 m thick was deposited by an ice flow to the east and east-southeast (Bergersen & Garnes, 1972, 1981; Garnes & Bergersen, 1977, 1980). Of the younger tills, C and D, C was correlated with the maximum extension of the ice sheet with an earlier ice-divide, and D with the deglaciation. The main directions of ice movements during the early phases of glaciation in Hardangervidda and Gudbrandsdalen show the same trend as the directions in central and northern Sweden, when the old clayey till was deposited.

It was previously concluded that the tills formed during the early part of the main Weichselian glaciation inside the limit of glaciation during isotope stage 3, as shown on the map in Fig. 8.4, are time-transgressive. As some of them may be younger than the early Middle Weichselian, the directions of ice flow presented in Fig. 10.2 were not included on the same map as the directions for the tills stratigraphically referred to as representing an early Middle Weichselian ice-advance during isotope stage 4 in the marginal areas of glaciation, outside the limit of the isotope 3 glaciation (see Fig. 8.4). Even if the age of some of these tills, as noted, is uncertain, their separation from Late Weichselian tills is based on the presence of nonglacial Middle Weichselian sediments between them. Thus, if early Middle Weichselian tills were preserved in the marginal parts of glaciation it is likely that at least some of the lodgment tills in the central parts are also from this time, as concluded in the studies referred to above. On the other hand, it is not likely that the lodgment tills, also presumably formed during the main Late Weichselian extensive glaciation, would be less well preserved than those formed during the early Middle Weichselian Substage.

In contrast to these tills that – on the whole – were laid down at times when the ice sheet was extensive and when the directions of ice flow were rather uniform over larger areas, as seen in Fig. 10.2, the tills laid down during the final phases of the Late Weichselian glaciation are different. They have often been dumped on top of the underlying deposits during the final stages of the decay of the ice sheet, near the retreating ice margin as subglacial or supraglacial melt-out tills, flow-tills, or as waterlain melt-out tills. When they show a preferred orientation of the clasts, the directions have great regional variations because the last movements were perpendicular to the lobe-shaped moraines formed during the retreat from the outermost position of the Late Weichselian ice sheet. There is a wealth of detailed information, from a great number of sites in Scandinavia, about the properties and formation of the youngest Late Weichselian tills, depicting the glaciological conditions during the withdrawal of the last ice sheet. Thus, they show, for instance, whether the retreating ice margin was grounded or afloat, and, further, if it was active or stagnant. In this way deductions about the manner in which the final retreat of the ice took place and about the formations of marginal landforms can be made. But the lithostratigraphical studies of these tills, however, give no additional evidence as to the chronology of the Late Weichselian Substage established with the help of other sediments.

The time-transgressive Late Weichselian tills have often preferred stone orientations similar to the directions of ice movements shown by the striae and interpreted as having been formed during the main advance of the Late Weichselian ice sheet, directions schematically shown in Fig. 10.2 according to the more detailed maps, such as those published by Glückert (1974, 1987). The directions for the tills referred to earlier as being partly early Middle Weichselian differ, on the other hand, from these directions, particularly in central Sweden and southwestern Finland. And the directions for the youngest Late Weichselian tills, which can generally be classified as being ablation tills, as, for instance, in Finnish Lapland (Hirvas, 1991) also differ from

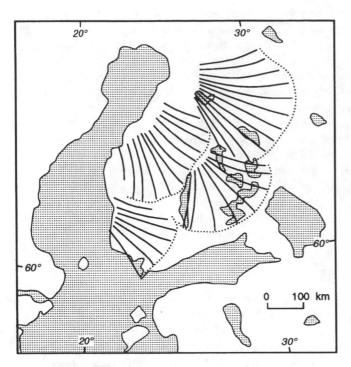

Figure 10.3. Time-transgressive ice lobes formed in Finland during last Weichselian deglaciation.

Figure 10.4. Rogen moraines in the type area near Lake Rogen, central Sweden. (Photo J. Lundqvist)

the main Late Weichselian directions. Their clasts have, as mentioned above, an orientation, if any, similar to the youngest striae, which show great regional variations depending on the shape of the retreating ice margin, as seen from the map schematically showing the pattern of the youngest striae produced by the Scandinavian ice sheet in Finland (Fig. 10.3). The general influence of the ice lobes on the youngest ice movements was already demonstrated on the Quaternary map of Finland presented by Hyyppä (1960), and has also been shown in detailed studies of smaller areas, such as in the investigation of northern Karelia by Repo (1957) and of the previously mentioned Koillismaa in northeastern Finland by R. Aario and Forsström (1979). The pattern with a lobe-shaped ice margin was particularly striking in Finland, in Denmark, and in areas south of the Baltic Sea, compared with areas closer to the Scandinavian mountains, as shown by the map of the glacigenic longitudinal lineations by Boulton et al. (1985). The Norwegian coast, on the other hand, shows variations on a smaller scale due to its irregular deep fjords and archipelago. But as a result of a more extensive Late Weichselian ice advance, compared with the earlier Weichselian advances, far beyond the present Norwegian coast, both in the north into the Barents Sea (T. O. Vorren & Kristoffersen, 1986) and in western Norway into the Norwegian Sea (Larsen & Sejrup, 1990), thick Late Weichselian tills were laid down on top of older Weichselian and in places Eemian sediments, as at Fjøsanger near Bergen (Mangerud et al., 1981b). There is also evidence from studies of ice movements that a dome of the ice sheet at times during the Late Weichselian and also earlier was

situated over the coastal area of southwestern Norway, thus west of the highest mountains (Anundsen, 1990).

This pattern of ice movements during the final phases of the Late Weichselian glaciation is also reflected in the distribution and orientation of glacial landforms which were formed in the marginal zone of the retreating ice sheet or at the margin (Table 10.2). Of the landforms a variety of elongated ridges classified as drumlins, formed in the inner marginal zone, may in places have a varied structure, but on the whole they consist of till and occur in flat areas (Lundqvist, 1977). In central Finland there are large areas in which the drumlins have a fan-shaped orientation, similar to the youngest striae (Glückert, 1973). The fluting observed mainly with the help of aerial photographs is a common feature reflecting the ice movements (Lundqvist, 1977) and is also interpreted as having been formed in the inner marginal zone of the ice (R. Aario, 1990). Elongated ridges classified as radial moraines were formed in the outer marginal zone, in crevasse fillings as some other forms in this zone (Lundqvist, 1977). Of these the Veiki moraines in northernmost Sweden, as defined by Hoppe (1952), consist of raised till surfaces surrounded by rim ridges, which may partly have been formed supraglacially and been modified by flow-till (Lundqvist, 1977). These irregular landforms formed during the last phases of deglaciation are related to the ridges and hummocks of till named Pulju moraines and described from northwestern Finnish Lapland (Kujansuu, 1967). The Rogen moraines consist of irregular ridges formed subglacially at right angles to the movement by an active ice (Fig. 10.4), but preserved because of dead-ice conditions during deglaciation (Lundqvist, 1977). They

Table 10.2. *The relationship of some glacial landforms to the ice margin*

Land form	Inner marginal zone	Outer marginal zone	Ice margin
Drumlins	x		
Fluting	x		
Radial moraines		x	
Veiki moraines		x	
Pulju moraines		x	
Rogen moraines	x	x	
De Geer moraines		(x)	x

were formed away from the ice-divide in contrast to the Pulju moraines that were deposited close to the last ice-divide. The orientation of the small distinct ridge of till called De Geer moraines, common in the lowlands of Sweden and in the south of Finland as well as on the west coast, is generally that of the ice margin, and they were formed in a relatively short time. Several ridges could be formed in a zone corresponding to the annual recession of the ice, but in some areas the intervals between the ridges correspond to the annual recession (Lundqvist, 1977). In a detailed study of the De Geer moraines in Finland, Zilliacus (1987) concluded that they were formed subglacially in basal crevasses and that their origin has some similarities with the Rogen moraines. Their orientation, however, reflects the general direction of the ice margin during its retreat, as seen in the area between Helsinki and Lahti in southern Finland, and on the west coast in the archipelago of Vaasa even small differences in the direction of the ice margin can be detected (Aartolahti, 1972). Further, in Sweden they have been found to curve around eskers, demonstrating "the formation of calving-bays around the outlets of glaciofluvial melt-water streams" (Lundqvist, 1977). And in the Møre area in western Norway, where De Geer moraines are common, they were interpreted as having been formed outside the margin of retreating ice-lobes in the fjords (Larsen, Longva, & Follestad, 1991). It seems, therefore, that even if some of the landforms were formed subglacially inside the ice margin (Table 10.2) their orientation can be used to reconstruct in detail the shape of the ice margin during its final retreat.

The majority of these landforms were formed during the retreat of the Late Weichselian ice sheet but, as already noted, older Early Weichselian landforms were preserved in northern Sweden in spite of having been later glaciated in Middle and Late Weichselian time (Lagerbäck, 1988a). The common Veiki moraines in this area, and also the less common Pulju moraines, consisting of both flow-till and poorly sorted gravel, as well as laminated silt, all covered by a till bed, were interpreted as being such landforms. This led to the conclusion that the erosion during the main Weichselian glaciation was very small in northern Sweden, close to the Scandinavian mountains, a

conclusion, if correct, that has a bearing on the glaciation models for these areas, as stressed by Lagerbäck (1988a). Even if the interpretation of the formation of the Veiki and Pulju moraines, restricted to the areas close to the former ice-divide, has to be modified, the other landforms (Table 10.2) seem to be clearly related to the pattern of the last deglaciation.

In the discussion of the evidence of the Middle and Late Weichselian glaciation, divided into two glaciations in the marginal parts, till beds and landforms were used in outlining the advances and withdrawals of the ice sheet as presented in the glaciation curves (Fig. 10.1). A certain uniform pattern could be traced in the area dealt with, well inside outer margin of the Late Weichselian glaciation. Denmark and Scania in southernmost Sweden differ, however, from the rest of the area. It is the only area close to the outer margin of the Weichselian glaciation, situated in Jutland. It is an area that was affected by ice advances of lobes both from Norway in the north and Sweden in the northeast, as well as from the Baltic basin from the east. This affected the lithostratigraphy of the area that has been linked with the moraines formed during the Late Weichselian deglaciation. Although these end moraines are discussed later, in connection with the Late Weichselian retreat of the ice, the stratigraphical position of the till beds is dealt with here. In identifying and correlating tills, their petrographic composition and till fabric were used and, especially in Denmark, kinetostratigraphy, where glaciotectonic data are used as a stratigraphical tool in separating sedimentary units affected by particular directions of ice movement (Berthelsen, 1979; Sjørring, 1982).

Table 10.3 shows the correlation of the Late Weichselian till stratigraphy in southernmost Sweden, in Scania (Skåne), with that in Denmark, and the correlation with the ice-advances and moraines, as well as the chronostratigraphy, which, however, is discussed in greater detail later. The directions and extensions of the separate ice-advances are shown on the maps in Fig. 10.5. In southern Sweden individual till beds could be accumulated during successive ice-advances, during which the directions of the ice movements changed considerably, which resulted in differences in petrographic composition between the basal and upper parts of the tills. The two oldest ice-advances are the Norwegian advance from the north over Jutland and the northern parts of the island of Sjaelland in Denmark, which deposited the Norwegian till, and the Old Baltic advance from the southeast from the Baltic basin, which deposited the Old Baltic till in Denmark. This advance also affected southernmost Sweden and deposited tills rich in Palaeozoic limestones – in Scania the Korsaröd, Bocksbacke, and Kvarnby tills, according to Ringberg (1988, 1989), as seen from the correlation shown in Table 10.3. The Norwegian advance is considered to have preceded the Old Baltic advance but has also been held to be younger than this advance. Further, even if an early Late Weichselian age has been given this advance in Table 10.3, an age between that for the Eemian and 47 ka has also been suggested for it (Petersen, 1984a), thus possibly an Early Weichselian age. It was already noted that the age of some older Weichselian tills

Table 10.3. *Correlation of Late Weichselian tills in Scania (Skåne) in southern Sweden and Denmark*

Moraines and ice advances		Denmark	Island of Ven	S Sweden (Scania)	Late Weichselian chronostratigraphy	
E (YB 2)	Low Baltic / Baelthav (Belt Sea) advance (Young Baltic ice 2)	Young Baltic tills	Kyrkbacken till	Malmö & Kågeröd tills	Bølling Oldest Dryas	13.0 ka BP 13.2 ka BP
				clays & gravels	Lockarp	
						13.5 ka BP
			Laebrink till	Upper part of S Sallerup till		
D (YB 1)	East Jylland (Jutland) advance (Young Baltic ice 1)	East Jylland till		Upper parts of Eslöv & Hardeberga tills & middle part of S Sallerup till		
C	Main stationary line (Main Weichselian advance, Northern ice)	Mid Danish till	Västernäs till	Basal parts of Hässleholm, Eslöv, Hardeberga, & S Sallerup tills		
				clays, sands & gravels		
	Old Baltic advance	Old Baltic till[a]		Korsaröd, Boksbacke, & Kvarnby tills		
	Norwegian advance	Norwegian till				21.5 ka BP

[a]Possibly Middle Weichselian
Sources: Mainly according to Sjörring (1974, 1982) and Ringberg (1988, 1989).

in Denmark are difficult to date, but, on the other hand, there are no tills in southernmost Sweden that can be placed in the Early or Middle Weichselian (Ringberg, 1988). After these two ice advances Denmark was overrun by the Weichselian ice, first out to its marginal position in Jutland, to the Main Stationary Line (C), during an advance, the Main Weichselian advance, of the Northeast Ice, which, in northern Jutland, reached a position close to the earlier margin of the Norwegian advance. The Mid Danish till in Denmark, as well as the Västernäs till on the small Danish island of Ven between Sjaelland and Scania, are correlated with the Main Weichselian advance, and also the basal parts of some tills in Scania, which are separated from the underlying older tills by clays, sands, and gravel formed during an interval. The direction of the ice movement became more easterly after the retreat of the ice margin from the Main Stationary Line, during the East Jylland (Jutland) advance, also called the Young Baltic Ice 1, YB 1 (= line D). The East Jylland till (= lower Young Baltic till) and the slightly younger Laebrink till on the island of Ven have been correlated with several tills in Scania, as shown in Table 10.3. After an ice-free interval in southernmost Sweden called Lockarp, but not interpreted as an interstadial by Berglund and Lagerlund (1981), during which clays and gravels were deposited, an ice lobe curved round the southern Baltic basin and reached Denmark and the southwest coast of Scania from the southeast. This Low Baltic or Baelthav

(Belt Sea) advance, also called the Young Baltic Ice 2, YB 2 (= line E), deposited a chalky till, in Denmark represented by the upper parts of the Young Baltic tills (Fig. 10.6) and in the Scanian till stratigraphy correlated with the Malmö and Kågeröd tills. These, as seen from Table 10.3, were deposited during the final Late Weichselian retreat of the ice. In an earlier presentation of the Weichselian stratigraphy in South Sweden, before those by Ringberg (1988, 1989), Berglund and Lagerlund (1981) concluded that the area was not glaciated after the Eemian until the Late Weichselian, when the main tills, the Dalby and Bracke tills were deposited, during the time corresponding to the Young Baltic ice advances in Denmark. These tills are covered by the Lund and Jonstorp tills, in age corresponding to the Malmö till and the tills correlated with it (Table 10.3). In the study of the stratigraphy of the Alnarp Valley, U. Miller (1977) distinguished several Weichselian tills, but as at that time she used 13 ka BP and not 25 ka BP as the age for the beginning of the Late Weichselian, some of the tills dated as being formed after 21.5 ka BP were classified as being Middle Weichselian instead of Late Weichselian. Of the four Late Weichselian tills, according to the division used here, the two youngest are correlated with the Young Baltic Ice, the third from the top with an ice movement from the north and northeast and the fourth, the oldest, with a Baltic ice. The till stratigraphy is thus similar to that given in Table 10.3. As all these tills are younger

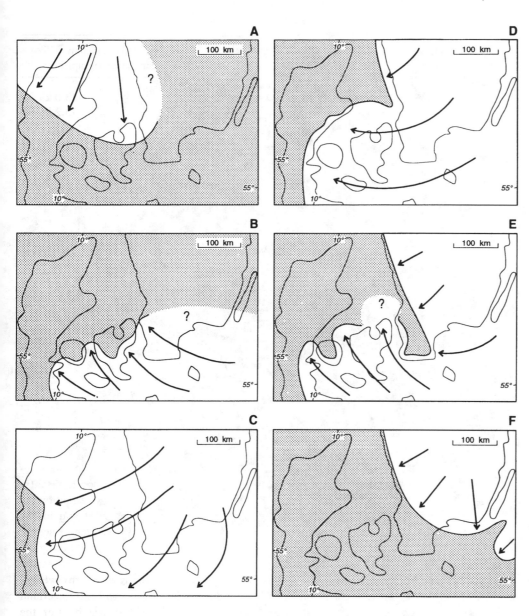

Figure 10.5. Late Weichselian glacier advances in Denmark and southernmost Sweden (after Ringberg, 1988, Fig. 11). A, Norwegian advance; B, Old Baltic advance; C, Main Stationary line; D, East Jutland advance; E, Baelthav (Belt Sea) advance; F, Youngest ice movement.

than the earlier mentioned Middle Weichselian Gärdslöv I Interstadial (see Fig. 9.1), they are all Late Weichselian, as are presumably most of the earlier mentioned tills listed in Table 10.3. Further, they can all be interpreted as having been deposited during ice advances from a single dome over the mainland with a changing flow pattern (Ringberg, 1988). A more complicated glaciation model with a separate ice dome in the southern parts of the Baltic basin, as suggested by Lagerlund (1983), has not been considered to be a likely alternative on the basis of the geological evidence available (Ringberg, 1988). The changes in the direction of ice movement were governed by the movement of the center of glaciation eastwards, away from the Scandinavian mountains, as a result of the buildup of the ice sheet. This resulted in ice advances from the east over Denmark and southernmost Sweden after ice movements from a northerly

direction, a shift of directions observed to have taken place also in the Hamburg area already in the Saalian, but also in the Weichselian (Ehlers, 1983; Ehlers, Meyer, & Stephan, 1984). This is in agreement with the previously mentioned model by Ljungner (1949) in which the ice-divide was displaced eastwards during the maximum extent of the ice sheet.

In its details the glaciation after the Early Weichselian and culminating during the Late Weichselian is, as presented, complex in southern Sweden and in Denmark, as well as along the Norwegian coast, and many of the correlations of the till beds are still tentative. It is particularly in the western marginal parts of the area of glaciation, with a stronger maritime influence, that several smaller ice advances, depositing separate till beds, succeeded one another, in contrast to the eastern, more continental, areas on the lee side of the Scandinavian mountains.

Figure 10.6. Cliff of Young Baltic till (see Table 10.3) with a great proportion of Baltic erratics, including about 50 percent Palaeozoic limestones. North coast of Sprogø in the Great Belt. (Photo K. S. Petersen)

On the other hand, the eastern areas discussed are further inside the outer limit of the Weichselian glaciation than the areas in the west, and therefore, possible smaller advances and recessions of the ice margin are not recorded in the stratigraphy of the former areas.

As a result of the complex nature of the ice advances in many of the areas reviewed, the need to use formal stratigraphical terms for the tills and other glacial sediments has increased. The difficulties in general correlations of the tills over large areas remain, however, and the problems are the same in areas with a great number of formal names for the tills, as in Scania and southern Sweden, as they are in areas where the classification of the tills is regionally based, as in Finnish Lapland.

10.2. Transport of erratics

The preferred orientation of the clasts in the tills, when undisturbed, together with the directions of the striae and some glacial landforms, have all, as mentioned in the previous chapter,

been used in separating the directions of ice flow during glaciations, particularly during the Late Weichselian glaciation and its recession. Additional evidence of the directions of flow can be obtained from the erratics, especially from the indicators, erratics for which the source area in the bedrock is known. The indicator fans show the variations in the direction of ice flow at the time during which the indicators were transported, and thus reflect the dynamics of the ice sheets that transported the erratics, whereas the indicator trains show the transport distances in a particular direction. The transport directions of the erratics, together with till fabric-studies, can directly be tied to till stratigraphy, whereas the directions of striae cannot alone be used as a stratigraphical tool.

In the central parts of the glaciation, east of the Scandinavian mountains, the lithology of the widespread Late Weichselian tills generally reflects the composition of the underlying bedrock. This is the case both for boulders and for finer fractions, such as the gravel fraction 2–6 mm studied under the microscope, as seen, for instance, in the lithological studies presented in the detailed descriptions of the Quaternary deposits of various counties in Sweden, such as of Värmland, Jämtland, and Västernorrland (Lundqvist, 1958, 1969a, 1987a). In addition, the lithology of the tills gives evidence of the underlying bedrock in areas poor in outcrops. Studies similar to those in Sweden have in Finland shown that the rock types represented in the tills are mostly local, as seen from the map of the percentages of different rocks in the 3–10 cm fraction in the tills of northern Karelia (Repo, 1957). The same can be seen on the maps of an area in northern Finland in which the percentages of different rock types identified in the tills are compared with the distribution of these rocks in the bedrock (Okko, 1941).

Even if there is a general correspondence between the lithology of the tills and the underlying bedrock, particularly apparent when presenting the results on small-scaled maps, the material was transported in the direction of the main ice movement, though most of it a very short distance. This is seen in the transects studied in the direction of the last movement of ice, across areas with distinct, and in stone counts, easily identifiable rock types. Thus Hellaakoski (1930), could show by studying clasts with a diameter >6 cm and 2–6 cm in a transect across a 26 km broad area of rapakivi in Laitila, southwestern Finland, that the relative amount of rapakivi increased sharply already 1.5 km from the proximal part of the outcrop in the direction of ice movement, in some samples to over 50 percent, and decreased again beyond the distal part, the percentage falling to 20 percent after 1–3.5 km. Similar results were obtained in later more detailed studies, in Finland by Virkkala (1969) in a transect across a 9 km broad area with gabbro in Hyvinkää in southern Finland, where the percentages dropped from 74 percent 1 km before the distal contact to 10 percent 4.5 km and to 4 percent after the contact. In a study of the central parts of southern Sweden, Gillberg (1965, 1967) showed how the percentages of indicators of Cambro–Silurian rocks in the tills decreased away from the outcrops in the direction of ice flow. In another investigation, Lindén (1975) studied two transects in the area west of Uppsala in central Sweden, across an area with an Archaean bedrock. In the study, in which different grade sizes were analyzed, it was concluded that of the whole material in the till fractions <20 mm about 60 percent had been transported less than 3.5 km, thus in agreement with the results from Finland. The finer fractions of the tills studied usually consist of minerals and cannot therefore be referred to any particular rock type.

The increase of the relative amount of till clasts of a particular rock type from the proximal contact of the source area of the bedrock, in the direction of ice movement, and the decrease after the distal contact, have been presented with exponential curves, in agreement with the observations made by Krumbein (1937). These have been presented as regression lines in diagrams drawn to a semilogarithmic scale, an approach tested and used in Sweden by Gillberg (1965, 1967) and used in Finland by Perttunen (1977). Krumbein, Gillberg, and Perttunen also used as one parameter the half-distance value, which is the distance at which the frequency of a particular rock type was halved (Bouchard & Salonen, 1990), specifically defined by Perttunen (1977) as the distance from the distal contact and not from any arbitrary point on the negative exponential curve. Thus, in two traverses of 24 km and 33 km across granitoids and basic volcanics in the Hämeenlinna area in southern Finland the half-distance values were 3.7 km and 4.7 km for the former rocks and 5.6 km and 4.2 km for the latter (Perttunen, 1977). Detailed studies of tills have shown that a number of factors affect the transport distances of the till matrix, such as resistance to abrasion and crushing during transport, and differences in transport distances have therefore been recorded for different-size fractions (Bouchard & Salonen, 1990). There is also a difference in the transport of the surface boulders as compared with boulders within the till beds. On the whole, however, these early studies already gave a good indication of the short transport distances of most of the material in the tills, a result confirmed and supplemented by later detailed studies. The exponential curves are only used in showing the overall local character of the till lithology. A different approach to the study of tills was introduced by Salonen (1986; Bouchard & Salonen, 1990) in his analysis of the glacial transport of surface boulders in Finland. Instead of using single rock types in the study of transport distances, he took into account all rock types represented in the boulder counts. The percentages and distances of each rock type to their source areas were plotted as "cumulative distance distribution curves," the method being called the "transport distance distribution" method. Thus, the distances for each boulder count were measured to the areas from which the boulders were transported, and not, as usually, from the area of a particular rock type in the direction of transport. Salonen also determined the half-distance values as done by Perttunen (1977). Salonen's (1986) results showed that the geometric means of the boulder transport distances varied between 0.4 km and 3.0 km and the average transport distances between 0.8 km and 10.0 km in cover moraine areas and 5.0 km and 17.0 in drumlin areas. In a later summary of the boulder transport in Finland, Salonen (1987) showed that almost 500 boulder fans in Finland are normally 1–5 km long and that the median length is 3.0 km. The half-distance values were found by Salonen (1986) to vary a great deal as result of differences in the deposition of the glacial deposits, but the arithmetic mean of 5.0 km and the median of 3.6 km are of the same order of magnitude as those quoted above and give additional evidence of the short transport of most of the material in the tills and of the surface boulders. The local character of the tills is also reflected on the geochemical maps of the till matrix used particularly in ore prospecting, on which the composition of the underlying bedrock is reflected. The good correspondence found between geochemical anomalies in the till and the variations in the underlying bedrock may in some cases indicate that the anomalies are due to postdepositional absorption and therefore show no influence of glacial transport (Kokkola, 1989). Apart from being important in prospecting, the geochemical anomalies can give additional evidence about glacial transport of finer fractions of tills.

The generalizations presented about the glacial transport of erratics in the tills are all, as mentioned, based on results from the areas east of the Scandinavian mountains and cannot be applied for the Norwegian coast west and north of these mountains. The glacial activity was quite different here because of the fjord landscape with great local differences in altitude. The deep fjords, such as Hardangerfjord in west Norway, have thick sediments including tills and turbidites (H. Holtedahl, 1975), whereas the accumulation of glacial deposits in the surroundings was smaller. Higher up in the mountains, on the other hand, the transport distances are more uniform, as on Hardangervidda, an area with a comparatively low relief consisting of Precambrian basement rocks, mostly granitic and granodioritic gneisses (T. O. Vorren, 1979). The transport distances of phyllite in the till were found to be shorter here than in the earlier mentioned areas; of 14 samples of till in which the phyllite in the 0.125–0.250 mm fraction had been transported 6–9 km 13 samples had less than 5 percent phyllite, compared with the mean value of about 25 percent at the starting point in the source area. This is partly due to phyllite being easily comminuted during glacial transport. In areas such as Denmark, where the bedrock consists of soft rocks, such as Cretaceous limestones, or even Tertiary sands and clays, the lithological character of tills cannot be studied in the same way as in areas with crystalline rocks. The transport distances are considerably smaller for the local sedimenatry rocks than for the far-traveled crystalline rocks (Marcussen, 1973).

In contrast to the transport of erratics in tills, where the material was carried by ice, the erratics in the glaciofluvial sediments of eskers have mainly been transported by water, even if the material originated from till. The possible part played by a direct erosion of the underlying bedrock by water cannot be determined (Gillberg, 1968). In his study of the glacial transport of erratics in the Laitila area in southwestern Finland, Hellaakoski (1930) also studied the changes of rapakivi erratics in the Laitila esker. The first erratics of rapakivi appeared 5–8 km from the proximal contact in the direction of ice movement and reached values of 80–90 percent in the 26 km broad area of rapakivi, percentages which are twice as high as those in tills. Similarly, there as a sharp fall in the amount of rapakivi erratics 5–8 km from the distal contact. But the local variations in the relative amount of erratics were found by Hellaakoski to be greater in the till than in the glaciofluvial material. In a study of the lithology of the material in the Hämeenlinna esker in southern Finland, Virkkala (1958) found that less than 40 percent of the erratics were transported more than 5 km, but that there were variations between different rock types. The effect of crushing was also found to be important by Gillberg (1968) in his study of eskers in Sweden. In the detailed study of the lithology of the 400 km long Badelunda esker in central Sweden and the Åre and Indal eskers further north in Jämtland, Lilliesköld (1990) obtained results similar to those of Hellaakoski in Finland. The first appearance of granite was 5–8 km from the proximal part, about 8 km for shales, and about 9 km for

gabbroic rocks that disintegrate easily. The grain-size fractions used in the study were 2–5.6 mm, 5.6–8 mm, and 8–16 mm. Most of the studies of the lithology of eskers have been concerned with eskers formed subglacially in subaquatic areas, that is, in areas submerged at the time of deglaciation. Less is known about the lithology of eskers in supra-aquatic areas, and about the feeding of material into the large marginal formations of glaciofluvial sediments, connected with the eskers but formed along the retreating ice margin.

In these studies of the relationship between the lithology of the tills and the underlying bedrock, in which various size fractions as well as surface boulders were used, the local character of the material was stressed. A small part of the till material was, however, transported further away from the source areas, and the indicators in this material have been used in drawing indicator fans, mostly boulder fans, of these erratics, commonly classified as far-traveled indicators. The indicator fans inside the outer limit of the Weichselian glaciation, when based on studies of the surface tills, reflect the movements of ice during the Late Weichselian glaciation. The indicators outside the limit were, on the other hand, transported earlier, chiefly during the more extensive Saalian glaciations.

The pattern of glacial transport is reflected on the map in Fig. 10.7, which shows some of the indicator fans of rocks with their source areas in Norway, Sweden, and Finland. In the north, in the area of the ice-divide of the last glaciation, the fans are short and broad, the directions showing the displacement of the ice-divide in a north–south direction, as already demonstrated in the till fabric studies. The two fans in Finnish Lapland shown in Fig. 10.7 are of Jatulian sandstone and conglomerate in Sodankylä and Kittilä, and of Nattanen granite (Tanner, 1914). The broad fan for the Umptek and Lujaur-Urt nepheline-syenite in the Kola peninsula further east (Flint, 1971) is similar to the fans in Lapland, but with an even wider spread of the erratics. Further south the indicator fan for the Lappajärvi impactite is the longest fan traced inside Finland, stretching from central Finland to the south coast (Kulonpalo, 1969). The impactite, which is suitable as an indicator because of its small source area and distinct composition, was earlier interpreted as a volcanic explosive breccia, a dacite called Kärnäite after the island in Lake Lappajärvi on which it was found. The indicator fans of rocks with their source areas in southern Finland are, however, longer than the fan mentioned earlier. Indicators of the Viipuri rapakivi granite from the massif in southeastern Finland have been traced in a broad area south of the Gulf of Finland, and of the Åland rapakivi granites in a wide area south of the Baltic Sea, in the west even beyond the outer limit of the Weichselian glaciation (Sederholm, 1911; Hausen, 1912; Flint, 1971; Viiding et al., 1971). In the spread of the Jotnian sandstone from Satakunta on the west coast of Finland, the changes in the directions of transport from this area during the retreat of the late Weichselian glaciation could be demonstrated, when compared with the transport of rapakivi indicators from the Vehmaa and Laitila massifs immediately south of the source area of Jotnia sandstone.

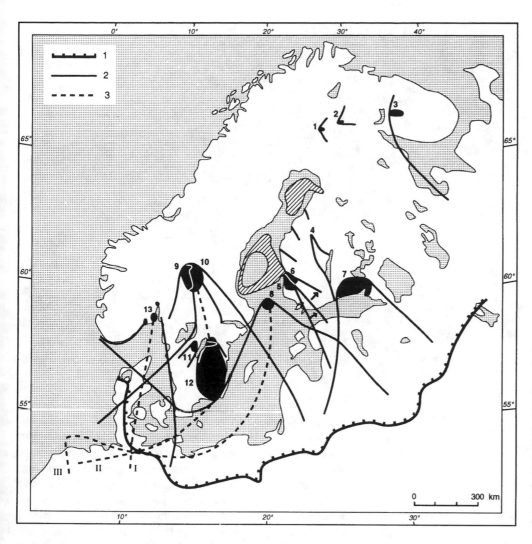

Figure 10.7. Indicator fans. Symbols: 1, outer limit of Weichselian glaciation; 2, lateral limit of fan; 3, path of dispersal (after Ehlers, 1983; Rappol, 1986). Source areas: 1, Jatulian sandstone and conglomerate; 2, Nattanen granite; 3, Umptek and Lujarv–Urt nepheline syenite; 4, Lappajärvi impactite; 5, Vehmaa and Laitila rapakivi granite; 6, Jotnian sandstone (Satakunta); 7, Viipuri rapakivi granite, 8, Åland rapakivi granite; 9, Jotnian (Dala) sandstone; 10, Dala porphyries; 11; Cambro–Silurian limestone; 12, Småland granite; 13, Oslo rhomb porphyries and larvikite.

After a direction of transport to the south and south-southeast the direction changed to a more easterly direction when the ice margin had retreated to northern Estonia, at the time of the formation of the Palivere moraine (Raukas, 1965). When the ice margin had retreated as far as southern Finland, to the first Salpausselkä moraine, the direction had become even more easterly, as shown by the spread of indicators of Jotnian sandstone (Donner, 1986). These indicators were, however, found in glaciofluvial material but their spread was determined by the general direction of the ice movement. As the source area of Jotnian sandstone continues into the Gulf of Bothnia the origin of the erratics along the Finnish west coast is likely to be outside the area on land in southwestern Finland indicated on the map (Salonen, 1991).

The shapes of the long indicator fans for the rocks that had their source areas in Sweden were influenced in the southwestern parts of the Baltic by the previously mentioned great changes in the direction of ice movement, as was to some extent the indicator fan for the Åland rapakivi. The shorter fans within Sweden are, however, more regular and show the same

variations in directions of the movement as the glacial striae. This can be seen in the dispersal of Jotnian sandstone from Dalarna (Dala sandstone) and of the Cambro–Silurian limestones further south (Overweel, 1977). Indicators or porphyries from Dalarna (Dala porphyries) have been traced in a broad area south of Sweden, beyond the outer limit of the Weichselian glaciation as far as The Netherlands in the southwest, and so have indicators of Småland granites and porphyries from southern Sweden (Overweel, 1977). Indicators of the Oslo romb porphyry have been found both to the south in Denmark and northern Germany as well as in The Netherlands and on the east coast of England (Flint, 1971). In England the romb porphyries from the Oslo area, as well as larvikites from the same area, have been found in the sediments of the North Sea Drift (Cromer Till) Formation, correlated with the continental Elsterian stage (Ehlers & Gibbard, 1991). But in The Netherlands the indicators from Oslo and those from Dalarna and Småland in Sweden, and from the Åland Islands in Finland, are associated with Saalian glacial sediments, from the time of the most extensive Saalian glaciation (Overweel, 1977), during stadial III (see Table 5.1).

Figure 10.8. Crystalline erratics from an area north of the Estonian coast at the eastern shore of Lake Vortsjärv, Estonia. (Photo J. Donner)

According to Ehlers (1983) a change in the direction of ice movement from Scandinavia took place during this Saalian glaciation in the Hamburg area, with an initial movement from the north bringing erratics from the Oslo area, followed by a movement over Sweden and ending with a movement from the east, which curved down the basin of the Baltic Sea and brought erratics from the Åland Islands, as a result of the eastward migration of the ice divide in Scandinavia (Fig. 10.7). The same succession, but with different transport directions, can be found in The Netherlands during the Saalian. An initial movement from the northeast was followed by one from the northwest or north-northwest bringing in material similar to that brought by the ice from the east in the Hamburg area (Rappol, 1986). This final northerly movement in The Netherlands was caused by a collision of the ice from Scandinavia with that of the British Isles, forcing them both to turn south in the North Sea basin (Fig. 10.7). A reverse order of Saalian ice advance directions has been mentioned by Ehlers (1990b) as an alternative.

In addition to the studies of the shapes and directions of fans for far-traveled indicators, the range of the source areas for the assemblages of the rock types found in tills have been investigated, in studies somewhat similar to the earlier mentioned study of the local erratics in Finland by Salonen. The proportion of erratics, particularly of crystalline rocks from the Fennoscandian Shield, is comparatively high in the areas with Palaeozoic or younger sedimentary rocks. In Denmark the proportion of crystalline rocks in glacial deposits is about 40–60 percent (H. W. Rasmussen, 1966), in deposits from the time of the Weichselian ice recession 30–40 percent (Hansen, 1965). The differences in the assemblages, caused by spatial differences in the direction of the earlier mentioned ice movements, were elucidated by the detailed study of indicators in Denmark by Milthers (1942), later used by Overweel (1977). Erratics of romb porphyries from the Oslo area are restricted to the northern parts of Jutland, the porphyries from Dalarna in Sweden to a zone further south, in the direction from northeast to southwest, and Baltic prophyries occur mainly in this zone and in the southeastern parts of the country. As the proportion of far-traveled erratics of crystalline rocks is comparatively high in Denmark, higher than would be expected from the studies of the transport distances in the more central areas of glaciation, Marcussen (1973) suggested that the erratics found in Denmark were carried by the ice step-by-step during several glaciations, at the beginning and end of each glaciation. The possibility of material transported by drift ice being incorporated in the tills was also taken into account by Marcussen. In Estonia, as in Denmark, the proportion of crystalline rocks is also high (Fig. 10.8), in glaciofluvial gravel on the island of Saaremaa (Ösel) 32 percent (Hausen, 1912), in the till at Viljandi 27 percent (Raukas, 1985), at Otepää 22–56 percent and at Haanja 26–39 percent (Raukas & Karukäpp, 1979), with, however, great differences within Estonia (Raukas, 1969). The source area of the erratics is foremost southern Finland, with erratics of rapakivi granite from the Åland Islands in the west and Viipuri rapakivi in the east, in places represented by 50–60 percent of all erratics of crystalline rocks. In a study of 961 erratics in Latvia further south Eskola (1933) found that the percentages of the main crystalline rocks, mainly granites and migmatites, corresponded to the percentages of these rocks in

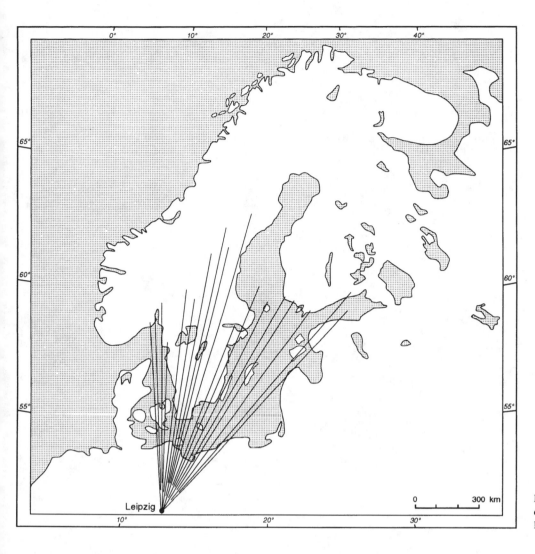

Figure 10.9. Origin of erratics of crystalline rocks in Leipzig (after Richter et al., 1986).

the bedrock of Finland, but that the erratics were mainly from southern Finland. The great number of large surface boulders is also striking in these marginal areas of glaciation, in Estonia particularly at the north coast. In Estonia, 1,900 erratic boulders with a diameter over 3 m are known, the largest with diameters of 56.5 and 49.6 m (Viiding, 1976, 1981). In eastern Germany the largest surface boulders are found in the northern parts near the coast; toward the marginal area of glaciation they clearly get smaller (Richter, Baudenbacher, & Eissmann, 1986). The range of the identified source areas for erratic boulders recorded in the Leipzig area during nearly 100 years is an example of the varied movements of the ice during the Elsterian and Saalian glaciations during which the erratics were carried to this area (Richter, Baudenbacher, & Eissmann, 1986). The erratics of sedimentary rocks came from Denmark, Sweden, Finland, and the area near St. Petersburg, the northernmost likely source being north of the Gulf of Bothnia, for a Lower Cambrian sandstone. The erratics of crystalline rocks (Fig. 10.9) have an almost equally wide source area, from the Oslo area over Sweden to eastern Finland. As Leipzig is north of the southern limits of the Elsterian and

Saalian glaciations, the erratics could have been transported into this area by the ice sheets during both of these cold stages. In The Netherlands all recorded erratics of Fennoscandian crystalline rocks were transported by a Saalian ice sheet, during one single glaciation. An inventory of these rocks showed some differences between the areas covered by the Saalian glaciation, differences caused by the previously mentioned changes in ice movement. The great number of indicators classified according to their provenance areas represent rock types from the Oslo area, large parts of Sweden, and the southern parts of Finland, like the erratics of rapakivi from Åland (Zandstra, 1986).

These detailed studies of assemblages of erratics confirm the conclusions about the wide distribution of indicators from the central parts of the glaciated areas, already shown by the few indicator fans shown in Fig. 10.7. Many indicators were transported over 500 km and some up to 1,200 km (Flint, 1971), although not always during one single glaciation. On the whole the transport distances, being dependent on the size of the ice sheets, were greater in the peripheral parts of the ice sheets, away from the area of the ice-divide. Most of the boulders have

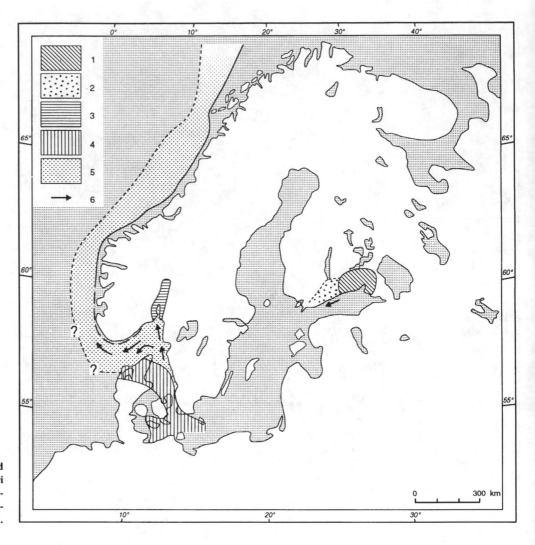

Figure 10.10. Dispersal of ice-rafted material. Source areas: 1, Viipuri rapakivi granite; 3, Oslo rhomb porphyries; 4, flint; 2 and 5, areas of dispersal (3–5 after Spjeldnaes, 1973).

been explained as having been eroded and transported during the retreat of the ice, resulting in short transport distances near the ice-divide where the ice became stagnant soon after eroding the underlying bedrock (Bouchard & Salonen, 1990). There were also differences in flow rates within an ice sheet which may account for some of the differences in transport distances found in the formerly glaciated area, some differences having been caused by a lobate flow of the receding ice, in some extreme cases by surges of the ice (Bouchard & Salonen, 1990).

In addition to subglacial and englacial transport of material a transport of erratics by ice-rafting took place in submerged areas with a calving ice-margin (Overweel, 1977). During the Late Weichselian deglaciation of southern Finland the coastal areas were submerged at the time of the ice-dammed Baltic Ice Lake, during the standstill of the ice-margin at the Salpausselkä end moraines. The icebergs from the calving ice-margin carried numerous boulders of Viipuri rapakivi from southeastern Finland westwards along the coast, in the general direction of the present coastal currents of the Gulf of Finland, to areas west of Helsinki (Espoo) as well as to Lahti and Hyvinkää along the outer

Salpausselkä end moraine (Hyyppä, 1950). The ice-rafting was not only restricted to these indicator boulders (Fig. 10.10) because dropstones, which have disturbed the underlying sediments, are frequently found in the finer sediments of the formerly submerged coastal areas, both in glaciofluvial sediments of eskers and end moraines and in glacial clays (Donner, 1986). The source areas for these erratics have, however, not been identified. Ice-rafting also took place along the western coast of Norway, also at the time of the Late Weichselian deglaciation. Rhomb porphyry from the Oslo area and flint from the source area in northern and eastern Denmark and southernmost Sweden (Fig. 10.10) were spread by icebergs along the Norwegian coast (Spjeldnaes, 1973). This agrees with the observations made by using side-scan sonar pictures that the floating icebergs left plough marks in the sediments on the shelf off the south coast as well as on the slopes of the Norwegian Trough (Belderson & Wilson, 1973). The ice-rafting was not only restricted to the boulders and smaller erratics. Several till-rafts have been reported from various parts of Sweden, which show that also larger bodies of sediment were carried by icebergs, such as a 56 m long and 0.4–0.7 m thick till-raft reported to have been found by G. De Geer at Bromma outside Stockholm (Overweel, 1977).

10.3. Glacial dynamics and thickness of ice

The glacial geological evidence, particularly from striae and the dispersal of erratics, led to the conclusion that the last glaciation in Scandinavia, like the previous ones, started in the mountains, that the ice-divide shifted to the east of the mountains as a result of the increased thickness of the ice during its maximum extent, and that the glaciation ended in the mountains (Hoppe & Liljequist, 1956; Flint, 1971). This is in agreement with the previously noted glaciation model by Ljungner (1949), showing the displacement of the ice-divide during the Weichselian (Lundqvist, 1986b). The evidence of ice movements in southwestern Norway has, however, been interpreted as showing that there was an initial ice-dome in the coastal area, on the western side of the highest mountains, at the onset of the glaciation and possibly also at its final stage (Anundsen, 1990). This conclusion would explain the observed great glacioisostatic subsidence of the Jaeren area as compared with other coastal areas of Norway. An early growth of the ice sheet over the fjord district is also in agreement with the distribution of the present precipitation, which is highest west of the mountains.

The climatic change which resulted in the growth of an ice sheet did not, according to the interpretations of the evidence in the field (Lundqvist, 1986b; Ehlers, Gibbard, & Rose, 1991), result in an "instantaneous glaciation" ("glacierization"). But the advance of the growing ice sheet must have been relatively rapid. On the basis of available dates it has been calculated that if southern Sweden was still ice-free 24 ka ago and if the ice margin reached the Hamburg area not later than 15 ka ago it would have advanced a distance of 450 km at a rate of about 50 m/year, and that if the Weichselian maximum extent of the ice was at about 20 ka BP the rate would have been about 75 m/year or more; during the growth from the central area of glaciation to the maximum extent, rates of 100–150 m/year would have been necessary (Ehlers, 1990a; Ehlers, Gibbard, & Rose, 1991).

When the ice sheet reached its maximum extent it resulted in a displacement of the cyclone tracks south of their present position and the area of maximum precipitation moved south (Liljequist, 1974). The strong katabatic winds from the glacial anticyclones above the ice sheet only affected a marginal area of less than 10 km outside the ice margin, where the winds close to the margin were easterly during the passage of cyclones from the west (Liljequist, 1974). In addition to the cyclones following close to the ice margin and coming from the west or north-northwest some, especially during the summer, moved up from the eastern Mediterranean area toward the northeast, resulting in comparatively warm summer temperatures and a steep temperature gradient near the margin of the ice sheet in the east, in eastern Russia (Hoppe & Liljequist, 1956). This relatively warm continental climate is reflected in the vegetational history during the warm stages as well as during the Late

Weichselian and early Holocene, in contrast to the vegetational history of western continental Europe.

With the knowledge of the extent of the ice sheets of the Scandinavian glaciations and the underlying topography it has been possible to make estimates of the surface forms of the past ice sheets, because they were governed by the same physical processes as the present-day ice sheets, with similar forms and thicknesses. In this way Robin (1964) estimated that the maximum thickness of the ice sheet was around 3,000 m during the last, Weichselian, glaciation and that it reached this thickness over the Baltic states during the most extensive Pleistocene glaciation. In a modeled transect of the European ice sheet from the west coast of Norway to Poland, Boulton et al. (1985) also estimated the maximum ice thickness to have been about 3,000 m during the last glaciation, a figure quoted by Svendsen and Mangerud (1987) in their analysis of the sea level history of western Norway during and after deglaciation. By using both longitudinal and transverse landforms, the flow patterns during various stages of the ice sheet decay were modeled by Boulton et al. (1985), and slightly modified models of ice thicknesses and flow lines during the Weichselian were presented by Ehlers (1990a, 1990b), one of which is shown in Fig. 10.11. The velocity distribution in the basal ice was also modelled by Boulton et al. Thus, during the Weichselian maximum the basal velocities would have been 300–400 m/year and near the ice divide 50 m/year or less. These figures are similar to those given by Paterson (1981) for an ice sheet in a steady state with a parabolic profile and a width of 2,000 km, similar to the Scandinavian ice during the Weichselian maximum, but with a thickness of about 4,700 m. The velocity would, according to Paterson, have been 135 m/year, for instance, 950 km from the ice-divide, in the area of southern Estonia. The higher velocities in the marginal parts of the ice sheet, when at its largest and during its early retreat, have a bearing on the observed transport distances of erratics. It was earlier noted how the longest indicator fans are connected with the times during which the ice sheets had reached their largest extent, both during the Saalian and the Weichselian cold stages. The observed pattern of the dispersal of erratics during the Weichselian is similar to the predicted maximum dispersal according to the model presented by Boulton et al., which is an example of how the model can be tested by field evidence. It can thus be seen that the total length of a glaciation and the basal velocities cited here put certain limits to the distances the erratics could have been transported during one single glaciation. These limits were taken into account in a comparison of the distance of transport of far-traveled indicators in Estonia with transport distances in Finland, closer to the central area of glaciation (Donner, 1989). Even if the model by Boulton et al. showing the thickness and directions of flow is generally applicable in the area of the Weichselian glaciation, it was modified by Ehlers (1990b) for southern Denmark and Schleswig-Holstein, to fit the evidence of ice movement directions in that area. There were also probably great regional variations in the velocities of ice movement within the area, with a comparatively rapid ice

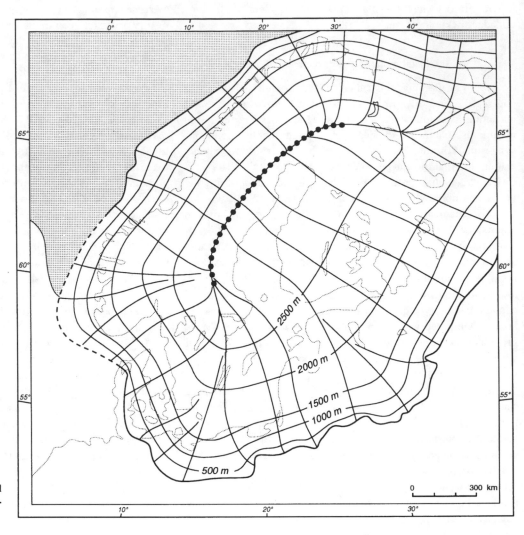

Figure 10.11. Thickness of ice and flow during the Weichselian glaciation (after Ehlers, 1990b).

movement in the southern part of the Baltic Sea basin curving westwards in Denmark and northern Germany, as shown by the previously mentioned studies of the tills and their erratics.

In areas with a varied succession of ice movements from different directions, such as the areas in Denmark and northern Germany, a detailed study of the glaciotectonic deformations have given additional evidence of the dynamics of the ice that deposited the tills and glacial sediments. The deformations caused by an active ice were formed both proglacially at the ice margin, referred to as trust-features, and subglacially beneath the ice (Hart & Boulton, 1991). In addition, deformations were also caused by a stagnant ice. The conclusions about the thermal conditions underneath the ice have mainly been based on theoretical calculations. These, together with field observations, have indicated that unfrozen sediments were commonly deformed under large ice sheets in addition to the frozen sediments near the ice margin (Hart & Boulton, 1991). The observations made of disturbed Quaternary sediments have been particularly detailed and numerous in the exposed cliffs of Jutland and the islands of Denmark, with thick sediments close to the outer

margin of the Weichselian glaciation, and, as noted earlier, with a varied history of glacial advances (Fig. 10.12). As examples from Denmark, the study of dislocated drift deposits in northwestern Sjaelland (Petersen, 1973) and the discussion of the use of glaciotectonic analysis in mapping (Petersen, 1978) may be mentioned. The stratigraphical studies in which the observations of lithology were combined with studies of directional elements, such as deformation structures and till fabric, led to a division into drift units, the previously mentioned kinetostratigraphic units that were defined as "deposited by an ice sheet or stream possessing a characteristic pattern and direction of movement" (Berthelsen, 1978). These units were, for instance, used in the study of a cliff profile on the island of Møn (Berthelsen, Konradi, & Petersen, 1977), Fig. 10.13, and later in the study of the Melsbjerg Hoved cliff and the cliffs at Viborg in northeastern Jutland (Rasmussen & Petersen, 1980) and in the study of the northern parts of the island of Samsø (Houmark-Nielsen & Berthelsen, 1981). The proglacial deformations so common in the marginal formations in Denmark (Houmark-Nielsen, 1988; Pedersen, Petersen, & Rasmussen, 1988) are also encountered in

Figure 10.12. Feggeklint in Limfjord, Jutland, with Eocene diatomite and dark ash layers dislocated by ice. (Photo K. S. Petersen)

Figure 10.13. Dislocated layers of till and calcareous sediment at Hvideklint on the island of Møn in Denmark, facing the Baltic. (Photo K. S. Petersen)

the marginal sediments in Norway, Sweden and Finland, particularly in the sediments of the major end moraines formed after readvances or during standstills of the retreating Late Weichselian or Early Flandrian ice margin. And, as observed earlier, the till-covered sediments from warm stages or non-glacial intervals during cold stages were also disturbed and displaced by the overriding ice. Whereas the conclusions about changes in the thermal conditions underneath an ice sheet during its growth, maximum extent, and decay are based on assumptions made when modeling the ice sheet, the deformations, when studied in detail, give direct evidence of the glacial dynamics at work at different times during a former glaciation. Of this evidence, that of the ice-flow directions is perhaps most useful. From what has been said it is clear that in stratigraphical studies of glacial sequences it is essential to include a study of possible disturbances of the sediments, in order to understand how a

sequence of sediments was formed and how it was later possibly changed.

In addition to the estimates of the extent and thickness of the Weichselian ice sheet at its maximum, arguments for and against the existence of ice-free areas, either as nunataks in the Scandinavian mountains or as coastal ice-free areas west of them, have been put forward. This problem centers around the question of whether or not the present distributions of some arctic–alpine plants in the Scandinavian mountains show that these areas or the neighboring coastal areas acted as refugia for these plants at the time of the most extensive Weichselian glaciation. The implication of the role played by possible ice-free refugia was different at the time over 100 years ago when this possibility was first discussed by Blytt and later by Sernander (Mangerud, 1973), because then the length of the last glaciation was considered to represent a much longer time than the relatively short Late Weichselian glaciation as it is now known, with its maximum about 18–20 ka ago. The botanical evidence used in support of refugia in the mountains or at the coast, in discussions restricted to the Weichselian glaciation, has been entirely based on the present distribution of some arctic–alpine plants, either to the mountains of Dovre and Jotunheimen in southern Scandinavia or to the area north of the arctic circle, to Troms and Finnmarken in northern Scandinavia, or to both areas. According to Gjaerevoll (Mangerud, 1973) there are about 30 species that occur in both areas and that have a bicentric distribution – about 10 species and subspecies with a southern unicentric distribution and 40 with a northern unicentric distribution. The number of bicentric species has, however, been somewhat reduced after a later critical assessment of these species, as mentioned by Mangerud (1973). Some of the species are endemic and have therefore been considered to show that there was a longer ice-free period in the mountains than the time after the last glaciation. Thus several subspecies of *Papaver radicatum* have been identified in both the southern and northern areas of its distribution (Nordhagen, 1933, 1936; E. Dahl, 1955). But some species have a wider area of distribution than was known at the time they were first studied. Thus, *Artemisia norvegica,* a unicentric species in Dovre in southern Norway, was known to grow outside this area only in the Ural mountains (Nordhagen, 1933), but was later found also at the coast of southwestern Norway and in northern Scotland (Mangerud, 1973). Even if the present distribution of some species can be reconciled with the idea of an immigration after the glaciation, this explanation for the distribution outside Norway of some arctic plants in Iceland, Greenland, and North America, and a few also in Novaya Zemlya and Spitsbergen, with an amphiatlantic distribution, has been difficult to accept; an immigration into Scandinavia after the last glaciation has therefore been considered improbable by E. Dahl (1955). He then concluded that much of the arctic–alpine flora had survived the last glaciation in refugia. These, as he had earlier shown (E. Dahl, 1947), would in most parts of Norway, where the high mountains are close to the shelf margin, have been mountain refugia, whereas in northernmost Norway they would have been coastal refugia, at a time when the ocean level was lower than at present.

Many factors, among them the average yearly maximum temperatures, determine the present distribution of the arctic–alpine species in the Scandinavian mountains (E. Dahl, 1955). Apart from the circumstantial evidence from their present area of distribution there is no biostratigraphical evidence of arctic–alpine plants having survived in the mountains or close to them during the Weichselian glaciation. On the other hand, there is strong evidence for a quick immigration after the glaciation, from the marginal areas of glaciation, a possibility already taken into account by Blytt in his early works before the concept of refugia became more accepted (Mangerud, 1973), and, in a number of palaeobotanical studies of Weichselian sediments in

the North Sea area, in which both macroscopic remains and pollen of arctic–alpine species have been identified, it has been shown that these species "wintered" in the periglacial areas during the glaciation and that they followed the retreating ice margin, as seen on the maps of the present distribution and fossil finds of *Salix herbacea* and *Dryas octopetala* by Hultén (1950). One important region for survival was the North Sea basin, with extensive icefree dry land areas at the time of the Weichselian glaciation. The immigration of some arctic–alpine species from the south of the Scandinavian mountains was, however, considered unlikely by E. Dahl (1955) because of the rapid retreat of the ice margin from central Sweden and southern Norway to the mountains during the beginning of the Flandrian, when the climate was comparatively warm. The climatic conditions at that time were considered to have been unfavorable for the arctic–alpine species to have survived near the ice margin. Later detailed investigations of Late Weichselian and Early Flandrian sediments in southeast Norway by Danielsen (1970, 1971), however, showed that a great many of the arctic–alpine species, such as *Dryas octopetala, Salix herbacea,* and *Koenigia islandica,* were present in this area at the time of the ice recession, even if only 31 species or about 13 percent of the about 230 Scandinavian arctic–alpine species of vascular plants have been identified as subfossils. Danielsen also suggested that the earlier mentioned Greenlandic–American species were dispersed into Scandinavia after the Weichselian glaciation, and concluded that there is no need for a theory of glacial survival in Scandinavia for the arctic–alpine species.

In addition to the lack of any biogeographical evidence there is, according to Hoppe (1959), no positive geological, geomorphological, or glaciological evidence of permanent ice-free areas in Norway during the Weichselian glaciation. Later studies have not proved this conclusion wrong, even if the significance of some of the geological evidence used has been discussed. R. Dahl (1972) has, for instance, put forward a hypothesis that there were ambulating refugia in Scandinavia as a result of shifting "ice culminations," a concept that is difficult to accept when taking into account what is known about the timing of the Weichselian glaciation and the signs of glacial activity in areas earlier taken as having been ice-free. But, as pointed out by Hoppe (1959), the areas which were first deglaciated after the extensive Weichselian glaciation largely correspond to the assumed areas of refugia. There are, according to O. Holtedahl (1960), no examples of steep nunataklike mountains being separated from the rounded land forms below by a sharp boundary, even if the glacial sculpturing of the valleys between the mountain summits has accentuated the difference, as in the mountain complex of the Syv Søstre (Seven Sisters) in Nordland (Ljungner, 1949; O. Holtedahl, 1960). In constructing theoretical ice-sheet profiles for the coast of western Norway, Nesje and Sejrup (1988) assumed that there was a low gradient on the profiles and that some summits in Møre at the coast and in the western part of Hallingskarvet in Hardangervidda rose above the ice (see also Ehlers, 1990b), an assumption which is at variance with the profiles produced by Anundsen (1990) for West Norway, with ice domes in the coastal areas. The distributions of erratics at high altitudes in Møre (H. Holtedahl, 1955), and similarly further north in the Lofoten Islands (Bergström, 1959), show that most of the high coastal mountains were covered by the ice sheet, and because of the unstable positions of the erratics as perched boulders they were, according to Bergström, most likely transported during the last Weichselian glaciation. Further evidence of a relatively recent glaciation of the summits is the preservation of striae at high altitudes; older striae would have been obliterated by weathering. The evidence of deeply weathered bedrock, earlier used as evidence for Weichselian ice-free areas, has later been questioned. As mentioned in the discussion of the weathering of the bedrock generally, weathered bedrock has been preserved at several localities throughout the area of glaciation. In the area of the Lofoten Islands in Norway, for instance, no time-dependent differences in weathering could be demonstrated in profiles from sea level up to the mountain summits (Bergström, 1959). There is a general increase in mechanical frost weathering with increasing altitude, a natural result of the difference in climate between lower and higher ground after the last glaciation (O. Holtedahl, 1960). In the study of the chemical weathering of the Quaternary sediments in the area of Møre, H. Holtedahl (1955), however, found no indication of age differences between samples from different altitudes, the highest being from about 700 m. This supports the conclusion that even the highest parts of this area were glaciated during the Weichselian, at a time when, according to H. Holtedahl, the ice margin was at least 40 km outside the present coastline.

11   Late Weichselian and Early Flandrian deglaciation

11.1. End moraines, tunnel valleys, and eskers

In Chapter 3, Section 2, in which the outer limits of the glaciations were discussed, the limit of the Weichselian glaciation, clearly more restricted than the previous Elsterian and Saalian glaciations (see Fig. 3.2), was also mentioned. It is in the Russian plain represented by the Bologoe moraine, in Poland by the Lezno moraine, and further west by the Brandenburg moraine, in one area being separated from the inner moraine called the Frankfurt–Poznan moraine (Nilsson, 1983; Kozarski, 1988). In Denmark, the Weichselian limit of glaciation (Fig. 11.1) is represented by the previously noted Main Stationary Line (C), which can be traced outside the coast of Jutland as a submarine ridge, the "Lille Fiskebank" moraine (B. Andersen, 1979). In earlier investigations it was concluded that the Weichselian ice sheet was joined to the British Devensian ice in the northern parts of the North Sea (B. Andersen, 1979; Jansen, van Weering & Eisma, 1979), but later detailed investigations of the sea floor with seismic profiles have shown that the ice sheets of Scandinavia and Britain were not connected during the Late Weichselian-Devensian glaciation (Ehlers, 1990b; Ehlers & Wingfield, 1991). The outer margin of the ice sheet was outside the coast of Norway, outside the numerous submarine moraines (B. Andersen, 1979, 1981), of which only some are shown on the map in Fig. 3.2.

Numerous end moraines were formed during the deglaciation of the Weichselian ice sheet, either as a result of standstills of the ice margin or as a result of readvances. According to the definition by Flint (1971) an end moraine is "a ridgelike accumulation of drift built along any part of the margin of an active glacier," whereas a complex of ridges forms an end moraine system. Most of the ridgelike features formed outside the retreating ice margin can be referred to as end moraines as defined by Flint; it is only in the valleys and fjords of Norway that terminal moraines were formed in front of glaciers. Both end and terminal moraines can be built up of till as well as of stratified drift, that is, of glaciofluvial material, as the term drift used in the definition by Flint implies, even if the term drift is no longer generally used. The morphological features identified as end or terminal moraines inside the outer margin of the Weichselian glaciation have been linked together and presented on maps as recession lines. In many areas the moraines can easily be followed in the field but in some areas they are discontinuous and their connection is therefore more difficult, as shown by alternative interpretations. This is, for instance, the case in some areas at the eastern margin of the retreating ice, where the connection of the end moraine features has been problematic.

The Weichselian recession lines and the local names used for the end moraines in Russia, Belarus, the three Baltic states, Poland, Germany, and Denmark presented in Fig. 3.2 were drawn on the basis of earlier published maps. For the northeastern areas, the maps by Lukashov (1982) and I. Ekman and Iljin (1991) were used; for eastern Europe, generally, those of Serebryanny and Raukas (1970), Raukas, Rähni, and Miidel (1971), and Faustova (1984) were used; but also the presentation of Poland by Kozarski (1988) has been taken into account. Many of the recession lines mentioned were already shown on the maps presented by B. Andersen (1981) and T. Nilsson (1983). For Denmark, the recession lines were drawn according to Sjörring (1974, 1982). South of the Baltic Sea basin, the main recession line after the outer limit of the Weichselian glaciation is represented by the Pomeranian moraine, also called the Inner Baltic moraine, the Frankfurt moraine being the Outer Baltic moraine. In the east, the Pomeranian moraine corresponds to the Vepsovo moraine, in Denmark to the East Jylland (Jutland) Line (D), which is the limit of the first Young Baltic ice advance, YB 1 (see Table 10.3).

For the end moraines formed during the recession from the Pomeranian moraine, and moraines correlated with it, to the moraines which can be referred to as belonging to the Fennoscandian moraines (Daly, 1934; W. B. Wright, 1937), recession lines cannot be drawn for the whole glaciated area. Some correlations have, however, been made between the moraines, particularly in the southeastern and eastern parts. In Germany two recession lines have been separated north of the Pomeranian moraine, the Velgast and North Rügen lines at the Baltic coast, of which the latter has been correlated with the moraines further east referred to the North Lithuanian–Haanja-

Figure 11.1. Section through outermost Weichselian moraine (Main Stationary Line) at Rovbjerg, west coast of Jutland. The section at the church of Trans shows the proximal part of the sandur. (Photo K. S. Petersen)

Luga line, which has been traced as far as the eastern shores of the White Sea. A younger line, Pandivere in Estonia, has been correlated with the Neva line, which has also been traced further east and north. It follows the south coast of Lake Ladoga, crosses Lake Onega, and curves around the eastern coast of the White Sea over to the Kola Peninsula, where it is represented by the Keiva I moraine. A local ice on the peninsula later formed the Keiva II moraine, which in many parts runs close to and parallel to the Keiva I moraine (Ekman & Iljin, 1991), even if the ice movements that formed the moraines were from different directions. In northwestern Estonia there is a younger recession line, Palivere, that has been connected with a moraine on the Karelian Isthmus called Väärämäenselkä (Raukas, 1977). The limit in Denmark of the Baelthav (Belt Sea) advance (Line E), also called the second Young Baltic ice advance (YB 2), with its continuation in southernmost Sweden, representing the Low Baltic advance (see Table 10.3), cannot directly be connected with the moraines representing the Velgast or North Rügen lines, of which the latter, however, is presumably younger than the Baelthav advance. Nor can the moraines in southern Sweden be directly connected across the Baltic Sea with the moraines in the Baltic states. The biostratigraphical studies in southern Sweden, discussed later, combined with radiocarbon dating, have led to a detailed scheme for the ice recession in this part, and the moraines have been fitted into this frame. The closely spaced end moraines along the west coast fan out toward the east where they are further apart. The Hönö Grötö Lines representing the Halland Coastal moraines (Lundqvist, 1986b) have been correlated with the above-mentioned Low Baltic readvance (Berglund, 1979). The Göteborg (Gothenburg) moraine, with its two large marginal formations, the "Fjärås bräcka" and "Svedas-

kogen" (Hillefors, 1979), is close to the west coast north of Gothenburg but curves inland further south. In the west the Berghem and Levene moraines are close to the Middle (Central) Swedish end moraines, which belong to the Fennoscandian moraines. Some of the main moraines at the outermost coast of Norway or on the shelf outside the coast (B. Andersen, 1979) are also included on the map in Fig. 3.2. The Lista moraine follows the southwestern coast; further north Skjoldryggen is the outermost clear moraine, in the northwest connected with the Egga I moraine with a slightly younger ridge, the Egga II moraine, inside it. In addition to these main ridges, several other submarine ridges have been mapped on the shelf (B. Andersen, 1979). The close spacing of the end moraines on the Norwegian shelf shows how narrow the area was into which the ice sheet could spread from the Scandinavian mountains as compared with the eastern and southeastern areas of the mountains, into which the ice could flow without being broken up by an irregular underlying relief and by deep water.

In many parts of the area of glaciation the Fennoscandian moraines are distinct landforms that can be traced over long distances. They represent the boundary between the area of Late Weichselian ice recession and the central area of early Flandrian deglaciation. The nature of the Fennoscandian moraines varies a great deal because of the different environments in which they were formed – on the eastern side of the Scandinavian mountains in an area with small differences in relief outside an ice margin terminating on dry land or in water, but along the coast of Norway at the margin of an irregular ice front in a fjord landscape (Fig. 11.2). The tracing of synchronous ice marginal positions has been particularly difficult in the latter area, and has required detailed studies in the field. Moreover, the reconstruc-

Figure 11.2. Fennoscandian end moraine at mouth of Ullsfjord in Norway. The shoreline corresponding to the formation of the end moraine is seen in the background. (Courtesy, B. Andersen)

tion of synchronous recession lines has in some areas been based more on radiocarbon ages of samples that can be related to the ice withdrawal than on morphological grounds. The datings have, on the other hand, also shown, as will be seen later, that there are age differences between the individual moraines of a particular substage. Because of this, and also because of the inaccuracies in the datings, the use of the name "Younger Dryas end moraines," referring to the Late Weichselian Younger Dryas Stadial, for the Fennoscoandian end moraines is misleading, as some of these moraines have been shown to be older and others younger than this stadial.

In Sweden, end moraines representing two recession lines have been referred to as the Middle (Central) Swedish moraines (Fig. 3.2). The northern moraines belong to the Billingen Line that passes Mount Billingen southeast of Lake Vänern, an area in which the recession of the ice margin regulated the outflow of the waters of the Baltic into the ocean, damming the Baltic into an ice lake when closing the gap. The moraines south of the Billingen Line represent the Skövde Line (Lundqvist, 1986b). In eastern Sweden the marginal formations have extensive glaciofluvial deposits, and a detailed mapping has resulted in the separation of several ice-marginal positions within the zone of the Middle Swedish moraines (Persson, 1983). In western Sweden, in Dalsland west of Lake Vänern, the moraines of the two recession lines of Billingen and Skövde are so close together that large formations of glaciofluvial material, such as the delta at Dals Ed, was formed during both standstills of the ice margin (Gillberg, 1961). But an older recession line, the Levene Line, can here be traced south of these two lines. In Central Sweden it curves around the southern end of Lake Vättern, but becomes difficult to trace further east (Berglund, 1979). In western

Sweden, in southwestern Värmland, the moraines are poorly developed and difficult to connect in detail with the moraines in the Oslofjord area or with those further east because of the irregular shape of the partly calving ice margin, particularly during the retreat from the moraines corresponding to the main Middle Swedish moraines (Lundqvist, 1988b). In the Oslofjord area several recession lines have been identified by connecting the moraines, of which the Ra moraine is morphologically the best developed. The main moraine ridge, in its eastern part consisting of two parallel ridges, has an older moraine outside it, which east of the fjord has been named the Onsøy moraine and west of it the Slagen moraine (Sørensen, 1979). The Ås, Ski, and Aker moraines are north of the Ra moraine. Of the above-mentioned moraines the Slagen-Onsøy moraine has been correlated with the Levene moraines in Sweden, the Ra moraine with the Skövde moraine, and the Ås moraine with the Billingen moraine (Table 11.1), not only on morphological grounds but largely on the basis of the dating of the deglaciation (Berglund, 1979; Sørensen, 1979). According to the correlation with Sweden, the Fennoscandian moraines, if this term is used, would be presented in the Oslo area by the Ra and Ås moraines. The Slagen–Onsøy moraine has also been called the Outer Ra moraine, but to avoid confusion in terminology it was suggested that this name should be abandoned (Sørensen, 1979).

Moraines correlated with the main Ra–Ås–Ski moraines in the Oslo fjord area have been traced all along the Norwegian coast up to the north and connected to represent an ice recession line, with small local age differences (B. Andersen, 1979). The moraines that represent this recession line were in many areas formed after a readvance of the ice margin, for example, the Herdla moraine in the Bergen area with a readvance of at least

Table 11.1. *Correlation of Fennoscandian moraines and moraines formed in the same zone of recession*

			Younger Dryas Fennoscandian moraines			
Norway			Tromsø– Lyngen (Main Substage) Tautra		Herdla	
	Slagen– Onsøy	Ra	Ås		Ski	Aker
Sweden		(Middle/Central Swedish)				
	Levene	Skövde	Billingen			
Finland		(Salpausselkäs)				
		Salpausselkä I Ss I	Salpausselkä II Ss II	Salpausselkä III Ss III		
		Selkäkangas– Paloharju	Uimaharju (Jaamankangas)			
Alternative:		⎧ Tuupovaara ⎩	Koitere (= Selkäkangas– Palokangas)	Pielisjärvi (Jaamankangas–⎫ Uimaharju) ⎭		
Russia **(Karelia)**		Rugozero	Kalevala			

40 km (Aarseth & Mangerud, 1974; Mangerud et al., 1979a; Mangerud, 1980; Anundsen & Fjeldskaar, 1983). At the west coast of central Norway the recession line has been traced in detail (Sollid & Sørbel, 1979) and includes the distinct moraine at Tautra near Tronheim (B. Andersen, 1979), and can be followed up the coast to the north coast of Finnmark, where the moraines represent the Tromsø–Lyngen moraine, locally referred to the Main substage in the Late Weichselian ice recession (B. Andersen, 1968, 1975, 1979; Sollid et al., 1973). As seen from the map in Fig. 3.2 there is only one line for the Fennoscandian moraines along the Norwegian coast, except in the Oslofjord area, in contrast to the areas east of the Scandinavian mountains. This reflects a difference in the nature of the ice recession between the two areas. On the lee side of the mountains the Fennoscandian moraines were formed in a zone with at least two parallel ridges, which in the east are comparatively far from each other.

The Salpausselkä end moraines in Finland form two distinct lobe-shaped arcs in southern Finland, from the coast in the southwest to eastern Finland. Salpausselkä I and II (Ss I and Ss II) form the main moraines, whereas Salpausselkä III (Ss III) can only be identified in the western part. As the ice margin in most parts of southern Finland terminated in the Baltic at a time during which it formed an ice-damned lake, the Baltic Ice Lake closed by the ice at Billingen in Sweden, marginal terraces, either deltas or sandur-deltas as defined by B. Andersen (1960), were deposited in addition to moraine ridges.

The Salpausselkä moraines are in fact predominantly glaciofluvial formations, in places over 1 km wide flat delta surfaces. This special character of the Salpausselkä moraines, in spite of their clear proximal ice-contact morphology, led to the conclusion that they cannot be called true end moraines (Flint, 1971). This undoubtedly stems from a misunderstanding of the word end moraine in local descriptions of the Salpausselkä ridges, because the term moraine ("moreeni") also includes in Finnish, as in Swedish, Norwegian, and Danish, the concept till. Thus, as this would imply that the formations classified as moraines have to be built up by till and the Salpausselkä ridges often consist of only glaciofluvial material, it led to the misconception. As the marginal deltas and sandur-deltas were build up to the water level they have been used in the reconstruction of the water level fluctuations during the formation of the Salpausselkä moraines. In this way it has been possible to link the recession with the changes in the Baltic, for a time from which there is very little other evidence of the land/sea-level changes. But this reconstruction is almost entirely based on morphological criteria and the interpretations have therefore varied. The main difficulty has been to determine if an even surface of a marginal formation represents a primary surface of a delta or sandur-delta, if it was formed below water level, or if it is a surface that was later eroded. Although the Late Weichselian and Flandrian land/sea-level changes will be discussed later, the evidence provided by the Salpausselkä moraines is given here, but without comparisons with areas outside Finland and without

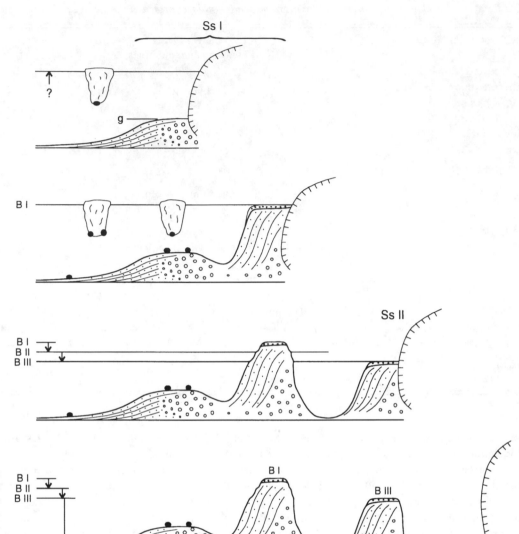

Figure 11.3. Relationship between marginal formations of Salpausselkä I and II and shorelines of Baltic Ice Lake and Yoldia Sea in the Salpausselkä zone of the Hyvinkää–Hämeenlinna area in southern Finland (after Donner, 1992).

a discussion of what events regulated these changes of level in the Baltic basin.

The fluctuations in water level during the formation of the Salpausselkä moraines have been explained as a result of the damming up of the waters of the Baltic Ice Lake and its final drainage (Donner, 1969a, 1978, 1982), a model that has been modified in later studies (Fyfe, 1990; Donner, 1992). The altitudes of the marginal deltas and sandur-deltas cannot directly be used in correlating them with one another, because the levels they represent have been tilted toward the southeast as a result of the differential land uplift during and after the deglaciation. The relative water-level changes during the formation of Salpausselkä I and II are schematically shown in Fig. 11.3. The highest relative level was reached during the formation of Salpausselkä I when the marginal terraces representing the B I level were formed. Among them is the approximately 1 km wide and 3 km long sandur-delta west of Lahti with well-preserved

melt-water channels from the proximal part to the distal edge, where they are closed by a shore bar at 150 m (Donner, 1951). It shows that the water level was at this level at the time when the ice margin retreated from Salpausselkä I at Lahti, when the flow of the melt-waters from the ice sheet had ceased to flow over the surface of the sandur-delta. Further west at Sairakkala in Kärkölä the B I level reaches its highest absolute level, 160 m, represented by a complex of a broad marginal terrace with kettle holes and some melt-water channels (M. Okko, 1962), including the braid delta of the Vesala fan (Fyfe, 1990). There are several marginal deltas accumulated to the B I level east of Lahti, in the southeastern part at an altitude of about 100 m, but southwest of Lahti only one at 150 m in Hikiä. In earlier studies it was assumed that before the water level had risen to the B I level, plateau-like marginal formations interpreted as deltas were formed 25 m below the B I level, either separately, as in the southwest at 112–125 m, or next to the deltas of the B I level, as at Utti in the southeast (Donner, 1982). The last-mentioned plateau forms an over 1 km wide and over 5 km long flat plain with its distal part at 95 m. Of the few surfaces representing this low level in the southwestern part of Salpausselkä I those at Lohja and Nummela are at 112 m. Further towards the southwest the end moraine is at a lower altitude and has no marginal terraces similar to the deltas. The low position of the water level in the Baltic, before it rose by 25 m, was taken to represent the level of the ocean at a time when there was a connection between the Baltic and the ocean, before the Baltic became an ice-dammed lake. The low level was named the g level (Sauramo, 1958), after a shoreline at the coast of the Arctic Ocean in northern Scandinavia (Tanner, 1930) assumed to be contemporaneous with the formation of the Salpausselkä I marginal deltas at the lower level. Even if this correlation is no longer valid (Donner, 1969a), the use of the term g level was later retained. In a detailed study of the southwestern part of Salpausselkä I, Glückert (1979), however, concluded that the g level represents a complex of surfaces from the time of a general rise of water level up to the B I level, whereas Synge (1980) accepted a single g level. In contrast to these studies, in which the low marginal terraces were interpreted as having been deposited up to the water level, the g level, the detailed study of the sedimentology of the southwestern arc of Salpausselkä I by Fyfe (1990) suggests that the low marginal terraces, such as the formation at 125 m in Hyvinkää, were deposited as overlapping fans below water level. This implies that there was not necessarily a rise of water level in the Salpausselkä zone before it stood at the B I level. The spatial distribution of marginal terraces at the B I level show that in the southwestern part of Salpausselkä I the water level was still at its highest position, the B I level, when the ice margin had withdrawn from the end moraine. The B I marginal terraces were therefore deposited somewhat northwest of the moraine as separate marginal deltas of eskers (Donner, 1978). In a reevaluation of the evidence of a transgression during the formation of Salpausselkä I (Donner, 1992) it was concluded that there is no conclusive proof of a transgression, and that the marginal plateaux of the lower g level were formed below water level as suggested by Fyfe (1990).

During the recession of the ice from Salpausselkä I to Salpausselkä II there was a 10 m drop of water level. The marginal terraces between the two moraines represent an intermediate level 5 m below the B I level and has been called the B II level. The marginal terraces of Salpausselkä II, at the B III level, are now, because of the subsequent uplift, at an altitude of over 160 m at their highest and at about 100 m in southeastern Finland. The final drainage of the ice-dammed lake in the Baltic took place when the ice margin stood at or just inside the Salpausselkä II moraine (see Fig. 11.3). There are several marginal deltas formed about 28 m below the level of the B III terraces at sites along the western arc of Salpausselkä II. The drop of water level, so clearly demonstrated morphologically by the occurrences of moraines with marginal terraces at two levels, was also important in the sedimentation of the glacial varved

clays used in dating the ice recession, as will be seen later. The drop of water level was the result of the final drainage of the Baltic Ice Lake through Central Sweden, resulting in a short-lived connection of the Baltic with the ocean, defined in the history of the Baltic as the Yoldian Sea stage. Salpausselkä III, inside Salpausselkä II in southwestern Finland, was formed at this time.

On the basis of the directions of striae north of the Salpausselkä moraines, and also on the basis of some glaciotectonic deformations of the proximal sediments of the moraines, including till-covered sediments, M. Okko (1962) concluded that there was a readvance of about 30 km of the ice-margin in the area west of Lahti before the formation of the Salpausselkä I moraine, after an interval that she called the Heinola deglaciation. Later, Rainio (1985) suggested that the readvance could be demonstrated throughout southern Finland, in the east called the Salpausselkä readvance, and that the ice margin had receded at least 80 km before its advance. This conclusion is also supported by the till stratigraphy, because north of the moraines a till, the "Salpausselkä phase till," in places overlies varved silts up to 65 km from the second Salpausselkä moraine (Hirvas & Nenonen, 1987), and by the orientation of the De Geer moraines south of the lobe-shaped moraines (Aartolahti, 1972). If the readvance took place before the formation of the Salpausselkä moraines, as also outlined by Lundqvist (1987b, 1988a), it was of the same magnitude as the readvance at the Norwegian coast. Even if there is strong evidence for an extensive readvance of the ice margin in southern Finland, no horizon that can be linked to such a readvance has been shown to exist in the glacial varved clays in the zone of the Salpausselkä moraines or south of them, even if there are gaps in the varve series (Niemelä, 1971). The greatest hiatus corresponds to over 1 ka.

In North Karelia in Finland, at the ends of the eastern arcs of the two Salpausselkä ridges Ss I and Ss II, the two moraines have been difficult to connect to the two moraines in eastern Karelia, that is, in Russia beyond the eastern border of Finland, as shown from the summary of the interpretations of the marginal formations of North Karelia by Eronen and Vesajoki (1988). There are essentially two main interpretations. As indicated on the map in Fig. 3.2 there is a northern end moraine, the Uimaharju moraine, which together with the extensive Jaamankangas marginal formation forms a lobe-shaped moraine deposited at the margin of an ice-lobe that advanced from the north and northwest, with Jaamankangas interpreted as an interlobate formation north of another southern lobe that formed Salpausselkä II. If the two lobes were contemporaneous, the Uimaharju moraine is a continuation of Salpausselkä II and Jaamankangas was formed between the lobes, as suggested by Ramsay (1891) and Rosberg (1892), and later by Repo (1957). This agrees with the interpretation of the moraines in eastern Karelia by Rosberg (1892) and that recently presented by I. Ekman and Iljin (1991). Accordingly, the southern moraines in North Karelia, primarily represented by the Selkäkangas–Palokangas marginal formation, would represent Salpausselkä I. The correlation is given in Table 11.1 and schematically on the map in Fig. 3.2. The other alternative, suggested by Rainio (1985, 1991) as a result of detailed studies of the ice-marginal formations in North Karelia, and supported in the studies by Hyvärinen (1973) and Eronen and Vesajoki (1988), is that Jaamankangas and the Uimaharju moraine, together representing the Pielisjärvi end moraine, are younger than Salpausselkä II and that the Selkäkangas-Palokangas moraine, representing the Koitere end moraine, should be linked with Salpausselkä II. Salpausselkä I would in this correlation dwindle in the east and be represented by formations of the Tuupovaara end moraine southeast of Salpausselkä II. This alternative is also given in Table 11.1. If only the evidence from Finland alone is taken into account it is not at variance with the connection of the other moraines in the zone of the Salpausselkäs. Jaamankangas in the east would in this scheme be broadly of the same age as Salpausselkä III in the southwest.

The correlation of the moraines of North Karelia with those in eastern Karelia used on the map in Fig. 3.2 agrees with the latest interpretation of the moraines in the later area. Of the two moraines, mainly formed outside an ice-margin resting on dry land, the case in most parts of North Karelia, the outer is the Rugozero moraine and the inner the Kalevala moraine (Table 11.1), correlated with Salpausselkä I and Salpausselkä II respectively (I. Ekman & Iljin, 1991). Near the White Sea, they are further apart than in earlier interpretations (Lukashov, 1982), but further north, in the northern part of the Kola peninsula, they are again as close together as near the Finnish border in the south. There is only a small gap to the Tromsø–Lyngen moraine in the north. Compared with the Fennoscandian moraines in Norway, Sweden, and Finland the dating of the moraines in eastern Karelia is still tentative, but the general positions of the ice-margin can be held to be well established and agree with the biostratigraphical evidence discussed later.

The final withdrawal of the ice was comparatively rapid from the zone of the Fennoscandian moraines and on the whole without major standstills or readvances of the ice margin. The earlier mentioned Ski and Aker moraines in the Oslo area are close to the moraines correlated with the Fennoscandian moraines (Table 11.1), as is the Salpausselkä III moraine in Finland. The Jaamankangas–Uimaharju moraine in eastern Finland may similarly be younger than the main Salpausselkä moraines I and II, an alternative correlation that, as has been seen, has a bearing on the age of the moraines in eastern Karelia. In central Finland there is an approximately 250 km long marginal formation clearly away from the zone of the Fenno-scandian moraines, with two arcs formed by a lobe-shaped ice-margin (see Fig. 3.2). The westernmost part, the Hämeenkangas moraine, belongs to the western arc called the Näsijärvi marginal formation (Virkkala, 1963). The eastern continuation is the arc formed by the Jyväskylä marginal formation, the arcs together named the Central Finnish formation (Repo & Tynni, 1971) or the Näsijärvi–Jyväskylä end moraine (Aartolahti, 1972; Glückert, 1974), or on some maps also the Central Finland ice-marginal formation.

The end moraines of the Weichselian ice sheet, formed at its maximum extent and during its retreat, were built up of glaciofluvial sediments in many areas. The influence of the meltwaters was even more pronounced during the deglaciation from these moraines. A characteristic feature in the marginal areas of glaciation are the tunnel valleys formed by strong subglacial meltwater streams. These landforms up to 0.5 km broad are common in North Germany and in Denmark inside the outer margin of the Weichselian glaciation and they cut through the landscape, in places even uphill toward the moraines as a result of the great force of the subglacial meltwater streams (H. W. Rasmussen, 1966). The tunnel valleys are not only erosional features, because there was also a deposition of a sheet of glaciofluvial sands and gravels at the time of deglaciation. Furthermore, esker ridges were formed in the valleys, either in tunnels or in open channels in dead ice in the marginal zone of the ice sheet, which finally in places deposited till on top of the gravels and sands – this without a readvance of the ice. In the more central parts of the glaciation, particularly on the eastern side of the Scandinavian mountains, the eskers reflect the pattern of deglaciation, where separate fan-shaped ice lobes formed the arcs of the major moraines, such as the Salpausselkä moraines in Finland. The direction of the eskers is usually perpendicular to the moraines, except in areas lying between lobes with different directions of ice movement, as seen from the schematic map in Fig. 11.4. In many parts of Sweden, however, the eskers and associated gravels and sands were deposited in the river valleys on the southeastern slope of the Scandinavian mountains and therefore all have the same general direction.

The formation of eskers has varied from place to place. In subaquatic areas, such as southern Finland (Fig. 11.5) and the coastal areas of Sweden, the material was deposited either in subglacial tunnels or at the mouths of these tunnels, in places forming deltas similar to the marginal terraces of the moraines. In the detailed study by K. G. Eriksson (1960) of sections in the Stockholm esker in Sweden the subglacial material was separated from that deposited later outside the mouth of the tunnel, on top of the esker core, in places divided into two parallel ridges. As the glaciofluvial material in eskers was transported through tunnels to the moraines, the eskers linked to the moraines have been described as feeding eskers. Their material was thus deposited in moraines together with more locally derived till. Some of the till in these moraines, as well as till found on top of eskers, was deposited at the final stage of deglaciation, as a subglacial melt-out till or some form of ablation till.

Eskers are characteristic landforms of the last Weichselian glaciation. But the coastal area of Ostrobothnia on the west coast of Finland has, as mentioned earlier, preserved till-covered eskers formed during the retreat of the Saalian glaciation before the Eemian warm stage. The well-consolidated gravels and sands of the Saalian eskers are clearly different from the sediments from the last deglaciation and the till that covers the Ostro-bothian eskers is a lodgment till and not a loose till found

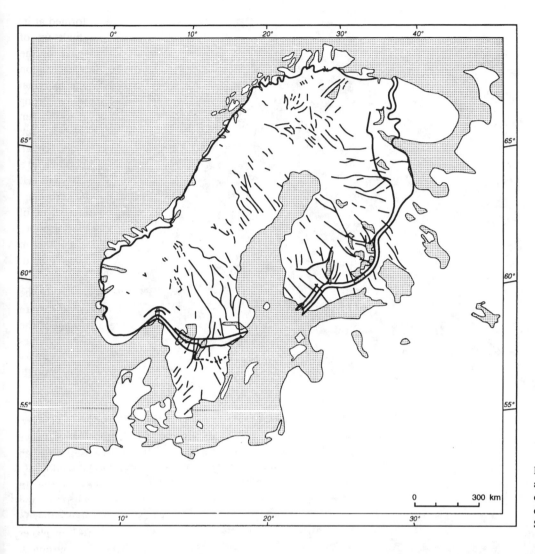

Figure 11.4. Fennoscandian moraines (see Figure 3.2) and main eskers of Scandinavia (compiled from general Quaternary maps of Norway, Sweden, and Finland).

elsewhere on top of the slopes of eskers or the proximal parts of moraines.

In areas where the ice margin terminated on dry land and where the ice melted as dead ice, eskers were often deposited in open channels. In the higher areas of the Scandinavian mountains as well as in Finnish Lapland the eskers form distinct narrow ridges that have not been affected by later changes by wave action as have the eskers in the subaquatic areas. In Norway eskers occur both in the higher mountain regions and in valleys. In the valleys, however, there are also remnants of thick glaciofluvial accumulations, later cut by the rivers and now forming conspicuous terraces (O. Holtedahl, 1960). Similar valley trains were formed in northern Scandinavia in the valleys of rivers flowing north, and in some valleys in northeastern Finland, such as in the Oulanka river valley flowing east towards the White Sea. Some of the deposits, particularly in Norway, were accumulated into short-lived local ice-dammed lakes in the valleys. The nature of the glaciofluvial deposition thus differed markedly from that in most parts of Sweden and Finland, with their network of long sinuous or beaded eskers.

The landforms described above, formed in a comparatively short time during the deglaciation and close to the retreating ice margin, cannot as such be used as stratigraphic units, even if the use of a morphostratigraphic classification has been suggested (Mangerud et al., 1974). This referred particularly to some of the major moraines, such as the Ra moraine. Direct evidence of the rate of retreat of the ice margin in some areas in Swedish and Finnish Lapland has, however, been obtained from the spacing of lateral drainage channels formed by glaciofluvial erosion on the slopes of valleys (Fig. 11.6). The channels were cut into the sediments on the slopes by the meltwaters released each warm season from the melting of the glaciers in the valleys. Even if they were not always formed annually the series of these channels can be used to estimate the rate at which the glaciers melted. In western Finnish Lapland the average vertical distance between the channels was estimated in one area at 2 m in a series with 120 channels, their gradient being 2.5–5: 100 (Kujansuu, 1967). In another area the distance was 3.0–3.7 m and the gradient 1.6–3: 100. On the basis of the lateral drainage channels, Kujansuu concluded that the withdrawal of the ice

Figure 11.5. The esker of Punkaharju in southeastern Finland. (Courtesy, the Finnish Travel Association)

Figure 11.6. Lateral drainage channels at Mt. Sonfjället, central Sweden, formed during the last deglaciation. (Photo J. Lundqvist)

margin in western Finnish Lapland toward the southwest took about 800 years. The average horizontal rate of retreat of the ice margin in the area studied, nearly 200 km in the direction of withdrawal, would then have been about 250 m/year, but with local variations. It is a fast rate compared with the earlier quoted rates of ice advance in the beginning of the Weichselian, but similar to the average rate of retreat, 260 m/year, of the ice margin from the Salpausselkä moraines in southern Finland, as determined with the help of varved clays (Sauramo, 1918, 1929). Thus, the study of the lateral drainage channels in Finnish Lapland is an example of how the annual discharge of meltwaters of a retreating ice margin on dry land eroded chronologically important channels that can be used in the study of the retreat of the ice margin, whereas the meltwaters from the ice deposited annual varved clays in the submerged areas. The glaciofluvial sands and gravels have, on the other hand, no chronological importance as such.

11.2. Ice-dammed lakes

During the Late Weichselian and Early Flandrian deglaciation lakes were formed between the ice margin and higher ground outside it. The areas in the central parts of the glaciation covered by these lakes classified as ice-dammed lakes are shown in Fig. 11.7. Many of these relatively short-lived lakes had complex histories with fluctuating water levels and changes in their extent. Furthermore, they could contain remnants of dead ice in addition to icebergs, and therefore had smaller expanses of open water than would be expected (Penttilä, 1963).

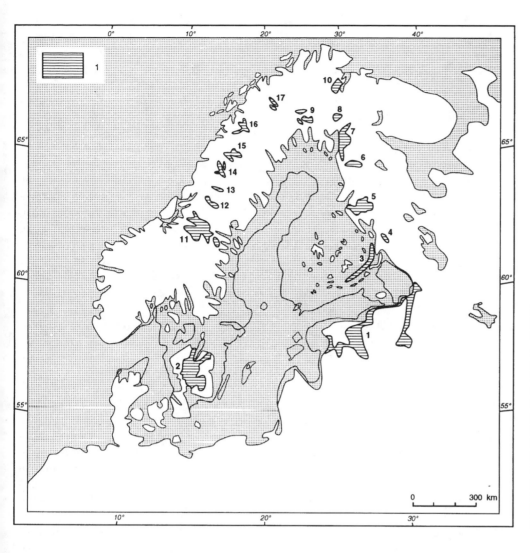

Figure 11.7. Late Weichselian and Early Flandrian subaquatic areas, including areas covered by ice-dammed lakes (symbol 1). 1, ice-dammed lake outside Pandivere–Neva moraine; 2, ice lake complex in southern Sweden; 3, ice lake north of Salpausselkä II in southeastern Finland; 4, Ilomantsi ice lake; 5, Sotkamo ice lake; 6, Posio–Kuusamo ice lake; 7, Salla ice lake; 8, Laanila ice lake; 9, ice lakes in western Finnish Lapland; 10, Inari ice lake. Ice lakes in Sweden: 11, central Jämtland; 12, Frostviken; 13, Borga; 14, Gäuta–Tärna; 15, Nasa–Jårrom; 16, Sitas; 17, northernmost Swedish Lapland.

They can be traced with the help of preserved raised beaches, and in many of them fine-grained glacial sediments were deposited that give additional evidence about their nature.

During the Late Weichselian retreat of the ice-margin from the areas south and southeast of the present Baltic Sea a series of ice-dammed lakes connected with one another were formed, changing in shape and extent as the ice withdrew further north. Of the ice-dammed lakes formed in eastern Europe (Kvasov, 1979) the one outside the previously mentioned Pandivere–Neva moraine is shown in Fig. 11.7, for the most part as presented by Raukas (1988). These early ice-dammed lakes were important for the spread into Scandinavia of freshwater glacial relics, mostly crustaceans, that survived the Weichselian glaciation in lakes further east, as demonstrated particularly by Segerstråle (1957, 1976, 1982). They could invade northern Europe along the proglacial ice-dammed lakes and later immigrate into the central area of glaciation, in which they now are found in lakes isolated from the Baltic basin. Their present distribution is clearly limited to lakes in the areas submerged after glaciation, as seen from the distribution maps presented

by Segerstråle (1957). During the Flandrian climatic optimum the relict species died out in most parts of the Baltic Sea because of an increased salinity, but later they again spread as far as their salinity tolerance permitted. *Mesidotea* and *Pontoporeia* could spread over the whole Baltic Sea area, whereas *Gammarocanthus* only survived in smaller lakes. The other faunal changes, especially of vertebrates, after the Weichselian glaciation will be discussed later.

As a result of the withdrawal of the ice margin from the Pandivere moraine in northern Estonia the ice-dammed lakes east of the Pandivere uplands were joined with the ice-dammed waters in the southern parts of the Baltic to form a large lake that in some studies has been suggested to represent the beginning of the formation of the Baltic Ice Lake (Donner & Raukas, 1989), which, as mentioned earlier, came to an end after the formation of the Fennoscandian moraines. Because the southern parts of the Baltic basin had an ice-dammed lake with successive outlets to the west as the ice-margin retreated north, the beginning of the formation of the Baltic Ice Lake has been defined in Sweden on the basis of the development in southern Sweden (E. Nilsson,

1953). The beginning was placed shortly before a drainage of the Baltic through Tyringe in southernmost Sweden, which presumably took place before the withdrawal of the ice margin from the Pandivere moraine in Estonia (Donner & Raukas, 1989). The area covered by the Baltic Ice Lake is not shown on the map in Fig. 11.7 because the evolution of the main Baltic Sea will be discussed later. The extent of the ice lake complex in southern Sweden, however, is shown on the map. It had several outlets toward the southwest and covered an area between the coastal areas submerged by the ocean and those by the Baltic during deglaciation.

In addition to the Late Weichselian ice-dammed lakes, proglacial lakes were dammed up against higher ground during the Early Flandrian deglaciation. One such short-lived lake was formed in southeastern Finland against Salpausselkä II, with two successive positions of water level, H IV and H V, above that of the Baltic after the drainage of the Baltic Ice Lake (Saarnisto, 1970). A separate ice-dammed lake was formed further east outside the marginal terrace of Selkäkangas (Table 11.1), south of Lake Koitere. This Ilomantsi Ice Lake was already recognized by Sauramo (1929), but later studied in detail by Hyvärinen (1971b). There were several ice-dammed lakes further north along the eastern border of Finland, of which the main ones are shown in Fig. 11.7, which for Finland is largely based on a map of supra- and subaquatic areas by Eronen and Haila (1981). The southernmost of them is a large lake in the Sotkamo area east of Kajaani. Its varved glacial clay series, connected to date the local deglaciation, show that the sedimentation in the Sotkamo ice-dammed lake lasted over 300 years (Kilpi, 1937). The drainage of the ice lake resulted in the formation of a few exceptionally thick, sandy, drainage varves. Further north the extent of an ice-dammed lake in the Posio–Kuusamo area was traced with the help of its former highest shoreline, which is now at about 260 m a.s.l. in the western part and just over 250 m in the eastern part, thus tilted toward the east due to differential uplift (Kurimo, 1979). The Posio–Kuusamo ice-dammed lake was an isolated lake, separated from the area submerged during deglaciation further west, as was the Ilomantsi lake. The large area in the Salla region further north that formed an irregular ice-dammed lake is, on the other hand, connected with the areas formerly submerged by the Baltic. The raised beaches of shorelines at about 250 m a.s.l. were earlier interpreted as representing the Baltic Ice Lake, which would have had an outlet east to the White Sea through Aapajärvi, with a threshold at 211 m, and through the basins of Kitkajärvi and Paanajärvi (Hyyppä, 1936). The interpretation that the Baltic Ice Lake extended up to northeastern Finland as a result of a Late Weichselian deglaciation of the areas along the eastern border of Finland was also adopted by Sauramo (1958), but later studies of the deglaciation showed that it took place in Early Flandrian time, as seen from results summarized by Hyvärinen (1973, 1975b; see also Donner & Raukas, 1989). These results led to the conclusion that the high positions of water level along the eastern Finnish border, as in the Salla region, represent local ice-dammed lakes. This was

taken into account on the earlier mentioned map produced by Eronen and Haila (1981), which is essentially similar to a map by Sauramo (1929), published before the study of the deglaciation history of eastern Finland by Hyyppä (1936).

In the eastern parts of Finnish Lapland the ice margin retreated toward the west and short-lived irregular ice-dammed lakes were formed in the depressions and valleys, such as the lake in the Laanila region (Sauramo, 1929), in which an area around Raututunturit was later studied in detail by Penttilä (1963). Similar lakes were formed in the valleys of western Finnish Lapland, where the ice margin retreated toward the southwest (Kujansuu, 1967). The extent of three of the main ice-dammed lakes is shown in Fig. 11.7. There were, in addition, numerous others, as shown by the map published by Kujansuu, all north of the area into which the Baltic Sea penetrated immediately after deglaciation. The sediments deposited into the ice-dammed lakes in Finnish Lapland consisted mainly of glaciofluvial sands and gravels from the meltwater streams in the valleys.

In northernmost Finnish Lapland the northeastern part of the Lake Inari basin was dammed by the ice-margin when it had retreated southwest from the Tromsø–Lyngen moraine at the Norwegian coast south of Varangerfjord (Tanner, 1930; Alhonen, 1969; Synge, 1969). The shoreline of the ice-dammed lake reaches an altitude of 130 m and tilts toward the northeast. The lake ceased to exist when an outlet was formed at Virtaniemi on the eastern side of Lake Inari and the water level dropped about 5 m. After a short connection of the lake basin with the ocean the lake was isolated from the sea and the independent Flandrian history as a lake started. Because of the strong outflow of freshwater from the lake, salt water was never able to reach it when it was connected with the sea.

During the final stage of the Early Flandrian deglaciation the ice sheet was situated east of the northern parts of the Scandinavian mountains. As a result a series of ice-dammed lakes were formed between the mountains and the ice, with the outflows mostly to the ocean. The final drainages of the lakes took place when the ice sheet broke up and melted on the slope east of these lakes. The major ice-dammed lakes schematically shown on the map in Fig. 11.7 have on the general Quaternary maps of Sweden (1: 100,000, 1958) been given the following names, from Jämtland in the south to Lapland in the north: Central Jämtland, Frostviken, Borga, Gäuta-Tärna, Nasa-Jårrom, and Sitas ice lakes. In addition, two ice lakes in northernmost Sweden are shown on the map. Some of the ice-dammed lakes were open lakes, others were filled with dead ice (Lundqvist, 1972, 1973), which, as mentioned earlier, is characteristic for ice-dammed lakes generally. The sediments deposited in the Swedish ice lakes varied due to their depths and shapes. Thick fine-grained sediments were only deposited in deep, open, ice lakes, in the Central Jämtland ice lake represented by thick glacial varved clays. Otherwise, sands and silts, with signs of current bedding, were deposited in some areas, and coarse glaciofluvial and littoral sediments in shallow-water environments (Lundqvist, 1972). The former shorelines can be traced with the help of preserved beaches, even if they were not well developed along all shores of the ice-dammed lakes.

11.3. Periglacial features

During the maximum extent of the Weichselian ice sheet there was a periglacial zone outside the ice margin. In this zone, features formed under cold conditions have been preserved, such as the patterned ground typical for the present arctic areas with permafrost, in sections seen as ice-wedge casts with infills of sand and gravel, features that come under the general heading of thermal contraction cracks. In glaciated areas these features have seldom survived the erosion during subsequent glaciations and therefore they are only occasionally found in sections, but when they occur they give additional evidence about the climatic conditions during the nonglacial intervals. Thus, Early Weichselian ice-wedge casts were recorded in the earlier mentioned sediments of the Dösebacka and Ellesbo drumlins in western Sweden (Hillefors, 1974) and in the esker sediments at Harrinkangas in Ostrobothnia in western Finland (Gibbard et al., 1989). In Finnish Lapland ice-wedge casts were found between the two uppermost Weichselian tills, II and III, postdating the formation of the Early Weichselian interstadial sediments (Hirvas, 1991), and in till-covered glaciofluvial material in the Tornio river valley, formed during an undefined Weichselian interval but possibly of the same age as the above-mentioned ice-wedge casts (Ristiluoma, 1974).

In addition to being common outside the outermost moraines of the Weichselian glaciation, periglacial features are also found at the surface of sediments deposited during the deglaciation in the Early Weichselian. The distribution of these features shows, when compared with the previously mentioned moraines, the area in which a cold periglacial climate with permafrost persisted outside the retreating ice margin. The clearest evidence comes from the distribution of patterned ground structures, which require a mean annual air temperature of less than $-6°$ C for their formation (West, 1977), a figure mainly based on the present distribution of these structures in Alaska, but valid also for Greenland. Additional evidence of a periglacial environment with bare ground can be obtained from the distribution of dunes and cover sands, and from signs of solifluction, but estimates of the mean annual temperatures cannot be based on this evidence alone. Permafrost, however, may at present already occur in areas with a mean annual temperature of about $-1°$ C to $-2°$ C (Péwé, Church, & Andersen, 1969).

By using low-level aerial photography, including the use of infrared pictures, large-scale patterned ground structures representing fossil ice-wedge polygons were observed as crop marks in fields in Denmark (Christensen, 1973a, 1973b, 1974). They occur both outside the outermost Weichselian moraine represented by the Main Stationary Line in Jutland and inside the moraine, in areas deglaciated in Early Weichselian time, as shown by areas studied in detail in southwestern Jutland and northern Jutland

west of Fredrikshavn, and they have also been observed in northwestern Sjaelland (Fig. 11.8). The pattern of polygons has been deformed in some places as a result of solifluction. In a study of the infilling of the former thermal-contraction cracks in southwestern Jutland, Kolstrup (1986) demonstrated that sand infilling in wedges does not necessarily have to be interpreted as showing arid conditions during their formation and that there is a great variety of wedge casts in the same area due to the types of sediments in which they occur (Fig. 11.9). In southern Jutland there are well-sorted aeolian sediments classified as coversands (Kolstrup & Jørgensen, 1982). They represent a stratigraphy similar to the coversands in Belgium, The Netherlands, and northern Germany, with an Older Coversand unit, resting on faceted pebbles and small stones on top of sand and gravel, representing the coldest part of the Weichselian, and a slightly coarser Younger Coversand representing the time of the Late Weichselian deglaciation, in The Netherlands separated into two units by the Allerød Interstadial. But, as is seen, this coversand stratigraphy is restricted to areas outside the boundary of the Weichselian glaciation and is a reflection of the conditions in a former, comparatively broad, periglacial zone.

In southern Sweden ice-wedge casts connected with patterned ground are particularly frequent, especially on top of the glaciofluvial marginal terraces of the moraines, such as the Göteborg (Gothenburg) moraine and older moraines in the county of Halland and further north around Gothenburg (H. Svensson, 1964; Hillefors, 1966, 1969). Even if remnants of patterned-ground features also occur further inland they are most frequent in the area along the west coast and in the counties of Skåne (Scania) and Blekinge and in the southern parts of the county of Småland, in areas that were not submerged (Johnsson, 1956, 1981, 1986). In this area, fossil ice-wedge casts, 1.5–6 m deep, are as frequent as in the older periglacial areas of central Europe. The ice-wedge casts further north, formed successively during the deglaciation, are more isolated and only 1–2 m deep, and occur only in areas south of the Skövde moraine (Johnsson, 1986), the older of the two Fennoscandian moraines in Sweden (Table 11.1). This distribution in Sweden shows that a severe periglacial climate persisted in Sweden only until the formation of the Fennoscandian moraines and that a zone with permafrost near the ice margin ceased to exist during the withdrawal of the ice from these moraines. A relationship between the distribution of the fossil-patterned ground with ice-wedge casts and moraines similar to that found in Sweden has also been observed in northernmost Scandinavia. On the map of the occurrences of fossil patterned ground in Finnmark drawn by Sollid et al. (1973), used for Fig. 11.8 and in which earlier observations by H. Svensson (1965) and H. Svensson et al. (1967) were also taken into account, it can be seen that their distribution, with a few exceptions, is restricted to the areas north of the Tromsø–Lyngen moraine. The polygons occur on raised marine deltas and glaciofluvial terraces. In contrast to these northern areas no ice-wedge casts have been recorded along the western coast of Norway (Larsen et al., 1984). In addition to having had a cold

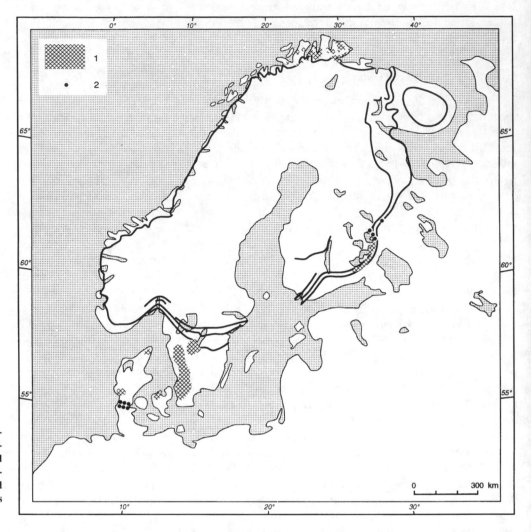

Figure 11.8. Distribution of Weichselian periglacial features in Denmark, Norway, Sweden, and Finland in relation to Fennoscandian moraines. 1, areas with fossil-patterned ground with ice-wedge casts; 2, areas with coversand.

periglacial climate the north coast had strong winds near the ice margin, which led to the formation of blow-outs and parabolic dunes at the coast of the Varanger peninsula (Sollid et al., 1973). The fossil patterned ground was particularly confined to these areas that had a thin snow cover as compared with areas less exposed to wind action. The influence of strong winds still persisted in the northern areas during the retreat of the ice margin from the Tromsø–Lyngen moraine, as seen from the distribution of aeolian sediments in areas with suitable source material, as in areas in Finnish Lapland with well-developed dunes on glaciofluvial sediments (Seppälä, 1971).

In Finland the area in which clear indications of a periglacial climate occur is restricted to the southeastern parts of the country, to the area of the Salpausselkä moraines and the area just outside them. Southern Finland was almost entirely submerged during deglaciation and eastern Finland was, as mentioned, deglaciated after the formation of the Fennoscandian moraines. Ice-wedge casts, some reaching a depth of 3 m, connected with a pattern of polygons, have only been recorded at sites on top of the marginal terraces of Salpausselkä I

southeast of Lake Saimaa and on Salpausselkä II further northeast, in the area in which the moraine curves northeastward toward Joensuu (Donner, Lappalainen, & West, 1968; Aartolahti, 1970). Some of these features are rather narrow thermal contraction cracks (Fig. 11.10) and could have been formed in a relatively short time, as the features described as load wedges on top of Salpausselkä I, on Joutsenonkangas, south of Lake Saimaa (Saarnisto, 1977), but others are well-developed ice-wedge casts similar to the earlier mentioned casts in Sweden and Denmark. They can therefore be used as evidence for a period of permafrost during the retreat of the ice margin from southeastern Finland (Fig. 11.8). In addition to these ice-wedge casts, which developed on the higher marginal terraces of Salpausselkä I and II built up to the B levels of the Baltic Ice Lake, wedge-shaped features interpreted as ice-wedge casts have also been recorded at two sites in southern Finland (Aartolahti, 1970) in the glaciofluvial material of Salpausselkä I, at altitudes corresponding to the lower marginal terraces of the g level. They were used to show that the terraces at the g level are primary marginal deltas built up to the water level, before the rise of the water

Figure 11.9. Ice wedge cast through till into Tertiary (Miocene) sand. Coastal cliff at Hvidbjerg, Vejle Fjord. (Photo K. S. Petersen)

Figure 11.10. Thermal contraction crack in outwash material of the Salpausselkä I moraine. Joutsenonkangas, southeastern Finland. (Photo J. Donner)

level to the B levels, an argument supported by other earlier mentioned evidence, foremost the presence of drop-stones on top of these terraces (Donner, 1982). Later observations, however, showed that the features in this area described as ice-wedge casts, being broad and covered by undisturbed layers of sand and gravel, are slump structures. Observations of these features in esker sands and gravels south of Salpausselkä I in Hyvinkää, not far from the previous sites, showed that in an area with slump structures and kettle holes caused by the melting of separate lumps of ice in the glaciofluvial material, some of them resemble ice-wedge casts, but differ from them in detail (Donner, 1992). They have coarse sand and gravel infillings sharply separated from the surrounding sediments, without a downward curling of their layers toward the wedges as in the wedges in southeastern Finland (Donner, Lappalainen, & West, 1968). If the features at the g level earlier described as ice-wedge casts are in fact slump structures, the suggestion by Fyfe (1990) that the marginal terraces were deposited as fans below water level is not contradicted by the presence of these features.

In addition to the ice-wedge casts in southeastern Finland, various periglacial structures have been recorded from the area in northern Karelia bounded by the Salpausselkä moraines and their continuations northeastwards (Table 11.1). The formation of these structures caused by frost action did not necessarily require permafrost. In Ilomantsi narrow thermal-contraction cracks occur in esker material below a stone horizon with ventifacts, covered by aeolian silt (Vesajoki, 1985). On the Jaamankangas moraine, on the formations of Kruununkangas

and Kulokangas, there is similarly a thin bed of aeolian cover sand on top of wind-polished stones (Markuse & Vesajoki, 1985). Further south at Tohmajärvi periglacial involutions were found to be covered by fine, probably aeolian sand (Gibbard & Saarnisto, 1977). These areas are all shown on the map in Fig. 11.8. The cover sand postdates the frost structures. The strong winds that deposited the sand in a periglacial environment also accumulated transverse, parabolic, longitudinal, and barchanoid dunes on top of the marginal terraces of Salpausselkä II and Jaamankangas, as well as east of them, as in the Tohmajärvi area (Lindroos, 1972). The prevailing winds that formed the periglacial dunes blew from between the northeast and the northwest, in agreement with the earlier mentioned conclusion about the winds close to the ice margin during the Weichselian glaciation (Liljequist, 1974). These wind directions were different from the more westerly winds that later, during and after the retreat of the ice margin, formed the shore dunes in the area.

Similar dunes were formed elsewhere in Finland throughout the deglaciation and emergence of the dry land. As long as there was an ice sheet the prevailing winds in Finland were from the northwest, from the high pressure above the ice, but later the winds were westerly and southwesterly, a change that is especially clearly recorded in the coastal zone of western Finland (Aartolahti, 1977).

When the distribution of the described areas with periglacial features, particularly that of patterned ground showing former areas with permafrost, is compared with the Fennoscandian moraines (Fig. 11.8), it can be seen that a severely cold periglacial climate only persisted until about the formation of these moraines. During the retreat of the ice margin from these moraines the climate had already changed so much that permafrost was not generally formed in the freshly exposed areas, but some indications of a narrow periglacial zone have been recorded (Lundqvist, 1962). The distribution of the areas with fossil periglacial features must also be compared with the map showing the formerly submerged areas (Fig. 11.7), because in Sweden and Finland the areas of dry land in which periglacial features could have developed was relatively small, especially immediately inside the Fennoscandian moraines.

The mean annual air temperature of less than $-6°$ C before and during the formation of the Fennoscandian moraines was in southern Finland, for instance, $10°-13°$ C less than the present annual temperature. There are at present in Norway, Sweden, and Finland no areas with continuous permafrost, but some of the northernmost parts lie within the area of discontinuous permafrost with a mean annual temperature of $-1°$ C to $-2°$ C, even if the observations of perennially frozen ground are few (Lundqvist, 1962). On the other hand, there are areas of patterned ground and other periglacial features in the higher parts of the Scandinavian mountains, especially in the north, as a result of a climate with periodic freeze–thaw cycles. Such conditions are found particularly in the zone below the snow line and above the tree line. The various forms of patterned ground and related frost features have each their own areas of distribution, some of them also being common below the tree line, as shown by Lundqvist (1962) in his comprehensive review of these features from south to north in Sweden, with a comparison with the occurrences of fossil patterned ground in southern Sweden. On the Swedish side of the Caledonides the vertical distributions of sorted circles, including stone pits, nonsorted circles, sorted and unsorted nets, polygons, steps, stripes, and cracks are chiefly restricted to the treeless zone above the tree line and concentrated below the lower limit of the boulder fields resting directly on the bedrock in the higher mountains, but also occurring above this limit. The occurrences of talus and boulder depressions are more difficult to relate to the present climatic conditions as they may have been entirely formed earlier under conditions other than those prevailing now. The conditions resulting in talus slopes in northern Finland were, for instance, present throughout the Flandrian (Söderman, 1980). On the islands of Öland and Gotland, nets of small polygons have developed on the flat surfaces of the Ordovician and Silurian limestones with only a thin soil cover and a sparse vegetation (*alvar* in Swedish), in a comparatively mild climate (Lundqvist, 1962). The frost phenomena in the mires will be mentioned later, as they can be related to the biostratigraphy of the peats.

11.4. Varve chronology

By systematically counting the annual varves of the glacial clays or clayey silts and measuring the variations of varve thicknesses, a method introduced by Gerard De Geer at the end of the last century, an independent way of dating the deglaciation in Sweden and Finland was established. In principle the withdrawal of the ice can be dated with an accuracy of one year and, as seen from the revisions of the varve chronology in Sweden, this accuracy has already been reached for some areas. A continuous varve chronology can only be established for those areas surrounding the present Baltic Sea which were submerged during deglaciation (Fig. 11.7), in an environment with fresh or brackish water suitable for the formation of annual varves. The varve chronology thus covers southern Sweden, the eastern parts of central and northern Sweden and southern Finland. The time scale based on the glacial varves formed during deglaciation was correlated with the postglacial time scale in northern Sweden, where thin varves continued to be deposited at the mouths of the rivers flowing into the Baltic, during the isostatic emergence of the coast following deglaciation. In this area the time scale could be brought up to the present time.

The Swedish varve chronology, the "Swedish Time Scale" established by De Geer (1940), covered the retreat of the ice from the area of the Middle Swedish moraines northwards and was connected with the chronology by Lidén (1913, 1938) of the younger varves in the river Ångermanälven, which brought the time scale close to the present time. The later revisions of this chronology extended it to southernmost Sweden. The details of the revisions of the chronology, with a mention of the investigators involved, were given by Fromm (1970) and later by Strömberg (1985a), and of the younger time scale by Cato (1985, 1987). In the following discussion, only the main outlines of the revised Swedish chronology will be discussed, as well as the Finnish varve chronology and its connection with the Swedish chronology. The main outlines of the varve chronologies in Sweden and Finland are given in Fig. 11.11; the dating of the deglaciation is shown on the map in Fig. 11.12.

Having measured and counted over 10,000 varves, Cato (1985, 1987) succeeded in connecting the varve chronology in the river Ångermanälven in Norrland with the present, using A.D. 1950 as a starting point. This chronology Cato connected with the zero year in the Swedish time scale, a thick varve at Döviken in the river valley of Indalsälven southwest of Ångermanälven, a varve interpreted as a thick varve formed during the drainage of an ice-dammed lake and separating the earlier formed glacial varves from later postglacial varves. Even if this transition in the varves

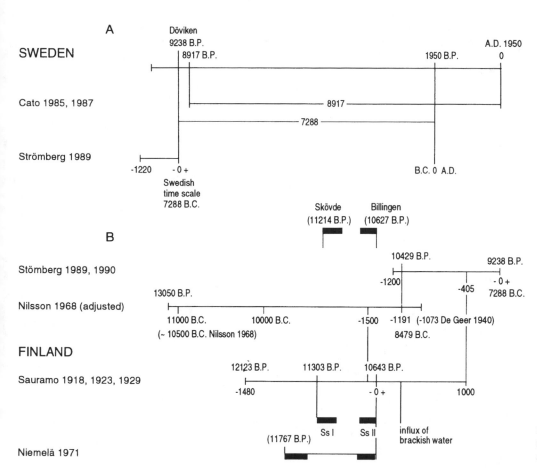

Figure 11.11. Main outlines of Swedish and Finnish varve chronologies and their connections.

is not a synchronous horizon in the varved clays in the area (Strömberg, 1989) it has, as it is well defined, been used in the varve chronology in Sweden. It was dated by Cato (1985, 1987) at 9238 B.P. (7288 B.C.) and from this zero year the older varve years are counted backwards (Fig. 11.11). The zero year differs from that originally used for the river valley of Ångermanälven by Lidén (1913, 1938), which is now dated at 8917 B.P. The younger glacial varves in the valley of Indalsälven were used in the detailed study of the ice recession from the coast and tied to the Swedish zero year by Borell and Offerberg (1955) and Hörnsten (1964), and these results were taken into account by Strömberg (1989) in his revised time scale for the ice recession from the Stockholm area north along the coast to Indalsälven, an area earlier covered by Järnefors (1963).

The revisions by Cato and Strömberg cover a period of over 10,000 years, as seen from Fig. 11.11, and bring the chronology back to the zone of the Middle Swedish moraines south of Stockholm. De Geer's (1940) chronology for this zone and for southern Sweden, as extended and revised by E. Nilsson (1968), gives the broad outlines of the ice recession. But it has been revised in its details, by Strömberg (1985b) in the area east of Mount Billingen in Västergötland, by Kristiansson (1982, 1986) from the area south of the Middle Swedish moraines to the area west of Kalmar opposite Öland, by Ringberg and Rudmark

(1985) south of Kalmar, and by Ringberg (1971, 1979, 1991) in Blekinge in southernmost Sweden. In constructing the map for the ice recession (Fig. 11.12) the lines drawn by E. Nilsson were used (see also Tauber, 1970), but the chronology was adjusted according to the corrections given by Strömberg (1989, 1990). The year −1073 in De Geer's chronology, used by E. Nilsson to mark the arrival of salt water into the Baltic when the ice margin was just south of Stockholm, and dated at 8015 B.C., was correlated by Strömberg with the year −1191 in the revised time scale, which gives it the date of 10,429 B.P. (8479 B.C.). Similarly, the corrected date for E. Nilsson's varve of 8213 B.C., correlated by him with the drainage of the Baltic Ice Lake after the formation of the Billingen moraine, becomes 10,627 B.P. (8677 B.C.), and the varve of 8800 B.C. marking the beginning of the standstill at the Skövde moraine becomes 11,214 B.P. (9264 B.C.), both being approximate ages. The dating of the drainage of the Baltic Ice Lake, "The Billingen Event," was based on the assumption that two or three thick varves were formed during the drainage, the uppermost of them being interpreted as the "catastrophe varve" (E. Nilsson, 1968). This correlation has later been questioned and reinterpreted. Strömberg (1990) suggested that a change in varve color from reddish or brownish clay to gray clay, found in a number of varve series in Sweden and dated at −1445 to −1500 in the revised time scale, corresponds to the

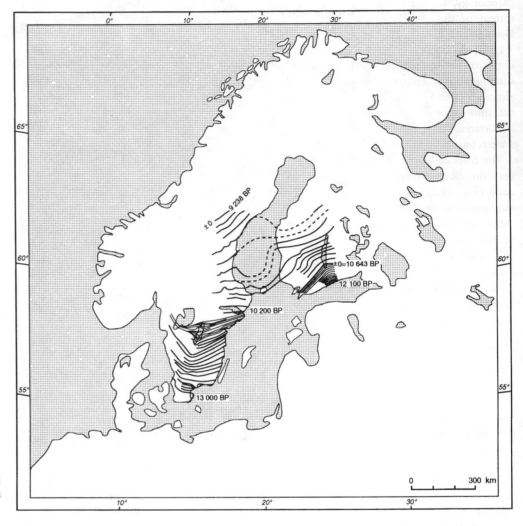

Figure 11.12. Late Weichselian and Early Flandrian ice recession as dated with varved clays.

drainage of the Baltic Ice Lake at Billingen. This correlation gives the drainage an age of about 10,700 B.P. (10,683 B.P. and 10,738 B.P., or 8733 B.C. and 8788 B.C.). These ages are slightly older than those given for the end of the formation of the Billingen moraine in Fig. 11.11. The recession line corresponding to the color change in the varves is in Östergötland 5–10 km north of the moraine. The opening up of the channel westwards at Billingen, causing the drainage of the Baltic Ice Lake, has been recorded in the thick clays in Gothenburg at the Swedish west coast (Cato, 1982). A sudden influx of freshwater with a marked increase of suspended clay material and freshwater diatoms marks the drainage into Skagerrak, a drainage of about 10,000 square km of water in about 400 years (Cato, 1982). The recorded change is about 250 km away from Billingen and forms a horizon between glacial sediments and overlying marine sediments.

If these revised ages of the varve chronology in Sweden are also taken into account for the recession lines drawn for southern Sweden south of the Middle Swedish moraines, the age of the southernmost line is about 13,000 B.P. – about 11,000 B.C.

instead of 10,600 B.C. as on E. Nilsson's map. The clays deposited in ice-dammed lakes in various parts of Denmark could not be used for an extension of the varve chronology to cover the Early Weichselian ice recession. The number of varves in the individual lakes, being far apart, was at most about 50. Furthermore, the varves were in many cases shown not to be annual varves but represent shorter time intervals, possibly daily variations in deposition (Hansen, 1940, 1965). The varves in Denmark are considerably thicker, often 20–40 cm, than the varves in Sweden and Finland. Even if the varved clays in Denmark could not directly be used for a time scale for the recession, Hansen (1940) estimated that the withdrawal of the ice from the Main Stationary Line in Jutland to northeastern Scania took at most 2,000–3,000 years. If these years are added to the age in southernmost Sweden of 13,000 B.P. the recession in Jutland would have started about 15,000 to 16,000 years ago, an estimated age which is a few thousand years younger than the age of 18,000 to 20,000 years B.P. used for the Main Stationary Line, but older than the age of 14,000 to 15,000 B.P. used for the East Jutland Line (Lundqvist, 1986b). In the general varve chronology, De Geer used the term Daniglacial time for the retreat of the ice margin in Denmark to southern Sweden, Gothiglacial time for the retreat to the line of −1074 years, now −1191 (10,429 B.P.), in the Swedish time scale, Finiglacial time to the zero year in the Swedish time scale, and Postglacial for the subsequent time up to the present. This division was also used by Sauramo (1929), but in later investigations of the ice recession the time units of the varve chronology have been of lesser importance than those used in biostratigraphy, where the time scale is based on radiocarbon ages.

The Swedish varve chronology extends further back in time than the chronology for Finland, which starts from the south coast (Fig. 11.12). The independent Finnish varve chronology established by Sauramo (1918, 1923) covers over 2,000 years and dates the withdrawal of the ice margin from the southern and western parts of the country, including the standstills at the Salpausselkä moraines. Sauramo chose as a zero year, ±0, a thick varve that he correlated with the drainage of the Baltic Ice Lake, seen as the previously mentioned drop from the B III level to a level about 28 m lower, at the time of the formation of Salpausselkä II or immediately after it (Fig. 11.13). The varve chronology was later extended to include Ostrobothnia in western Finland, as shown by the recession lines drawn by Sauramo (1929). The separate varve chronology for the earlier mentioned Sotkamo ice-dammed lake, covering over 300 years, was linked by Kilpi (1937) with Sauramo's chronology and given the years +700 to +1020, that is, years after the retreat of the ice-margin from Salpausselkä II. The formation of the two first Salpausselkä moraines, I and II, representing the Fennoscandian moraines in southern Finland (Table 11.1), took over 600 years according to Sauramo, Salpausselkä I having been formed between the years −660 and −443 and Salpausselkä II between −183 and ±0. The change from diatactic to symmict clay at +292 in Sauramo's chronology was interpreted as showing that there was an influx of brackish water after the drainage of the Baltic Ice Lake.

Before a direct connection could be established between the Swedish varve chronology and the floating Finnish chronology, various correlations were proposed. In many comparisons a correlation between Sauramo's zero year and the drainage varve used by E. Nilsson (1968) seemed natural. This correlation was, for instance, used by Niemelä (1971) in his study of the varved clays in southern Finland, in which he gave an age of 10,163 B.P. (= E. Nilsson's 8213 B.C.) to the Finnish zero year. Earlier Sauramo (1940) had correlated that year +292 in the Finnish time scale with the varve in Sweden showing the arrival of salt water in the Stockholm region, varve −1073 in De Geer's (1940) chronology, which was also used by E. Nilsson (1968). This was later, as mentioned, revised to −1191 (10,429 B.P.). The correlation by Sauramo was based on the assumption that the change at this time from diatactic to symmict clay, recorded both in Sweden and Finland, reflected a synchronous influx of brackish water. As a result of the revision of the Swedish time scale north of the Middle Swedish moraines (Cato, 1987;

Figure 11.13. Varved clays at Jokela, southern Finland. Thick varve below thin varves denotes the zero year in the Finnish varve chronology dated at 10,643 B.P. (see Fig. 11.11). (Photo J. Donner)

Strömberg, 1989), and new investigations of varve series from Åland and from southwestern Finland (Strömberg, 1990), a direct connection between the Swedish and Finnish varve chronologies was established. This connection was furthered by the presence of a stratigraphical horizon that could be identified both in Sweden and in Finland, including Åland. It was called the "spot zone" by Strömberg (1989, 1990) and consists of varved clays with fragments of Ordovician limestone deposited by icebergs and derived from the bottom of the southern Bothnian Sea northwest of Åland off the Swedish coast. The lower limit of the spot zone is, according to Strömberg, synchronous but the number of varves within it vary. On the Swedish coast there are 90–100 varves in this zone, on Åland 40–50 varves, and in southwestern Finland about 30. Supported by the position of the spot zone, Strömberg (1990) could connect the revised Swedish time scale with the Finnish varve series, particularly with the varves between +500 and +650 in Sauramo's time scale. Thus +500 in this scale was connected with −905 in the Swedish time scale, the whole comparison given by Strömberg (1990) covering the time period from −100 to +1000 in Sauramo's chronology. From this it follows that the Finnish ±0 year corresponds to the year −1405 in the Swedish time scale, which gives it an age of 10,643 B.P. (8693 B.C.). If this age is used in a revision of Sauramo's chronology for the retreat of the ice from the coastal area of southern Finland, the ice-marginal position of the year −1480 at the coast is dated at 12,123 B.P., and the beginning of the formation of Salpausselkä I at 11,303 B.P. The ages obtained in this way for the formation of the two Salpausselkä moraines are thus very close to the ages obtained for the formation of the Middle Swedish Skövde and Billingen moraines, as seen in Fig. 11.11. Further, the lengths of the periods during which each moraine was formed were of the same magnitude. There is therefore a close similarity between the revised Swedish varve chronology and Sauramo's Finnish chronology according to the connection presented by Strömberg (1990). The varve chronology presented by Niemelä (1971) for southern Finland, covering 2,500 years, in which he interpreted the earlier varve chronology by Sauramo as having gaps and therefore being too short, gives older ages for the Salpausselkä moraines. If Niemelä's time scale is used for the retreat before the Finnish zero year the formation of Salpausselkä I would have already started at 11,767 B.P. (Fig. 11.11). As Niemelä's chronology is mainly based on varve series south of the Salpausselkä I moraine their connection with Sauramo's ±0 year is only tentative. The accuracy of the extended time scale suggested by Niemelä must therefore still be questioned (Strömberg, 1990).

Some earlier correlations of the Swedish and Finnish varve chronologies were, as mentioned, based partly on changes in the varves from diatactic to symmict clays, but the connection by Strömberg (1990) was based mainly on varve graphs. The age of the Finnish ±0 year was in this way dated as 10,643 B.P. But as direct comparisons of the varves were made with the Finnish varve sequences between the varves +500 and +650 there are possible errors in the Finnish time scale between these varves and the varve ±0, particularly between varves +310 and ±0 that represent very thin varves (Strömberg, 1990). There is also some uncertainty about the linkage between the varve ±0 and the drainage of the Baltic Ice Lake, as well as the reason for the color changes found in the varve series, whether they were synchronous or not. There is still not an exact correlation between the Swedish and Finnish time scales as regards the time of the drainage of the Baltic Ice Lake and the subsequent short influx of salt or brackish water into the Baltic basin. If the correlations presented in Fig. 11.11 of the ages of the formation of the Salpausselkä moraines and the Middle Swedish moraines are used, and the drainage of the Baltic Ice Lake is dated at just over 10,600 B.P., the influx of ocean water took place about 100 years earlier in the Stockholm area than in western Finland, assuming, as earlier mentioned, that the changeover from diatactic to symmict clay in the latter are was caused by this influx. Thes conclusions, however, cannot be verified until the Finnish varve chronology is fully revised.

In addition to the widespread varved clays in Finland, which have been used in dating the ice recession, there are separate areas with the varved clays south of the Gulf of Finland, in Estonia, Latvia, and Lithuania. The varve series measured by Sauramo (1925a) were, however, comparatively short, the longest being only 115 years, and they could therefore not be used for dating. But further east, in the area of Lake Ladoga, thicker varved clays occur. In Helylä near Sortavala in Karelia, north of Lake Ladoga, 1682 varves were counted in a 6 m thick section. The formation of the varved clays ceased here, according to Bakhmutov, I. Ekman, and Zagnyi (1987), at the time of the drainage of the Baltic Ice Lake. If the date of about 10,700 B.P. is used for the drainage the oldes varves in the Helylä section were formed at about 12,400 B.P., an age close to the date of 12,100 B.P. obtained for the ice-marginal position at the south coast of Finland. These examples show that the time scale based on varved clays can perhaps be extended further back in time in the areas southeast of Finland. This is supported by the observations made by Sauramo (1925b) in the Neva valley near St. Petersburg, where up to 400 varves were found in 2–8 m thick varved clays.

The time scales in these Swedish and Finnish varve chronologies were given with one year's accuracy, as given in the detailed studies from which they were quoted, but the margins of error for different parts of the time scales mentioned in these studies were not included. The accuracy of dating has, however, reached a level where the assumed errors are of the order of tens of years, in some areas possibly a couple of hundred years. Thus the varve studies give a comparatively accurate picture of the ice recession in Sweden and Finland. It can be seen from the map in Fig. 11.12 that the Late Weichselian retreat of the ice margin in southern Sweden and southernmost Finland slowed down before the standstills at the Fennoscandian moraines, which were formed during two main standstills of about 200 years each, with an interval of about 200 years between, and which ended at about 10,600 to 10,700 B.P. at the time of the drainage of the Baltic Ice Lake. This was followed at least in the Stockholm area by an influx of salt water at about 10,400 B.P. If these varve years are compared with the radiocarbon age of 10,000 B.P. used in defining the boundary between the Late Weichselian and the Flandrian (see Tables 1.1 and 7.1), it can be seen that the varve years would indicate that the formation of the Fennoscandian moraines came to an end well before the beginning of the Flandrian (Holocene). The boundary of 10,000 radiocarbon years B.P. is, however, a chronostratigraphic boundary and is therefore of minor importance in describing the dynamics of the deglaciation. After the formation of the Fennoscandian moraines the withdrawal of the ice clearly quickened, at the Swedish east coast to 200–300 m/year (Strömberg, 1989), in Finland at an average rate of 260 m/year, but in some areas up to 500 m/year, after having had an average rate of 60 m/year south of the Salpausselkä moraines (Sauramo, 1924, 1940). Compared with the earlier mentioned rates of advance of the ice during the Weichselian glaciation, from 50 m/year up to 100–150 m/year, the rates of retreat are of the same order of magnitude.

The pronounced change in the rate of the retreat of the ice margin after the slow Late Weichselian retreat and the standstills at the Fennoscandian moraines is a reflection of the climatic changes that governed the end of the Weichselian deglaciation. The slowing down of the deglaciation to a halt at the Fennoscandian moraines was the result of a climatic deterioration, followed by a rapid amelioration resulting in the final melting of the ice sheet. The pattern of climatic changes reflected in the changeover from the slow Late Weichselian rate of retreat to the subsequent faster rate agrees with the pattern found in the distribution of periglacial features described earlier. The Fennoscandian moraines form a dividing line between areas in which a cold periglacial climate with permafrost persisted during deglaciation and areas inside these moraines with generally no traces of former permafrost. The evidence from the varved clays and from the periglacial features of the climatic conditions is corroborated by the biostratigraphical evidence, which will be discussed next.

11.5. Biostratigraphy

The pronounced climatic oscillations toward the end of the Late Weichselian have biostratigraphically been recorded throughout northwestern Europe and the identified interstadials and stadials have been widely used in studies of the time of deglaciation. In the zonation introduced by Jessen (1935) for the Danish pollen diagrams, the late-glacial period was divided into three zones, Older Dryas, I; Allerød, II; and Younger Dryas, III – Allerød representing a warmer oscillation between the cold Older and Younger Dryas periods and, as such, an interstadial followed by the Younger Dryas Stadial (Table 11.2). The name of the interstadial derived from the site in northern Sjaelland where sediments representing the Allerød oscillation had been found already at the beginning of this century. After the introduction of Jessen's zonation, Iversen (1942) found an older interstadial below Allerød that he called the Bølling oscillation after the lake in Jutland where it was first identified. This led to a subdivision of Jessen's zone I, Bølling representing zone Ib between zone Ia, Oldest Dryas, and zone Ic, Older Dryas, which is between Bølling and Allerød (Iversen, 1947, 1954). In pollen diagrams, particularly in that from Bølling, in which these late-glacial interstadials were separated, the lowermost zone, Ia, marks a transition period with the first signs of a climatic amelioration after a full-glacial arctic climate, but with a treeless tundra vegetation. There is in zone Ia a marked increase of *Artemisia* pollen and the first occurrences of *Hippophaë,* with a general dominance of nonarboreal pollen as throughout the late-glacial, showing that closed forests did not spread until the beginning of the postglacial period in Jessen's division, which corresponds to the Flandrian. The Bølling Interstadial is characterized by an increase of *Betula* pollen, the vegetation having changed to an open park-tundra. During the Older Dryas Stadial, Ic, there was again a treeless tundra vegetation, followed during the Allerød Interstadial by a new immigration and spread of *Betula* and with low percentages of *Pinus*. The climatic deterioration during the

Younger Dryas Stadial, zone III, caused an increase of *Artemisia,* grasses, sedges, and other nonarboreal pollen, with a corresponding decline of the arboreal pollen. The transition to the first postglacial (Flandrian) period, zone IV, shows the reaction of the vegetation to the quickly rising temperature, with, for instance, first a rise of *Empetrum* followed by a peak of *Juniperus* before the general spreat of *Betula,* marking the initial spread of the forest (Iversen, 1954, 1973). By comparing the late-glacial sites in Denmark with one another Iversen (1947) was able to show the spatial differences in vegetational history. During the Allerød Interstadial, northern Jutland had a rather open vegetation with *Betula,* a park-tundra, whereas further southeast, and on the island of Bornholm in the Baltic, the forest was more closed and *Pinus* was already present. During the Younger Dryas Stadial a tundra vegetation took over in most parts of Denmark, with remains of *Betula* stands only in the northernmost parts of the country. The climatic changes of the late-glacial period were also recorded in the sediments, in most areas in the lakes as freshwater muds of the Allerød Interstadial between clays deposited during the cold stadials, as a result of solifluction in a periglacial environment. Throughout the late-glacial period, organic lake sediments were formed only in southernmost Denmark (Iversen, 1947). The main outlines of the vegetational history in Denmark, as summarized by Iversen, were later outlined in more detail in the posthumously published general account of the environmental changes in Denmark after the last glaciation (Iversen, 1973); the Late Weichselian flora had been treated in detail already earlier (Iversen, 1954).

The detailed pollen stratigraphical division of the late-glacial zones by Jessen and Iversen described above was slightly modified later. Because of the variations within zone II, Krog (1954) divided it into three subzones, a, b, and c, in the study of the Allerød section of Ruds Vedby in western Sjaelland, one of the first sites in which the organic sediments of the Allerød Interstadial were successfully radiocarbon-dated (Fig. 11.14), even if some of the sediments consisted of lake marl (Iversen, 1953; Krog, 1954). The division of zone II by Krog was adopted by S. T. Andersen (1980) in his study of the Late Weichselian birch assemblages, in which he separated pollen of dwarf birch, *Betula nana,* from those of the tree birch *B. pubescens* and showed that the former was more common during the cold stadials, during which tree birches were virtually absent, as already mentioned. Similar results were obtained by Fredskild (1975) in his study of an Allerød site on the Island of Langeland in southern Denmark.

The palynological evidence, together with the stratigraphical evidence, reflect the pronounced climatic fluctuations during the Late Weichselian, the Younger Dryas being a cold oscillation during the general amelioration of climate, similar but larger than the Older Dryas. The interstadials have also been taken to represent oscillations during the Late Weichselian, as seen from the studies in which they were first identified. What is important, however, is that because of the recorded fluctuations the interstadials and stadials can be used as stratigraphical units over a wide area, the pollen diagrams reflecting a general parallelism of the

Table 11.2. *Late Weichselian biostratigraphy in Denmark and Sweden compared with age of Fennoscandian moraines*

Pollen assemblage zones in Denmark[a]	Chronozones[b]	Conventional radiocarbon ages B.P.	Chronozones[c]	Suggested calibration of radiocarbon ages[d]	Biostratigraphical evidence from Mt. Billingen area[e]	Formation of Fennoscandian moraines in Finland and Sweden acc. to varve chronology
		——— 10 ka ———			Final drainage of Baltic Ice Lake	
LW 3	b / a	Younger Dryas	Younger Dryas	10.6 ka 10.4–10.5 ka		10.6 ka
						Middle Swedish and Salpausselkä (I & moraines
		——— 11 ka ———	——— 11.45 ka		First drainage of Baltic Ice Lake	11.3 ka
LW 2	c / b / a	Allerød	Allerød			
		——— 11.8 ka				
	c	Older Dryas				
		——— 12 ka ———	——— 12.05 ka			
LW 1	b	Bølling	Older Dryas ——— 12.2 ka Bølling	12.2 ka		
	a	(Oldest Dryas)				
		——— 13 ka ———				

Sources: [a]S. T. Andersen, 1980 [d]Björck, Sandegren, & Holmqvist, 1987
[b]Mangerud et al., 1974 [e]Björck & Digerfeldt, 1982a, 1982b
[c]Berglund & Rapp, 1988

vegetational history, with the maximum and minima of the combined tree pollen curves in the diagrams being synchronous throughout Central Europe, including Denmark (Iversen, 1973). But the boundaries between the pollen analytical zones vary from place to place and are time-transgressive. These differences cannot be identified on the basis of pollen analysis alone. Some features in the pollen diagrams seem to be characteristic over large areas, such as the previously mentioned strong representation of *Artemisia* in zone III and the *Empetrum–Juniperus* sequence at the transition into the Flandrian, the latter feature not necessarily being synchronous over a wide area.

In the chronostratigraphical division of the Late Weichselian suggested by Mangerud et al. (1974) and, as stated, used here, the boundaries between the interstadials and stadials were given ages based on conventional radiocarbon ages obtained from southern Scandinavia, and for the older boundaries on ages from Central Europe, from noncalcareous freshwater sediments. Thus the area used for this division comprises mainly Denmark, southern Sweden and southern Norway. In presenting the division for the Late Weichselian, Mangerud et al. (1974) used 31 radiocarbon ages from biostratigraphically well-documented sites. The age for the lower boundary of Bølling was placed at 13 ka BP, which means that Bølling covers both Oldest Dryas, zone Ia, and Bølling, zone Ib, in Iversen's (1954) division, as seen in Table 11.2, in which the pollen assemblage zones defined by S. T. Andersen (1980) are compared with the chronozones by Mangerud et al. (1974). Andersen's Late Weichselian pollen zones LW 1, LW 2, and LW 3, in which 13 ka B.P. was used as the lower boundary, correspond to zones I as used by Iversen (1954), and II and III in Jensen's zonation. In the chronostratigraphical division the Bølling/Older Dryas boundary was placed at 12 ka B.P., the Older Dryas/Allerød boundary at 11.8 ka B.P., the

Figure 11.14. Allerød sediments with radiocarbon ages (listed by Krog, 1954). Brickworks at Ruds Vedby in western Sjaelland, Denmark. (Photo J. Donner)

Allerød/Younger Dryas boundary at 11 ka B.P., and the upper boundary of Younger Dryas at 10 ka B.P., which is also the Weichselian/Flandrian boundary. The stratigraphical change from an open vegetation to closed birch forest is, however, slightly older in southern Scandinavia, as pointed out by Mangerud et al. (1974) when they presented their division. The ages of the chronozones presented in Table 11.2 differ somewhat from those used by Iversen (1973) for Denmark in which his definitions of the interstadials and stadials were more strictly based on the biostratigraphical changes. Thus, the age of Bølling was given an age of 12.5 ka to 12 ka B.P., in accordance with the previously mentioned correlation of his zone Ib with Bølling. Iversen gave an age of 10.3 ka B.P. for the upper boundary of the Younger Dryas, which is more in keeping with the change in the vegetational history and with environment changes generally, as will be seen in the discussion of results from detailed studies of the Late Weichselian in Sweden.

In Chapter 7, on the application of radicarbon dating in the

study of the Early and Middle Weichselian interstadials, the range of the method was discussed. The use of radiocarbon dates in the study of the Late Weichselian interstadials is fraught with difficulties of another nature, one being purely technical, that is, caused by the different ways in which the radiocarbon ages have been calculated. The ages given by Mangerud et al. (1974) were conventional radiocarbon ages based on the Libby half-life of 5568 ± 30 B.P. or 5570 ± 30 B.P. When the $\delta^{13}C$ values have been measured the dates have been corrected for isotopic fractionation, normalized to $\delta^{13}C = -25.0‰$. This correction, which is now common practice, was not always applied during the early stages of radiocarbon dating. In addition to the corrections based on the $\delta^{13}C$, using the mean value of $-25.0‰$ for terrestrial plants, corrections have been made for marine shells or plants based on the apparent age of recent shells. This reservoir effect of the ocean water varies and has been determined for different coasts, the value used for the Norwegian coast being 440 years (Mangerud & Gulliksen, 1975), close to the average of 450 ± 40 reported earlier (Mangerud, 1972). The value used by Fredén (1988) for the Swedish west coast was 400 ± 25, and – further inland in Sweden – Björck and Digerfeldt (1982b) used the value of 300 years for marine-brackish sediments. These apparent ages are for dates normalized to $\delta^{13}C = -25‰$. If they were normalized to $\delta^{13}C = 0‰$, which is the approximate average for marine shells, the influence of the apparent age of 400–450 years is negligible, and therefore the ages for marine shells from northern Norway were neither normalized nor corrected for the apparent age by Donner, Eronen, and Jungner (1977).

After new determinations the Libby half-life of ^{14}C was found to be 5730 ± 40 years, but the use of the old half-life value was recommended (Godwin, 1962). As the new half-life was considered to be the best value available, ages based on this value have also been reported, together with conventional ages. The latter have been converted by multiplying them by 1.03. In the present discussion of the Late Weichselian all dates are as a rule given as conventional ages B.P. in agreement with the recommendation of Godwin.

These differences in calculating the radiocarbon ages do not cause problems in correlations as long as it is clearly stated what the ages mean. In most studies the ages are given as conventional radiocarbon years. A more serious problem is the difficulty in obtaining reliable dates for the Late Weichselian sediments poor in organic matter. These sediments were, in the area dealt with here, formed mainly in the periglacial zone close to the retreating ice margin, and if the radiocarbon ages from them are used in dating the deglaciation possible errors in the dates result in erroneous conclusions. Several examples of too old radiocarbon dates for Late Weichselian and Early Flandrian sediments poor in organic matter in Finland were given by Donner and Jungner (1974). Many dates were too old because the basal silty muds or peats contained redeposited older organic material, as could be seen already from the presence of redeposited pollen. Thus ages of up to 20 ka were obtained for about 10 ka-old sediments, in addition to the extreme ages for the clay at

Somero, where the mixture of Eemian and younger material gave an age of over 30 ka. Another source of error is the presence of graphite redeposited from the Precambrian bedrock, which in Central Sweden gave ages for basal samples about 1,500 years too old (Lundqvist, 1973) and which similarly was demonstrated to have affected the ages of the basal peat at Varrassuo near Lahti in Finland, the ages there being up to 2,000 years too old. In order to obtain more reliable dates, different fractions of sediments poor in organic matter have been dated separately. Thus, the fraction soluble in NaOH and the insoluble fraction of lake sediments in Sweden were dated and the results implied that the former fraction gave more reliable dates (Olsson, 1979).

It is often difficult to estimate how much the radiocabon ages in sediments poor in organic matter have been affected by the redeposited organic material or graphite, whereas the influence of the hard-water effect is easier to estimate by comparing the ages from calcareous sediments with ages from corresponding noncalcareous sediments from nearby sites. The hard-water effect, which in Flandrian sediments in northeastern Finland caused the ages to be about 1,000 to 3,000 years too old, is on the whole not a problem in most parts of Norway, Sweden, and Finland, but in Denmark it hampers the exact dating of some Late Weichselian sediments. In addition to difficulties encountered in dealing with radiocarbon ages that are too old, the dates may be correct and lead to a reinterpretation of the sedimentary sequence. The Allerød site at Nørre Lyngby in northern Jutland, for instance, had to be reinterpreted as the radiocarbon dates showed that most of the dated material had been redeposited and disturbed in the studied freshwater basin (Krog, 1978). The dates seemed at first to be too young.

This discussion shows how difficult it is to base the detailed division of the Late Weichselian entirely on radiocarbon dates, as reliable dates cannot be obtained for many sites for which the biostratigraphical evidence is good. A number of new radiocarbon dates have, however, been reported after those 31 ages used by Mangerud et al. (1974) in their chronostratigraphical division of the Late Weichselian. But there are still very few reliable dates for pre-Allerød sediments and its subdivision is therefore difficult. Results from southernmost Sweden show that the short Older Dryas is older than 12 ka and is difficult to date. Its usefulness as a chronozone has therefore been questioned by Björck and Håkansson (1982), even if a short stadial just before 12 ka can be identified at some sites. As the most practical solution Björck (1984) suggested that the time between 13 ka and 11 ka B.P. should be treated as one interstadial. Berglund and Rapp (1988) took into consideration the new radiocarbon ages, retained the main chronostratigraphical division by Mangerud et al., but made one revision. They used the dates 12,200 B.P. and 12,050 B.P. for the boundaries of the Older Dryas chronozone, as shown in Table 11.2.

Those radiocarbon dates that have been considered to be reliable have been used for an independent radiocarbon time scale for the Late Weichselian. If the spatial disturbance of sites with identified Late Weichselian sediments is compared with the deglaciation, zones of ice retreat corresponding to the interstadials and stadials can be identified. But in doing so, the radiocarbon time scale cannot directly be used in dating the Late Weichselian deglaciation because it differs from the varve chronology. The radiocarbon time scale cannot directly be calibrated to calendar years as the Bristlecone Pine tree-ring series does not extend into the Late Weichselian. In addition, the varve chronology beyond 10 ka B.P. is, as earlier pointed out, still open to revision. In an extension of the calibration of the radiocarbon time scale Stuvier et al. (1986), in a work in which they used the south German oak chronology, the Lake of Clouds varve series from North America, and the Swedish varves, concluded that the conventional radiocarbon years between 10 ka B.P. and 13 ka B.P. are at least 1,000 years too young, not taking into account the short oscillations over the centuries. This estimate exceeds that suggested by Tauber (1970) in a comparison of the varve chronology, before revision, with the radiocarbon time scale. Tauber suggested that the Late Weichselian radiocarbon ages are about 350 years too young. In a later comparison of the time scales, backed by magnetostratigraphic studies of varved clays and radiocarbon dated sediments, and taking into account the revised varve chronology, Björck, Sandegren, and Holmqvist (1987) suggested that the difference in years between the radiocarbon and varve time scales increased between 12 ka and 11 ka B.P. According to this comparison the years would correspond to each other in the two time scales somewhere between 12 ka and 13 ka B.P., after which the radiocarbon ages would be about 200 years too young at 12 ka B.P. and about 600 years too young at 10 ka B.P. In order to give an idea of the influence of the approximate calibrations of the Late Weichselian radiocarbon time scale, the calibrated ages suggested by Björck, Sandegren, and Holmqvist were given in Table 11.2. A reliable direct radiocarbon dating of the organic material in varved clays has not been possible because of the low organic content of the clays. This was demonstrated by Hörnsten and Olsson (1964) in the dating of varves +56 to +82 at Lugnvik in the Valley of Ångermanälven, which for the soluble fraction gave an age of 9,000 + 1,400 − 1,200. Even if the date is close to the age of the varves – 9156–9182 B.P. in the revised chronology – the errors of the radiocarbon are too great for the age to be of use for a calibration of the radiocarbon time scale.

The general outlines of the vegetational history of the Late Weichselian were known already soon after the Bølling and Allerød Interstadials had been identified, as seen from the summary by Iversen (1947) of the evidence from Denmark. A number of sites were also known throughout Europe from the British Isles and the Continent, as well as from sites in southwestern Norway and southern Sweden. The distribution of the known sites with Late Weichselian interstadial sediments was shown on the maps presented by Gams (1950) and Gross (1954). They concluded that the Younger Dryas period corresponds to the time of the formation of the Fennoscandian moraines in Norway, Sweden, and Finland. This conclusion was, for instance,

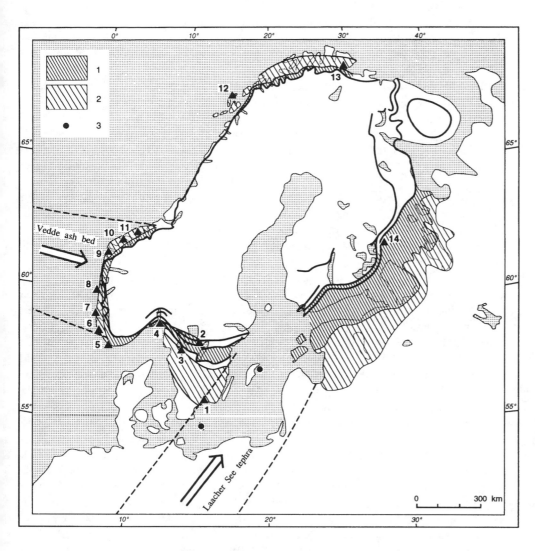

Figure 11.15. Late Weichselian chronozones and fans of Laacher See tephra (11 ka B.P.) and Vedde ash (10.6 ka B.P.) compared with ice recession and Fennoscandian moraines. Symbols: 1, area deglaciation in Allerød time; 2, in Bølling time; 3, sites with Laacher See tephra. Black triangles show sites listed in Table 11.3.

already reached by Mohrén (1938) when discussing Late Weichselian sites in Sweden, and by Donner (1951) in a study of sites in southern Finland and in Karelia, with comparisons with sites further south. Many of the sites in Finland and elsewhere have later been reinterpreted when studied in more detail, and the correlation of the Late Weichselian interstadials and stadials with the ice recession has become more accurate, even if the main outlines of the correlations have remained the same. A number of Late Weichselian sites with Allerød deposits have been recorded from Sweden as far north as the zone of the Middle Swedish moraines. The investigation by Fries (1951) of the west coast area and by Berglund (1966a) of Blekinge in southeastern Sweden, and their comparisons with sites in southern Sweden generally, already showed how *Betula* forests with some *Pinus* only reached southernmost Sweden in Allerød time. A park-tundra with *Betula*, similar to that in northern Jutland, spread into a belt north of this forest, including the west coast, and between this belt and the Middle Swedish moraines there was a rather broad belt of tundra vegetation. During the subsequent Younger Dryas period most parts of southern Sweden had a

tundra vegetation; only the southernmost areas had park-tundra. The vegetation thus is a reflection of the cold climatic conditions of the periglacial zone close to the ice margin. By using a number of radiocarbon dates from the Swedish Late Weichselian sites and the ages for the chronozones given by Berglund and Rapp (1988), as shown in Table 11.2, the pollen-analysis chronozones were compared with the Swedish varve chronology in which the earlier described revised varve ages were taken into account, as also the most recent varve investigations in southern Sweden. In this way the correlation of the chronozones with ice-marginal zones developed into the scheme presented in Fig. 11.15, through a number of investigations dealing with this problem (Berglund, 1979; Björck & Håkansson, 1982; Björck & Möller, 1987; Berglund & Rapp, 1988; Björck, Berglund, & Digerfeldt, 1988; Lemdahl, 1988). According to this scheme the position of the ice margin at 13 ka B.P., corresponding to the beginning of the Bølling Interstadial, crossed over from southeastern Sweden northwestwards to the west coast. It should be noted that the recession lines in southern Sweden have been revised and therefore differ somewhat from those given in Fig. 11.12, which were based on the map constructed by E. Nilsson (1968). The recession lines for the beginning and the end of the Older Dryas Stadial, at 12 ka B.P. and 11.8 ka B.P. according to ages used by Mangerud et al. (1974), cross southern Sweden southwest of the lakes Vänern and Vättern, the line for the end of the stadial following the Levene moraine (Berglund, 1979). If, however, the radiocarbon ages (Table 11.2) suggested by Björck, Sandegren, and Holmqvist (1987) were used for Older Dryas, the recession lines would be somewhat further south, clearly south of the Levene moraine. The ice margin retreated to the Middle Swedish moraines during the Allerød Interstadial, at a time when the retreat of the ice margin was slowing down before the standstill or readvance at the moraines, as seen from the recession lines based on the varve chronology (Fig. 11.12).

The investigations of several lake basins in the zone of the Middle Swedish moraines were undertaken in order to elucidate in detail the pattern of deglaciation as well as the shoreline displacement in connection with the drainage of the Baltic Ice Lake at the time of the withdrawal of the ice margin from Mt. Billingen. The presence of Younger Dryas minerogenous sediments with a strong representation of nonarboreal pollen, including *Artemisia,* and a diatom flora of the Baltic Ice Lake, were recorded from the zone of the Middle Swedish moraines at Kolmården, southwest of Stockholm (Florin, 1979), but the main studies were undertaken further west at Hunneberg south of Lake Vänern, just south of the Levene moraine (Björck & Digerfeldt, 1982a, 1982b), and in the area of Mt. Billingen (Björck & Digerfeldt, 1984, 1986). Fourteen lake basins were studied in the latter area, which is in the zone of the Middle Swedish moraines and crucial for the study of the ice recession resulting in the drainage of the Baltic Ice Lake (Fig. 11.15). The regional pollen-assemblage zones were correlated with the Allerød and Younger Dryas chronozones and the first Flandrian chronozone, with the help of radiocarbon dates. The Late Weichselian sediments of clay and clay-mud, and the pollen diagrams with a strong representation of nonarboreal pollen, reflect the conditions close to the ice margin, but the presence of sediments from the end of the Allerød zone could be demonstrated. As a result of the studies in the area of Mt. Billingen it was shown that the ice margin retreated northwards beyond Mt. Billingen in the later part of Allerød and the Baltic Ice Lake was drained (Björck & Digerfeldt, 1984). Then followed a readvance during Younger Dryas time, reaching its maximum extent at 10.5–10.6 ka B.P., when the Baltic Ice Lake was dammed, until the drainage at 10.4–10.5 ka B.P., lowering the water level 20–30 m 10.3–10.4 ka B.P. (Table 11.2) when the outlet to the sea north of Mt. Billingen was already over 5 km wide. With the help of the basins studied from various altitudes a curve could be drawn for the shoreline displacement west of Mt. Billingen, with a rapid regression at the end of Younger Dryas. A shoreline displacement curve was earlier obtained for the Hunneberg area, where fifteen lake basins were investigated (Björck & Digerfeldt, 1982a, 1982b), in which a reservoir effect of 300 years was taken into account for the marine-brackish sediments, as already mentioned. In the

Hunneberg area, as it was south of the Middle Swedish moraines, the older sediments represented the whole Allerød chronozone and, in one lake basin, possibly older sediments also were present.

These studies in Sweden show that the ice margin retreated north of the Middle Swedish moraines in late Allerød time in the area west of Lake Vättern and was followed by a readvance in Younger Dryas time to Mt. Billingen, again damming the Baltic Ice Lake during the formation of the Billingen moraine, a sequence of events already suggested in earlier studies of the deglaciation of Sweden (Berglund, 1979; Björck, 1979). The final drainage of the Baltic Ice Lake as a result of the retreat of the ice margin from the Mt. Billingen area, as dated in the studies mentioned above, was biostratigraphically recorded in the detailed studies in Blekinge in southeastern Sweden, on the Baltic side of the outlet of the ice-dammed lake, as well as the earlier history of this lake from the time southeastern Sweden was deglaciated (Björck, 1979, 1981). As Blekinge was deglaciated prior to 12 ka B.P. the shoreline displacement could be studied from the time of the Bølling Interstadial to the time of the drainage of the Baltic Ice Lake. The curve constructed by Björck shows that the relative regression of water level had a rapid lowering at the end of Allerød as a result of the drainage caused by the withdrawal of the ice margin north of Mt. Billingen. In Younger Dryas time there was a slow transgression of about 5 m as a result of the readvance of the ice margin to Mt. Billingen, until the final drainage of the Baltic Ice Lake, in Blekinge recorded as a drop of water level from 30 m a.s.l. to at least 4 m (Björck, 1981), thus at least 26 m, a figure in agreement with observations elsewhere in the Baltic Sea area. The biostratigraphical results from Blekinge have a bearing on the interpretation of the land/sea-level changes in the Salpausselkä zone in Finland based on morphological criteria (see Chapter 11, section 1). If the transgression of the level of the Baltic Ice Lake from the end of Allerød to the end of Younger Dryas was only a few meters, as demonstrated in Blekinge, it supports the suggestion by Fyfe (1990) that the low marginal terraces at the g level in Finland were formed below water level and that there was no major transgression in the Salpausselkä zone up to the B I level (Donner, 1992). This explanation makes it unnecessary to assume that there were clear differences in the isostatic uplift between the two areas.

According to the varve chronology in Sweden and Finland, as presented in Fig. 11.11, the Middle Swedish moraines and the two first Salpausselkä moraines in Finland, I and II, were formed between about 11.3 ka and 10.6 ka B.P. If these dates are compared with the biostratigraphical evidence from the Mt. Billingen area, in which conventional radiocarbon ages were used, there is a close correspondence between the two chronologies (see Table 11.2). The approximate calibration of the radiocarbon ages by adding 600 years at 10,000 years B.P., as suggested by Björck, Sandegren, and Holmqvist (1987), is not compatible with the correlation presented in Table 11.2. What seems, however, to be established from the varve studies and the biostratigraphical investigations is that the Fennoscandian moraines in Sweden and Finland started to form at a time corresponding to the end of the Allerød Chronozone and that their formation ended in the middle of the Younger Dryas, just before the drainage of the Baltic Ice Lake. The biostratigraphical evidence from terrestrial sediments in southern Sweden, as summarized by Berglund and Rapp (1988), shows that a climatically cold phase with arctic conditions began at about 11 ka B.P., corresponding to the lower limit of Younger Dryas, but that it had already ended at about 10.5 ka B.P., close to the earlier mentioned age of 10.3 used by Iversen (1973) for the upper boundary of Younger Dryas. In Sweden the cold phase was followed by a subarctic phase until 10.2 ka B.P. and after that a warmer temperate climate. The age used by Iversen thus falls within the transitional phase between 10.5 ka and 10.2 ka B.P. Both the evidence from Denmark and from southern Sweden clearly show that the marked amelioration of climate took place before 10 ka B.P., which is used as the Pleistocene–Holocene boundary. The age bracket for the formation of the Fennoscandian moraines in Sweden and Finland, as determined by the varve chronology, is, as seen in Table 11.2 and also indicated in Table 11.1, in agreement with the biostratigraphical evidence, placing the formation of the moraines in the coldest time of Younger Dryas.

In addition to the studies of the Late Weichselian development of southern Sweden as outlined above, there is a detailed investigation by Fredén (1988) of the marine faunal remains, particularly of molluskan assemblages combined with their dating, in southwestern Sweden, particularly in the Uddevalla region in Bohuslän and the Vänern basin. The oldest arctic to arctic–boreal faunas were found on the west coast and were dated at 13 ka to 10.2 ka B.P. using, as mentioned earlier, a value of 400 years for the reservoir effect, and the youngest faunas in the western part of the Vänern basin, dated at 9.9 ka to 9.6 ka B.P. An increase in water temperature at about 10.2 ka B.P. resulted in the almost total disappearance of the arctic–boreal fauna on the west coast, a change recorded in several marine sequences along the coast and dated at 10.2 ka to 10.3 ka B.P. (Fredén, 1988), which is in agreement with the results from terrestrial sequences. After the final withdrawal of the ice margin from Mt. Billingen and the broadening of the relatively short-lived connecting sound between the Baltic and the ocean, the Närke Strait, brackish water was able to penetrate some way into the Baltic. Shells of *Portlandia (Yoldia) arctica* were recorded in the varved clays of the Mälaren valley in the Stockholm region in clays representing a period of at least 90 years (Fredén, 1988). This molluskan evidence has been connected with the evidence from varved clays in the Stockholm area of an influx of brackish water after the drainage of the Baltic Ice Lake.

In Norway the biostratigraphical evidence of the Late Weichselian has been obtained from sites in the narrow area outside the Fennoscandian moraines and, because these moraines were at least in some areas formed after a readvance of the ice margin, also partly from sites in the area inside the moraine that was deglaciated before this readvance. In the Oslo area no Late Weichselian fresh water sediments have been found because most of it was submerged during the formation of the Ra moraines and the moraines north of them. The detailed pollen stratigraphical investigations in the inner part of the Oslofjord area (Hafsten, 1956) and in the lowland area east of the fjord, in Østfold (Danielsen, 1970), both showed that the oldest freshwater sediments are from early Flandrian time. But they also demonstrated that the Oslo area was an important gateway for the arctic–alpine species to immigrate into Norway during the deglaciation. With the help of radiocabon dates of marine shells from sediments associated with the moraines in the Oslo area the ice marginal positions could be dated in detail (Sørensen, 1979). In this way the ages for the Ra, Ås, and Ski moraines all fell within the time interval of 11 ka and 10 ka BP, thus corresponding to the Younger Dryas Chronozone. The oldest of the moraines, Slagen–Onsøy, has, as mentioned earlier, been connected with the Levene moraine in Sweden and therefore been placed in the Allerød Chronozone (B. Andersen, 1979). There are some discrepancies in the exact correlation between the moraines in the Oslo area with those in Sweden, as seen in a correlation between the Oslo area with Värmland (Lundqvist, 1988b). It seems, however, clear that the two moraines Ra and Ås correspond to the two moraines of the Fennoscandian moraines further east (Tables 11.1 and 11.2), but in addition to these the Ski moraine, correlated with the Salpausselkä III moraine, was formed toward the end of the Younger Dryas time, just before 10 ka B.P. The age of the Aker moraine is younger, formed at about 9.8 ka B.P. (Sørensen, 1979). This chronological scheme for the moraines in the Oslo area is an example of how they can be dated by using radiocarbon ages of marine shells, but an exact correlation of this scheme with that for Sweden based on extensive biostratigraphical evidence is not possible.

In contrast to the Oslofjord area (Table 11.3, Fig. 11.15) the coastal strip in the rest of Norway, outside the moraines correlated with the Ra, Ås, and Ski moraines, was to a large extent already dry land during the Late Weichselian. There are therefore a number of sites along the coast with sediments from this time. From many of these sites there is palynological evidence of the vegetational history throughout the Late Weichselian, starting with the Bølling Interstadial, backed by a number of radiocarbon dates that are consistent with the biostratigraphical evidence. As the area was close to the ice margin, the vegetation during the interstadials was an open birch forest in the south and an arctic–subarctic vegetation in the north, while the vegetation during the stadials was dominated by herbs in the whole area, with *Artemisia* as a typical element particularly in the Younger Dryas, but also earlier. At the site of Høylandsmyr on the Lista peninsula in South Norway clayey nekron-mud of Allerød age lies underneath Younger Dryas clay and Flandrian mud (Hafsten, 1963), which is a familiar stratigraphical sequence from the sites in Denmark. The palynological evidence from Lista of the Allerød Interstadial between two stadials is very

Table 11.3. *Areas and sites of special interest in dating the Late Weichselian development of the Baltic Ice Lake and the formation of the Fennoscandian moraines as shown in Figure 11.15*

Sweden
1. Blekinge (Björck, 1979, 1981)
2. Mt. Billingen (Björck & Digerfeldt, 1984, 1986)
3. Hunneberg (Björck & Digerfeldt, 1982a, 1982b)

Norway
4. Oslo (Sørensen, 1979)
5. Lista (Hafsten, 1963)
6. Jaeren, Brøndmyra (Chanda, 1965)
7. Karmøy, Liastemmen (Paus, 1989)
8. Bergen (Mangerud, 1970, 1977; Krzywinski & Stabell, 1984)
9. Nordfjord (Larsen et al., 1984), Vedde ash bed
10. Sunnmøre, Ålesund (Kristiansen, Mangerud, & Lømo, 1988), Vedde ash bed, Sunnmøre, Bergsøy, Gurskøy and Leinøy (Svendsen & Mangerud, 1990), Vedde ash bed
11. Nordvestlandet, Tingvollhalvøya (Johansen, Henningsmoen, & Sollid, 1985), Vedde ash bed
12. Andøya (K.-D. Vorren, 1978, 1982)
13. Varanger peninsula (Prentice, 1981, 1982)

Finland
14. Southeastern Finland (Hyvärinen, 1971a, 1973; Donner, 1971, 1978)

clear and similar to that earlier described by Faegri (1940) from Jaeren further north. In a reinvestigation of the fresh water sediments of Brøndmyra, Chanda (1965) was able to demonstrate the presence of the Bølling Interstadial, supported by radiocarbon dates, but in the pollen diagram, as in some other diagrams from further north, the Older Dryas is weakly developed. Discussing the detailed palynological results from the Liastemmen bog on Karmøy, Paus (1989) suggested that the time between 13 ka and 11 ka B.P. should be characterized as one single interstadial, in accordance with results elsewhere in northwestern Europe, where the short-lived Older Dryas climatic deterioration cannot always be traced in the pollen diagrams. The Older Dryas Chronozone can, however, when dated be placed in a chronological scheme, as done by Paus, even if it is not registered in the pollen diagram. In a study of the lake sediments in several basins in the Yrkje area, on the mainland near Karmøy, mainly dealing with the land/sea-level changes, the radiocarbon dates showed that even this area had, at least partly, become ice-free already in Bølling time (Anundsen & Fjeldskaar, 1983).

Studies of the Late Weichselian have been undertaken in the Bergen area both at the outer coast and close to the main moraine, the Herdla moraine, which, as earlier mentioned, was formed after a 40 km readvance of the ice. Twenty-eight basins were studied on the island of Sotra in Hordaland, west of

Bergen, by Kryzwinski and Stabell (1984) and the oldest recorded sediments were from Bølling time, the oldest radiocarbon date being 12,650 ± 110 (T-2634) from Hamravatn 29 m a.s.l. With the help of a number of dates the local pollen-assemblage zones could be correlated with the chronozones, and with the help of diatom analysis the times when the lake basins were isolated from the sea to become independent lakes could be determined. The relative regression of the sea as a result of the isostatic uplift of the area was interrupted by an over 10 m transgression starting in late Bølling time and ending in mid-Younger Dryas time, reaching up to 40 m a.s.l. This transgression was named the Bømlo transgression (Krzywinski & Stabell, 1984). The land/sea-level changes in Norway will, however, be treated in more detail later against the evidence from other areas. The results from Sotra show that the area became deglaciated already in early Bølling time and remained ice-free after that time. A readvance of the ice margin at the outer coast in late Bølling time, as suggested earlier on the basis of results from Blomøy north of Sotra (Mangerud, 1977), was not verified by the subsequent results from Sotra. Further inland from Sotra, marine faunas with radiocarbon ages placing them in the Allerød were already recorded from the Bergen and Os areas by H. Holtedahl (1964). Later radiocarbon dates of shells from marine beds, partly from inside the Herdla moraine, could be used to bracket the readvance during which the moraine was formed (Mangerud, 1970, 1977). The dates show that the ice advance that deposited the moraine reached its maximum just before 10 ka B.P. and stayed in this position only about 100 years before its rapid retreat in the beginning of the Flandrian. Because the readvance deposited a till on top of fossil-bearing marine sediments, from which the shells could be dated, the stratigraphical control of the oscillation of the ice margin is very good. In addition to being used for radiocarbon dating, the marine mollusks and the foraminifera could be used to record the temperature changes in the coastal waters from temperate conditions during Allerød to cooler conditions during Younger Dryas, changes parallel to those in the vegetation (Mangerud, 1977). The Herdla moraine in the Bergen area has an age similar to the Ski moraine in the Oslo area (Sørensen, 1979). A pronounced readvance following a retreat of the ice margin in Allerød time is, as seen from the radiocarbon dates from the coast of western Norway (Mangerud, 1970, 1977), particularly characteristic for the Bergen area and it was not until the end of the Younger Dryas time that the ice margin here reached its furthest position, even if the ice advance had started already in the beginning of Younger Dryas time, or perhaps even earlier. In the study of the Nordfjord area further north in western Norway, Larsen et al. (1984) concluded that the evidence from the coastal areas suggests that the greatest precipitation in Younger Dryas time was in the Bergen–Nordfjord region, with more continental conditions both north and south of this region. This conclusion is in keeping with the evidence of the Late Weichselian behavior of the ice margin. It can also be noted that the Herdla moraine in the Bergen area reached the outer coast at Bergen and north of it, whereas further south and north the marginal positions correlated with the Herdla moraine are further inland.

In the study of the Nordfjord area, at Stad and Vågsøy at the outer coast and at some distance from the main moraine, local cirque glaciers became active during the Younger Dryas climatic deterioration, as witnessed by cirque moraines at several localities (Larsen et al., 1984). The pollen stratigraphy in a lake basin just outside the cirque moraine at Kråkenes on Vågsøy represents the time from Bølling to the beginning of the Flandrian, with a radiocarbon date of 12,320 ± 120 (T-2534) from the basal silty mud, in addition to six other dates higher up in the sequence. The Younger Dryas laminated silts with a predominance of *Salix* and herbs, covering the interstadial muds of Bølling and Allerød, are comparatively thick, about 70 cm, and contain a thin layer of volcanic ash, as some other Younger Dryas sediments in Norway. Their dating and occurrence will be discussed in Section 6, together with other Late Weichselian ash layers.

In three lake basins in the Ålesund area of Sunnmøre, with marine basal sediments covered by lacustrine sediments, the pollen stratigraphy and the radiocarbon dates show that the sequences represent the time from Bølling to the beginning of the Flandrian, as in the Nordfjord area further southwest, and that the main vegetational history was broadly similar in both areas (Kristiansen, Mangerud, & Lømo, 1988). Bølling and Allerød sediments were similarly encountered in eight lake basins on the islands of Bergsøy, Leinøy, and Gurskøy in Sunnmøre, immediately southwest of Ålesund, in a study of the land/sea-level changes (Svendsen & Mangerud, 1990). There is also evidence of the activity of Younger Dryas cirque glaciers in the Ålesund area, as in the Nordfjord area. Ålesund is in an area close to the northern limit of open birch forests during the preceding Allerød Interstadial, which is also reflected in the sediments where the organic content of the Allerød sediments is smaller than further south in Norway and in Denmark.

In the study of seven sites from the archipelago of Nordvestlandet, mainly from the peninsula of Tingvollhalvøya, along a transect from the outer coast inland, it was shown that the archipelago was deglaciated in late Bølling time and that the areas further inland were ice free in Allerød time (Johansen, Henningsmoen, & Sollid, 1985). The conclusions about the early deglaciation were based on radiocarbon dates from the sediments in which the pre-Allerød time was poorly represented in the pollen stratigraphy. Allerød and Younger Dryas were, however, clearly recorded, the latter as a *Salix–Artemisia*–NAP (nonarboreal pollen) assemblage zone.

In addition to this evidence of the Late Weichselian biostratigraphy and deglaciation of the southern parts of coastal Norway, the northern parts have areas from which very old sediments have been recorded. The oldest sediments from Lake Endlevatn on Andøya, from a basin about 35 m a.s.l., were dated at about 18 ka B.P. (K.-D. Vorren, 1978; T. O. Vorren, 1982). From these sediments a pollen stratigraphical record was obtained that showed the early Late Weichselian dry arctic vegetation domi-

nated by grasses, being replaced by alpine and subalpine species in Bølling time. A vegetation with a strong representation of *Salix* was established in Allerød time, interrupted by the harsh conditions with *Artemisia* in Younger Dryas. A sequence of sediments equally as old as that from Endlevatn was obtained from the nearby Lake Aeråsvatn (T. O. Vorren et al., 1988), but there the pollen stratigraphical evidence was less complete, even if the same pattern of vegetational history as that in the Endlevatn diagram can be seen. The evidence from Andøya, together with the stratigraphical evidence and 50 radiocarbon dates older than 10 ka B.P. from the coast of Troms northeast of Andøya, and from the shelf outside, were used to reconstruct the deglaciation of this area after the maximum of the Late Weichselian glaciation (Rokoengen, Bugge, & Løfaldi, 1979; T. O. Vorren & K.-D. Vorren, 1979; T. O. Vorren, 1982; T. O. Vorren et al., 1988). As is seen from the radiocarbon ages these areas were deglaciated shortly after the maximum extent of the Late Weichselian glaciation. The moraine complexes formed during the retreat of the ice have been divided into groups, of which the earlier mentioned Egga moraines belong to the oldest. The exact correlation of the older moraines is still tentative, whereas the position of the Tromsø–Lyngen moraine along the coast is well known. The radiocarbon ages related to this moraine show that it was formed in Younger Dryas time, having an age similar to the Ra and Ås moraines in the Oslo area (B. Andersen, 1975), but some ages indicate that they may partly have been formed already in Allerød time (B. Andersen, 1979).

As a result of the stratigraphical studies of glaciomarine sediments from the shelf it was possible to compare the pollen stratigraphical evidence from Andøya with the changes in the foraminifera and mollusk faunas in the dated marine sediments. There was thus a Mid (High) Arctic marine environment at the coast until about 13 ka B.P. when it changed to a Low Arctic, and finally to a High (Mid) Boreal environment at about 10 ka B.P., at the beginning of the Flandrian (T. O. Vorren, 1982). The major macrofaunal change in the shelf sediments of northern Norway at 10 ka B.P., from an Arctic to a Boreal fauna, was caused by a rise in temperature and salinity, as well as an increase in nutrient supply, as the Norwegian Current came into these coastal waters (Thomsen & T. O. Vorren, 1986). The displacement of the faunal zones since the Late Weichselian was considerable as seen when the results are compared with the present distribution of the marine faunas (see Chapter 13, Section 6).

The extreme, cold climatic conditions in Finnmark in northernmost Norway deglaciated in the Late Weichselian, as already seen from the distribution of the fossil-patterned ground (Fig. 11.8), are also recorded in pollen diagrams from lake sediments on the Varanger peninsula. The oldest pollen assemblage zone at Østervatnet was correlated with the late Bølling Chronozone, in a correlation based on the biostratigraphical changes as the radiocarbon dates were 1 ka to 2 ka too old because of the hardwater effect (Prentice, 1981). The pollen diagram showed that there was a tundra vegetation throughout the Late Weichselian, with an *Artemisia*–grass steppe during Older and Younger

Dryas. It was not until the beginning of the Flandrian that tree birches immigrated into the area. In three other pollen diagrams from the Varanger peninsula two, those from Bergebyvatnet and Holmfjellvatnet start in Allerød, and one, from Stjernevatnet, in Younger Dryas. They could all be used for a detailed reconstruction of the open plant communities of the Late Weichselian (Prentice, 1982). The changeover from an *Artemisia* zone to a birch zone at the Weichselian–Flandrian boundary had already been recorded earlier from Bruvatnet, a lake basin ɛt the head of the Varangerfjord (Hyvärinen, 1975).

The distribution of the sites with Late Weichselian sediments shows that the outer coastal areas of northern Norway were deglaciated at least in Bølling time and that the area close to the Tromsø–Lyngen moraine was deglaciated in Allerød time, as schematically shown in Fig. 11.15. The readvance of the ice margin to the Tromsø–Lyngen moraine was not as pronounced as that recorded further south in Norway. But in the Andøya area the ice margin was already close to the Tromsø–Lyngen moraine in Allerød time, at a time called the Skarpnes event in the local deglaciation scheme (T. O. Vorren et al., 1988). In addition to the results from the freshwater sediments from Finnmarken a study of the Barents Sea suggested that the outermost coast here, similar to most parts of the shelf north of Finnmarken, was deglaciated between 16 ka B.P. and 13 ka B.P., before Bølling (T. O. Vorren, Hald, & Lebesbye, 1988). According to this suggested deglaciation pattern there would be a coastal strip in northernmost Norway that was deglaciated at a relatively early time, before 13 ka B.P., as compared to the coastal areas further south.

The general pattern of Late Weichselian climatic changes recorded biostratigraphically along the coast of Norway in sediments formed close to the retreating ice margin was also recorded in the marine sediments away from the coast, in the Norwegian Sea (Jansen et al., 1983). The analysis of the variations in the distribution of both planktonic and benthic foraminifera showed the same general changes as the changes found in the marine faunas in the shelf area outside Troms. The Norwegian Sea was in the Late Weichselian an environment with icebergs and pack-ice throughout the year until 13 ka B.P. when it became seasonally ice free, conditions that prevailed even during the Younger Dryas. The final rapid warming of the Norwegian Sea started at about 10 ka B.P. In Skagerrak, between Norway and Denmark, and closer to the mainlands, the sediments were influenced by terrigenous material, with large quantities of ice-rafted material, until about 10 ka B.P. when temperate water penetrated into Skagerrak. This change was recorded in an over 10 m long core from a water depth of 325 m, in which the shells, pollen, stable isotopes, foraminifera, radiolaria, and sediments were studied, a core from which the oldest radiocarbon date for shell material was 10,200 B.P. (Björklund et al., 1985).

As is seen from the previously described sites there were areas of dry land during the Late Weichselian deglaciation in Sweden and Norway, outside the zone of the Fennoscandian moraines. In Finland, on the other hand, there was only a narrow strip of dry

land in the southeastern part of the country outside these moraines, as shown in Fig. 11.15. But there are some sites in this area with freshwater sediments from the end of the Late Weichselian, consisting of fine-grained minerogenic sediments poor in organic matter and therefore difficult to date with the radiocarbon method. In the pollen diagrams there is lowermost an *Artemisia* zone, underneath a *Betula* zone that in its lower sections shows a transition from an open vegetation to closed birch forest (Hyvärinen, 1971a, 1973; Donner, 1971, 1978). A pollen composition similar to the *Artemisia* zone in southeastern Finland was also found in pollen diagrams of varved clays and silts in the Mäntsälä–Helsinki area northeast of Helsinki (Tynni, 1960, 1966). After the introduction of radiocarbon dating, and after taking into account the previously mentioned errors in this method, there are no freshwater sediments in Finland that can be held to represent Allerød. The radiocarbon age for the upper boundary of the *Artemisia* zone, marking the changeover to organic lake muds, was on the basis of four dates from four sites outside Salpausselkä II, studied by Repo and Tynni (1967), dated at about 10.1 ka to 10.0 ka B.P. (Donner, 1978). The change in vegetation is thus here, close to the ice margin, slightly older than the Late Weichselian–Flandrian boundary, but not as much as about 300 years older, as in Denmark. The change in vegetation from the open vegetation, with a strong representation of *Artemisia,* to the immigration of birch is as in other boundaries of pollen-assemblage zones time-transgressive, and in this case it was influenced by the proximity of the retreating ice margin. When the dating of the change in vegetation in southeastern Finland to 10.1 ka to 10.0 ka B.P. is compared with the dating of the formation of the Salpausselkä moraines, ending with the formation of the Salpausselkä II moraine at about 10.6 B.P. (Table 11.2), it can be seen that ice wedges had time to develop at the surface of this moraine, at a time with permafrost, before the climatic amelioration recorded in the pollen diagrams, when the ice margin had already retreated from the Salpausselkä II moraine.

Even if there are no sediments in southern and southeastern Finland that by pollen analysis can be referred to Allerød, it can on the basis of the varve dating of the ice recession be assumed that the areas south and southeast of the Salpausselkä moraines were ice-free in Allerød time. The ice margin had, as seen earlier, receded to the south coast of Finland already at about 12 ka B.P., and it has been assumed that the Gulf of Finland was also ice-free in Allerød as a result of a breaking up of stagnant ice. This would agree with the dating of the Pandivere moraine in northern Estonia and its continuation in the east, the Neva moraine, to about 12.2 ka to 12.0 ka B.P. (Serebryanny & Raukas, 1967; Serebryanny, Raukas, & Punning, 1970; Raukas, 1977). In Russian Karelia the pollen stratigraphical evidence of the Late Weichselian ice recession is better than in the area of the Gulf of Finland. There are a number of pollen diagrams from Karelia in which Allerød can be identified even if the influence of redeposited pollen is strong in the fine-grained minerogenous sediments (I. Ekman, 1987; Elina & Filimonova, 1987). The zone that was deglaciated in Allerød time broadly comprises southern and southeastern Finland, the Gulf of Finland, the Karelian Isthmus, Lake Ladoga, and the area between the Neva and Rugozero moraines in Russian Karelia (Fig. 11.15), in agreement with maps of the ice recession presented earlier (Hyvärinen, 1973; Donner, 1978; I. Ekman, 1987). As there was a readvance of the ice margin in southern Finland before the formation of the Salpausselkä I moraine, as noted earlier, the ice had probably already withdrawn north of this moraine in Allerød time. Whether this was also the case in Karelia is not known. According to the reconstructions of the deglaciation, the ice margin withdrew from the zone between the North Lithuanian–Haanja–Luga recession line, with an approximate age of about 13 ka B.P., to the Pandivere–Neva line during the Bølling Interstadial. This is only a tentative correlation as the exact dating of the moraines representing the recession lines is still open. If the patten of the Late Weichselian deglaciation as presented in Fig. 11.15 is correct the zone from which the ice margin retreated during Allerød is comparatively broad in the southeastern area of glaciation as compared with the areas in Sweden and Norway. This would agree with the pattern of Late Weichselian moraines, which are more widely spaced in the southeastern areas than in the coastal areas of Norway in the west.

Of all the Late Weichselian moraines the Fennoscandian moraines and moraines associated with them are the best known. The biostratigraphical evidence combined with datings, as presented in Tables 11.1 and 11.2, show that the moraines classified as belonging to the Fennoscandian moraines consist of two main moraines in the Oslo area, in Sweden, Finland, and Russia, but of mainly one moraine along most of the Norwegian coast. As there are smaller age differences between the moraines and some of the correlations between the moraines are still tentative, a morphostratigraphical classification valid for the whole area cannot be used. The local names mentioned in Table 11.1 should therefore be retained in detailed descriptions, as also the names of moraines closely associated with the Fennoscandian moraines. Because of the age differences encountered between the Fennoscandian moraines no exact age bracket can be given for their formation, but it seems to have started already in Allerød time and their formation ended before the end of Younger Dryas. Because the boundaries for the Younger Dryas Chronozone were defined as 11 ka and 10 ka B.P. the Fennoscandian moraines cannot be called the Younger Dryas moraines. Further, the cold climate characteristic for Younger Dryas came to an end 300 to 100 years before 10 ka B.P., depending on the closeness to the ice margin, as noted earlier.

From the preceding discussion it can be seen that a number of parallel moraines formed toward the end of the Late Weichselian and some still at the beginning of the Flandrian and that some of the main ones formed during Younger Dryas. Biostratigraphically only one cold oscillation, that of Younger Dryas, has been identified after Allerød. It follows that the standstills and readvances of the ice margin before its rapid Flandrian retreat have not been recorded as colder fluctuations in the biostrati-graphical sequences and that the mechanism of the jerky behavior of the ice has not been explained. In some areas, as in the Baltic Sea area, the fluctuations in water level probably affected the behavior of the ice margin, grounding it at times when the water level dropped and calving was reduced (Lund-qvist, 1987b). A rise in water level, on the other hand, would have caused an advance of the ice.

On the basis of the evidence from Norway, Mangerud (1987) estimated that the mean summer temperature was 8–10° C lower than at present during Younger Dryas and that it dropped by 5–6° C in less than two centuries at the transition from Allerød. Similarly, the end of the cold climate during Younger Dryas was abrupt (Dansgaard, White, & Johnsen, 1989). The Greenland ice cores show that there was a change in the North Atlantic region already at 10.7 ka B.P., with a warming of 7° C in South Greenland in about 50 years. These figures for Younger Dryas show the magnitude of the climatic oscillation and therefore clarify how such a fluctuation had a strong impact on the retreat of the ice sheet and how it can so plainly be recorded biostratigraphically.

11.6. Tephras

In the upper part of the Late Weichselian, in the time interval of 13 ka to 10 ka B.P. from which there is good biostratigraphical evidence of the previously discussed intersta-dials and stadials, there are in northwestern Europe two tephra horizons that have been used as marker beds for correlations. The older of the two consists of the volcanic ash with a phonolitic composition from the Laacher See in the Eifel mountains near Bonn in Germany. Its distribution is divided into three fans, one 600 km-long fan south over Switzerland, recorded at a number of sites (Wegmüller & Welten, 1973), and as far as Turin (Torino) in Italy, another 100 km-long fan west-southwest over part of Belgium, and the longest fan, 1100 km long, northeast as far as the Baltic Sea area, where it has been traced on the island of Bornholm in a profile of a bog, Vallensgård Mose, with sediments representing both Bølling and Allerød (Usinger, 1978), and in cores northeast of the island of Gotland (van der Bogaard & Schmincke, 1985; Lang, 1985). The eruption had three main phases with different compositions but they can all be referred to one explosive eruption, which deposited the tephra in less than 10 days. The thickness of the tephra is 50 m near the crater (van der Bogaard & Schmincke, 1985), and decreases further away from it, being between 5–10 mm at the south coast of the Baltic Sea and only a few mm on Bornholm. The differences in transport directions were due to differences in wind directions at the altitudes to which the eruption columns reached during the eruptive phases. The atmosphere had vertical zones with different wind directions at the time of the eruption.

Through pollen stratigraphy the Laacher See Tephra has been placed in the upper part of Allerød, but 16 radiocarbon ages dating the tephra have a mean of 10,920 B.P., with a likely age between 10,950 B.P. and 11,050 B.P. when the frequency distribution of the ages are taken into account (van der Bogaard & Schmincke, 1985). An age of about 11,000±50 B.P. is, however, usually quoted for the Laacher See Tephra. Such an age places it at the Allerød–Younger Dryas boundary in the chronostratigraphical division used here. The margins of the northeastern fan of the Laacher See Tephra, shown in Fig. 11.15 and drawn on the basis of the sites further southwest in Germany, indicate that it should be possible to trace this tephra in the sediments in the southern parts of the Baltic Sea as well as at the coasts of Lithuania and Latvia, at least as far northeast as the Bay of Riga.

The second Late Weichselian tephra horizon in northwestern Europe is the Vedde Ash Bed (initally described as the Sula Ash Bed) found in freshwater and marine sediments along the west coast of Norway (Fig. 11.15). The glass particles of the ash have a basaltic and rhyolitic composition that suggests an Icelandic origin, most likely from an eruption of Katla (Mangerud et al., 1984). The ash bed in the coastal sites, many of which are listed in Table 11.3, reaches a thickness of 23 cm but is sometimes very thin. The sites, including the Kråkenes profile in the Nordfjord area (Larsen et al., 1984), several basins in Nordvestlandet (Johansen, Henningsmoen, & Sollid, 1985), and a number of sites in Sunnmøre (Kristiansen, Mangerud, & Lømo, 1988; Svendsen & Mangerud, 1990), show that the fan is 500 km broad at the coast. The weighted mean for four radiocarbon dates used

by Mangerud et al. (1984) was 10,600 ±50 B.P., an age that has generally been accepted for the Vedde Ash Bed. It was thus deposited during the Younger Dryas and is a useful marker bed in the minerogenous sediments of that chronozone along the Norwegian coast, outside the previously described moraines formed at that time. The Vedde Ash Bed has also been identified in cores from the Norwegian Sea (Jansen et al., 1983) and been correlated with ash zone I found in the North Atlantic cores from a wide area (Mangerud et al., 1984; Kvamme et al., 1989). The ash horizon found in the sediments from the continental shelf east and northwest of northern Scotland has also been correlated with the Vedde Ash Bed (Long & Morton, 1987), and so has the older of two ash horizons in the sediments in the Rockall Trough and the Faeroe–Shetland Channel northwest of Scotland (Stoker et al., 1989).

As noted, the ages of the Laacher See Tephra and the Vedde Ash Bed are close to each other, but they can be separated on the basis of both their mineralogical composition and their distribution and are therefore important tephrochronological horizons that can be used as lithostratigraphical marker beds (Mangerud et al., 1984). In addition to the two Late Weichselian ash horizons some Flandrian ash falls have been identified and dated. They will, however, be mentioned in connection with the Flandrian biostratigraphy.

12 Flandrian biostratigraphy and climatic changes

12.1. Pollen stratigraphy

As a result of the rapid amelioration of climate shortly before the Weichselian–Flandrian boundary, placed at 10 ka B.P., in conventional radiocarbon years, which also marks the boundary between the Pleistocene and Holocene (Recent), purely organic freshwater sediments soon started to form in basins of the areas that were not submerged at that time. The deposition of lake sediments was followed by the growth of mires, and their peats spread laterally into the areas surrounding the lake basins. There is thus an abundance of preserved undisturbed organic sediments containing an often complete biostratigraphical record of the Flandrian that has been studied in great detail, especially after the introduction of pollen-analysis by Lennart von Post, as mentioned in Chapter 1. The Flandrian pollen stratigraphical record has also been used in comparisons with the record from earlier warm stages. There are, however, some limitations in this approach. First, the Flandrian has not yet ended and the upper part of the stratigraphical record is therefore missing. Second, the forest composition and flora have been increasingly influenced by the activities of man for the last few thousand years. The present vegetation is therefore in most parts not natural and cannot directly be used in a detailed comparison with the vegetation of earlier warm stages. The general outlines of forest history in the Flandrian pollen diagrams can, however, be used for comparisons and also to establish a pollen stratigraphical frame for the Flandrian.

In most modern palynological investigations the diagrams have been divided into local or regional pollen-assemblage zones and their boundaries given ages based on radiocarbon dates. In the most sophisticated studies the division of the assemblage zones is based on numerical zonation methods, as in the description of diagrams from southern Sweden by Birks and Berglund (1979). The computer-based methods make the handling of the large Flandrian material possible. The present account, however, has to be limited to the results based on conventional pollen stratigraphical divisions, used in outlining the main changes during the Flandrian. The pollen-assemblage zones were therefore compared with the chronozones for continental northwestern Europe suggested by Mangerud et al. (1974) for the Flandrian, as a continuation for the Late Weichselian zones. The ages of the boundaries between the chronozones are given in Table 12.1 together with the division into the three substages of Early, Middle, and Late Flandrian. The use of the Blytt–Sernander terms for the names of the chronozones no longer implies that they have any climatic connotations (Mangerud et al., 1974). In the Flandrian pollen diagrams the local or regional pollen-assemblage zones, with their well-defined boundaries, have been used as stratigraphical units with which environmental changes have been compared, in addition to their use in describing vegetational history, particularly that of the forests. This is similar to the use of pollen diagrams in the study of older warm stages. But because the changes in the Flandrian pollen diagrams can be dated with the radiocarbon method, the formal division into zones loses some of its importance; the changes can directly be referred to years B.P. The ages for the chronozones (Table 12.1) coincide with changes in the pollen diagrams only in a limited area, as they are based on averages of dates – mainly from southern Sweden.

In order to demonstrate the main features of the Flandrian vegetational history in the area south and east of the Scandinavian mountains, three schematic percentage diagrams of arboreal pollen diagrams from southern Sweden, southern Finland, and northern Finland are compared with one another in Fig. 12.1. In all three examples the local or regional zone boundaries have been radiocarbon-dated. Similar comparisons of diagrams have been made earlier, for larger areas of Europe particularly by Janssen (1974), including the schematic diagrams earlier presented for Finland (Donner, 1963) and for Sweden (Fries, 1965) in which three diagrams were used to show the south–northerly differences in forest history. With the increased use of radiocarbon dating these presentations were later revised, as seen from later comparisons of diagrams and zonations in Finland (Donner, 1971; Hyvärinen, 1972).

The first schematic diagram in Fig. 12.1 represents Scania and Blekinge in southernmost Sweden and is based on the detailed pollen diagram of Ageröd bog in Scania, from which there are a number of radiocarbon dates (T. Nilsson, 1964a). The ages for the

Table 12.1. *Flandrian chronozones for NW Europe as suggested by Mangerud et al. (1974) compared with local and regional zonations used for three schematic percentage diagrams in Figure 12.1*

Substages	Chronozones	Radiocarbon ages B.P.	Scania & Blekinge, S Sweden[a]	S Finland[b]		N Finland, Akuvaara[c]	
Late Flandrian	Subatlantic		Subatlantic	Spruce–pine	IX		
		2.5 ka	2.25 ka		2.5 ka	Pine–birch–spruce (Aku 3)	
							3.2 ka
Middle Flandrian {	Subboreal		Subboreal	Birch–alder– hazel–elm	VI–VIII	Pine–birch–alder (Aku 2)	
		5 ka	5.25 ka				
	Atlantic		Atlantic				7.5 ka
		8 ka	8.15 ka		8 ka		
						Birch (Aku 1)	
Early Flandrian {	Boreal		Boreal	Pine	V		
		9 ka	9.65 ka		8.9 ka		
	Preboreal		Preboreal	Birch	IV		
		10 ka					
			10.25 ka				

Sources: [a]Berglund, 1966b
[b]Donner, 1971
[c]Hyvärinen, 1975a

zone boundaries are the revised ages used by Berglund (1966b) in his comparison of the zone boundaries in Blekinge with those in Scania (Table 12.1). Scania and Blekinge are near the northern boundary of the present temperate vegetation zone (see Fig. 3.3) and the influence of the trees of this zone is clearly reflected in the pollen diagram. After the initial maximum of *Betula*, the spread of *Corylus* at the same time as that of *Pinus* is striking; the percentages of the former are here, as in northwestern Europe generally, based on the total tree-pollen sum but excluding *Corylus* as its pollen was very abundant. The Atlantic and Subboreal chronozones have a strong representation of *Alnus* and *Quercus,* in addition to continuous curves of *Tilia, Ulmus,* and *Fraxinus,* the boundary of the zones being marked by a decline in the *Ulmus* curve, a feature generally found in diagrams from northwestern Europe. The immigrations of *Fagus* and *Carpinus,* with their tree lines in southern Sweden north of the area, were late in Subboreal time, with the spread of *Fagus* in the Subatlantic. *Picea* has low percentages in this zone because of its late immigration from the east into the areas north of the temperate zone. The pollen-analytical zones for Scania and Blekinge, with the same names as those later suggested by Mangerud et al. (1974) for the chronozones, were divided on the basis of the actual changes in the pollen diagram from the Ageröd bog. Therefore, they do not coincide with the boundaries for the chronozones even if some are close to them in age. The most noticeable differences are in the ages of the Preboreal and Boreal zones which are older than the corresponding chronozones (Table 12.1). It is particularly in the early zones, based on the maxima of *Betula* and *Pinus,* that time differences can be demonstrated because of the

differences in the initial immigration of these trees shortly after the Flandrian climatic amelioration after deglaciation. These time differences can clearly be seen in the comparison of the diagrams in Fig. 12.1.

The second schematic diagram, that for southern Finland, is mainly based on the pollen diagrams from the raised bog Varrassuo and the adjoining lake Työtjärvi in Hollola west of Lahti, on Salpausselkä I, in the southern part of the present boreal vegetation zone (Donner et al., 1978). The diagram was divided into the regional pollen-assemblage zones of southwestern Finland and the ages for the zone boundaries on the radiocarbon ages from that area (Donner, 1971). The first two zones are close in age to the Preboreal and Boreal chronozones (Fig. 12.1), and the beginning of the spruce–pine zone coincides with the lower boundary of the Subatlantic chronozone. The birch–alder–hazel–elm zone, corresponding to the Atlantic and Subboreal chronozones, cannot with any certainty be divided in the diagrams for southern Finland, even if the time-transgressive beginning of the *Tilia* curve has been used in dividing the zone into two parts in earlier divisions. As the regional pollen zones for southwestern Finland correspond to the zones introduced by Sauramo (1949), which were based on the zonation by Firbas (1949) for Central Europe as adopted from the earlier division of Danish diagrams by Jessen (1935), they were also included in Table 12.1, as they are often still used. The schematic diagram for southern Finland shows how *Quercus,* with its present tree limit at the southwestern coast of Finland, and *Tilia, Ulmus,* and *Corylus,* with their tree limits in Central Finland, together with *Alnus* were relatively common elements in the Atlantic and Subboreal forests, repre-

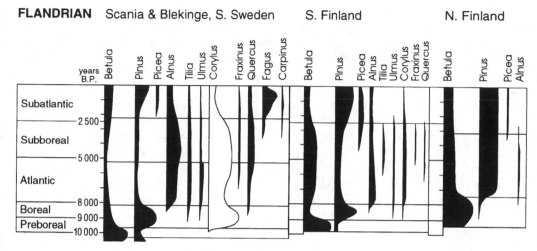

Figure 12.1. Schematic Flandrian pollen diagrams from Sweden and Finland with their zones compared with the Flandrian chronozones (the sites for the diagrams are shown in Figure 12.2). The diagram for Scania and Blekinge is based on a diagram from Ageröd bog (after T. Nilsson, 1964a), the diagram for southern Finland on diagrams of the bog Varrassuo and Lake Työtjärvi west of Lahti (after Donner et al., 1978), and the diagram for northern Finland on a diagram from a lake in Akuvaara (after Hyvärinen, 1975a).

senting a southern element in the area of the present boreal forest, until the onset of the Subatlantic, when they clearly withdrew south or became more sparsely represented. The changes show how the northern boundary of the hemiboreal vegetation zone with oak was situated in a more northerly position at the time of the Flandrian climatic optimum. *Picea* immigrated into southern Finland already in Subboreal time and became common in Subatlantic time, together with the relative increase of *Pinus*.

The third schematic diagram in Fig. 12.1 is from the northern boreal vegetation zone in Finland. It is based on the diagram from a small lake basin from Akuvaara, about 2 km northwest of the shore of Lake Inari, and the chronology is based on 5 radiocarbon dates showing a linear rate of sedimentation curve (Hyvärinen, 1975a). Of the local assemblage zones the lowermost birch zone (Aku 1), with its upper boundary at 7.5 ka B.P., was followed by a pine–birch–alder zone (Aku 2) with predominantly pine. In northernmost Finland the low percentages of *Alnus* are represented by *A. incana*. At 3.2 ka B.P. there is a slight change in the diagram as a result of the immigration at that time of *Picea,* at about the same time as the end of the continuous curve for alder. Otherwise the uppermost pine–birch–spruce zone (Aku 3) does not differ from the previous zone. Compared with the diagrams from further south the Akuvaara diagram shows a very uniform forest history after the first pioneer phase with a birch period followed by the spread of pine, both somewhat later than in southern Finland.

From the comparison of the three schematical pollen diagrams it can be seen that more stratigraphical units, zones, can be identified in diagrams from the areas of the present temperate and hemiboreal vegetation zones than in diagrams from the northern areas of the borcal zone, where there are fewer tree species. In diagrams from Denmark and southern Sweden and Norway there are at least five main pollen-assemblage zones, but only three in diagrams from northern Finland, as well as in northern Sweden, as seen in a diagram from, for instance, Leveäniemi (Lundqvist, 1971). The use of pollen diagrams alone as a stratigraphical tool in studies of the Flandrian biostratig-

raphy is thus limited in the northern areas without the control of radiocarbon dates. Further south, on the other hand, the regional pollen-assemblage zones can be successfully used in biostratigraphical studies even without the backing of radiocarbon dates from each individual site. The ages of the chronozones and of the local and regional pollen-assemblage zones (PAZ) are, however, based on conventional radiocarbon years that differ, as earlier pointed out, from sidereal years and, furthermore, are inaccurate because of the uncertainty inherent in the radiocarbon method. In the biostratigraphical studies of the Flandrian, as well as of the Weichselian, it is therefore essential to have a sound stratigraphical basis in combination with the radiocarbon chronology, which has so often been misleading on its own because its inaccuracies and errors have not been fully understood. The ages given for the regional pollen-assemblage zones are averages of a number of radiocarbon dates with a wide range, greater than the standard errors for the individual dates, as seen from the ages for the Atlantic–Subboreal boundary in Scania (T. Nilsson, 1964a) and for all Flandrian zone boundaries in Finland (K. Tolonen & Ruuhijärvi, 1976; Donner et al., 1978). The ages quoted in the schemes presented in Table 12.1 and Fig. 12.1 must therefore be viewed against this background. The ages used for the local assemblage zones for the Akuvaara diagram from northern Finland are, on the other hand, only those from the site itself and are therefore more accurate. In some profiles with varying sedimentation rates the ages based on the best-fitting curves for rates of sedimentation are less accurate.

In order to establish a more exact chronology for the lake sediments than that provided by radiocarbon dates, attempts have been made to use as a chronological frame the thin laminations found in many lake sediments and shown, especially with the help of the seasonal variations in the production of diatoms, to be annual varves. The thin varves, which can be identified with the help of X-ray radiographs of thin sections, have been useful in dating the youngest lake sediments that are otherwise difficult to date, as, for instance, demonstrated in the study of Järlasjön near Stockholm (Digerfeldt, Battarbee, & Bengtsson, 1975) and in the study of Lovojärvi in southern Finland (Saarnisto, Huttunen, & K. Tolonen, 1977). In lake basins ideal for the formation of laminations interpreted as annual varves, a record of practically the whole Flandrian sequence can be obtained, for example, from Lake Valkiajärvi in central Finland with nearly 3 m thick sediments from 25 m depth (Saarnisto, 1985). From the organic lake sediments, in which the varves had an average thickness of 0.3 mm, 8,262 varves were counted. Taking into account the basal, slightly thicker, varves it was estimated that the lake sediments, taken all in all, represent 9,500 years – nearly the total time after the deglaciation of the area. The varves can thus in some cases be used in dating the sedimentary record, but varve chronologies similar to those for glacial varves cannot be established as there is no way of connecting varve series from different lakes with one another merely on the basis of the varves. But the varves can, as was earlier noted for the Holsteinian and Eemian sediments in Germany, be used for estimates of the length in years of warm stages (Saarnisto, 1988). The independent chronology of lake varves, in principle representing sidereal years, has been difficult to compare with the chronology based on radiocarbon years, mainly because of difficulties in using the varve sediments for dating. A comparison of about 6,000 varves in Lake Ahvenainen in southern Finland with 13 radiocarbon dates showed, however, that there was the expected age difference in the lower part of the sequence, the radiocarbon dates being a few hundred years too young at about 4,000 to 5,000 radiocarbon years B.P. (M. Tolonen, 1978). In the upper part of the sequence the radiocarbon ages, being 240 to 600 years too old, were explained to have been affected by redeposited organic material washed into the lake basin during the deforestation of the catchment area. There is thus in the comparison of the two chronologies established for Lake Ahvenainen some uncertainty as to the real differences in years between them, and an accuracy similar to that obtained in comparisons between tree-ring chronologies and radiocarbon dates can hardly be achieved in any lake. A conversion of the Flandrian chronology based on radiocarbon dates to calendar years is therefore still best done by using the detailed calibration tables based on dendrochronology, especially since this chronology has been extended to the beginning of the Flandrian. The conventional radiocarbon years are, however, still generally used as a chronological tool in the study of the Flandrian.

In addition to the use of pollen diagrams in stratigraphical studies, their use in paleobotanical studies has led to more detailed and refined interpretations of the changes in vegetation, in which types other than conventional percentage diagrams have been constructed, as shown by the examples given in Chapter 3, Section 3. The potential of these methods has already been demonstrated but they have not yet been widely used in pollen stratigraphical studies, apart from purely paleobotanical studies mainly of the Flandrian. In the general account of the Flandrian vegetational history against the background of that of the older warm stages, their use is therefore still limited. One aspect of the Flandrian that, however, can be studied in detail is the immigration histories of tree species with the help of radiocarbon dates,

which is not possible for older warm stages. The time-transgressive character of the spread of *Betula* and *Pinus* after deglaciation is already reflected in the comparison of the three diagrams in Fig. 12.1, but the late Flandrian spread of *Picea abies* from the east over Finland into Sweden and Norway has been of special interest and is well documented by numerous radiocarbon dates. The presentations of its spread vary, however, because of slightly different interpretations of the levels in the pollen diagrams at which spruce could be considered to have immigrated to the surroundings of the investigated sites. The isolines constructed by Moe (1970) and used for the map in Fig. 12.2 for the distribution of of spruce at different times, were based on the dates for the rise of *Picea* pollen percentages to more than 15 percent, which by Moe was considered to show that spruce had a continuous distribution; small groups of spruce had previously spread into the areas in which it later became common. The area of its present distribution is also shown on the map. A slight change for the isoline of 5 ka B.P., as drawn by Moe for Finland, was made because the radiocarbon dates from this area show that spruce had already reached the eastern parts of south-central Finland by that time (K. Tolonen & Ruuhijärvi, 1976; Donner, 1979). Earlier maps for Finland were constructed by R. Aario (1965a) and Aartolahti (1966). The general pattern of the spread of spruce as presented in Fig. 12.2 agrees with later more detailed maps for smaller areas. The maps presented by Persson (1975) for Sweden also show the early rapid immigration into northern Sweden from Finland. The late spread of spruce into Norway has been reconstructed in detail with the use of a great number of radiocarbon dates (Hafsten, Hennisngsmoen, & Høeg, 1979; Hafsten, 1985). In discussing the general spread of spruce into Scandinavia, Tallantire (1972) suggested that the spread of spruce northwards probably took place stepwise, as indicated by the clustering of the radiocarbon ages, and that it is also probable that there were local stands of spruce in favorable areas hundreds of years before its general spread. The dating of the spread of the two alders *Alnus glutinosa* and *A. incana* showed that the former became established in large areas around 8 ka B.P. at the same time as the latter in the northern parts of Norway, Sweden, and Finland, even if alder was present in the southern areas of Scandinavia before that time (Tallantire, 1974).

The changes in the Flandrian pollen diagrams (Fig. 12.1), as with the changes of previous warm stages, show the general development toward a climax forest followed by the retrogressive phase, as described by Iversen (1958, 1964), changes that reflect a general amelioration of climate to its optimum followed by a deterioration. The translation of this evidence into exact figures in degrees is, however, difficult, in spite of the knowledge of the temperature requirements in mean temperature of the warmest and coldest months as well as the dependence on the amount of summer rainfall for the distribution of the tree species (see Fig. 3.6). But some general climatic changes have been deduced from the pollen diagrams of northern and northwestern Europe. According to the reconstruction of the changes in the European forest during the Flandrian by Huntley (1990), the

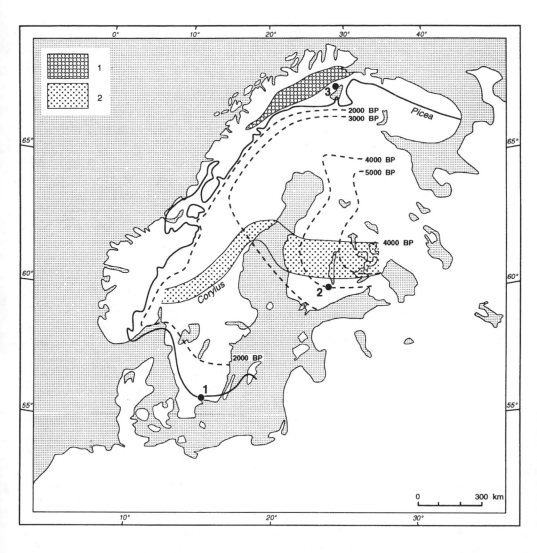

Figure 12.2. Locations of pollen diagrams in Figure 12.1 (1, 2, and 3), immigration of *Picea* from the east up to its present limit, northward spread of *Corylus* during the Flandrian climatic optimum and retreat to its present limit (symbol 2), and the distribution of palsas in the north (symbol 1).

replacement in northern Europe of *Betula* and *Pinus* by *Alnus* and *Picea* indicates a general increase in temperature. In the area of the temperate vegetation there first were relatively warm summer temperatures with moderate precipitation, at a time when *Pinus* and *Corylus* were common, followed by a time with lower precipitation, at a time when *Alnus* and *Tilia* spread, until 5 ka B.P. After that, and especially after the spread of *Fagus* and *Carpinus* in the last millennia, the summer temperatures decreased and the precipitation increased. In using the available pollen data, Huntley and Prentice (1988) concluded that the mean July temperatures were warmer than those at present in most parts of Europe, more than 2° C in northern Scandinavia and in southern Norway, southern Sweden, southwestern Finland, and in Denmark, but less in the areas between. They also showed that there were areas further south in Europe where the temperatures were 4° C higher. The figure of 2° C more for the July temperatures agrees with the analysis by Iversen (1944) of the Flandrian relative pollen frequencies of mistletoe, *Viscum album;* ivy, *Hedera helix;* and Holly, *Ilex aquifolium,* in Denmark, in which he concluded that the July temperatures

were about 2° C higher than at present, a conclusion also reached by Hafsten (1957) on the basis of the history of *Viscum* and *Hedera* in Norway. Iversen also concluded that the highest annual temperatures were reached in late Boreal and in Atlantic times, the highest summer temperatures at the boundary between the Atlantic and Subboreal times, the boundaries not exactly corresponding to the boundaries of the chronozones shown in Table 12.1. The temperature curve of July temperatures drawn by Iversen (1973) for the Late Weichselian and Flandrian in Denmark has a maximum at about 7 ka to 6 ka B.P. of just over 2° C higher temperatures than at present. If one assumes that the annual mean temperature 6 ka ago was about 1–2° C higher than at present it means that it had risen by at least 10° C from Younger Dryas times, which had annual mean temperatures of −6° C (Chapter 11, Section 3) in the areas outside the Fennoscandian moraines with fossil-patterned ground structures. The curve for the July temperatures in Denmark shows a rise of just over 6° C, but of over 8° C from pre-Bølling times (Iversen, 1973). These figures give a general idea of the magnitude of the changes in temperature after the last Weichselian glaciation.

These results, including the curve for the temperature changes, can be used as background information in discussing the climatic changes that can be deduced from the Flandrian macroscopic plant remains to be dealt with in the next chapter.

12.2. Macroscopic plant remains and stratigraphy of lakes and mires

The macroscopic plant remains in the lake sediments and in the peats of the mires reflect the natural plant successions of the hydroseres leading up to the climax communities of the Flandrian climatic optimum, as well as the Subatlantic climatic deterioration and fluctuations between wet and dry periods throughout the Flandrian, as seen from the bulk of evidence from stratigraphical studies of mires in northern and northwestern Europe (Godwin, 1975). The initial, natural plant successions in the studied basins took place during the Early Flandrian amelioration of climate, or had already done so in the Late Weichselian in areas deglaciated at that time. The pioneer phases of the vegetation during deglaciation were already discussed earlier, especially the immigration of arctic–alpine species into the Scandinavian mountains after the Weichselian glaciation. As mentioned, about 13 percent of the present Scandinavian mountain vascular plants have, according to Danielsen (1971), been recorded as Late Weichselian or Early Flandrian subfossils south or east of these mountains, *Dryas octopetala* and *Koenigia islandica* among them. The open vegetation was soon replaced by forest, with the initial immigration of *Betula* followed by that of *Pinus*, as seen in the pollen diagrams. The climax vegetation of the Flandrian was established mainly during the Atlantic, during the period described as the climatic optimum, but which has no clearly defined culmination identifiable in a large region. It was during the climatic optimum that many plants reached their northernmost range limit, most clearly witnessed by their macroscopic remains that show that these plants actually grew at the sites investigated, in contrast to the pollen evidence. The best records have been obtained for plants with easily recognizable macrofossils. It is therefore natural that the former range of hazel, *Corylus avellana*, is well known because of its subfossil nuts found in peats. Compared with its present distribution (Fig. 12.2) the finds of *Corylus* show that its limit was further north during the climatic optimum both in Sweden and in Finland, as demonstrated already in the beginning of the century and later verified in numerous investigations resulting from additional finds of nuts (Tallantire, 1981; B. Eriksson, Aalto, & Kankainen, 1991). The results from Finland show that finds of hazel – north of its present range – are from the beginning of the Subboreal chronozone, from about 5 ka to 4 ka B.P., but that its range maximum in Scandinavia as a whole was at about 6 ka to 4 ka B.P. It seems, as suggested by Tallantire (1981), that the climatic optimum reached its peak progressively later toward the north. The northward spread of *Corylus* during the Flandrian climatic optimum reflects the typical pattern for the deciduous trees with their present tree limits at or just north of the northern boundary

of the present hemiboral zone, but differs markedly from that for the immigration of *Picea* as reconstructed on the basis of pollen diagrams, as seen from the map in Fig. 12.2. The late and comparatively slow immigration of *Picea* from the east into the climax vegetation of the Flandrian climatic optimum accelerated after about 4 ka B.P. and particularly at the beginning of the climatic deterioration starting at 2.5 ka B.P.

In addition to the evidence from changes in the distribution of trees, macrofossils of water plants show the influence of the climatic optimum on the vegetation. The finds of fossil fruits of water chestnut, *Trapa natans*, in southern and central Sweden as well as in southern and central Finland (Hultén, 1950), far north of its present limit in Central Europe, is good evidence of formerly higher summer temperatures than at present (Hintikka, 1963). The finds in Finland are from the beginning of the Atlantic and from the Subboreal (Alhonen, 1964). Apart from *Trapa natans*, the macrofossils of, for instance, *Ceratophyllum submersum* (Backman, 1943), *Najas flexilis* (Backman, 1948), *N. marina* (Backman, 1941), *N. minor* (Backman, 1951), and *N. tenuissima* (Backman, 1950) all show how they had more northerly distributions during the climatic optimum. Many of the finds are from Ostrobothnia, the west coast of Finland. There their relative frequency was due to the exceptionally favorable environment for these water plants during the climatic optimum. Because of the isostatic uplift, several shallow lakes were formed along the flat Ostrobothnian coast – lakes with a rich eutrophic vegetation (Backman, 1950). Later these lakes became oligotrophic, before being overgrown with mires. The general increase of peat growth after the climatic optimum caused by the climatic deterioration thus resulted in the disappearance of suitable habitats for these water plants, which is reflected in the changes of their distribution and frequency.

Using the earlier quoted figures for the changes in temperature during the Late Weichselian and the Flandrian, the curve in Fig. 12.3 showing the general trend of the mean temperature for July in Denmark was drawn after the curve constructed by Iversen (1973), with its culmination at about 7 ka to 6 ka B.P. In addition to the evidence of the climatic optimum obtained from the former distribution of plants, the radiocarbon-dated fossil finds of pine wood, mainly preserved trunks or stumps, in the Scandinavian mountains and in northern Finland show the Flandrian changes in the altitude of the tree line for pine, changes that show the development after the climatic optimum. The three curves in Fig. 12.3 were drawn on the basis of the results from Hardangervidda in northern Norway (Moe, 1979), Jämtland further north in Sweden (Lundqvist, 1969a), and from Finnish Lapland (Eronen, 1979; Eronen & Huttunen, 1987). They all show the culmination after the spread of pine, with higher altitudes reached in the south than in Lapland. The highest altitudes were reached in Hardangervidda and Jämtland at about 8 ka B.P., but in Finnish Lapland between 6 ka and 5 ka B.P. This delay partly reflects the later immigration of pine into the northern areas, as seen in the pollen diagrams, but may also reflect the previously mentioned progressive shift northward of

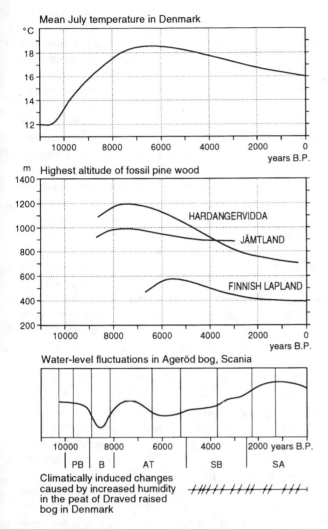

Figure 12.3. Schematic curves of mean July temperature in Denmark (after Iversen, 1973); highest altitudes of fossil pine in Hardangervidda (after Moe, 1979), Jämtland (after Lundqvist, 1969a), and Finnish Lapland (after Eronen & Huttunen, 1987); water-level fluctuations in Ageröd bog, Scania (after T. Nilsson, 1964b), and shifts in peat formation in Draved raised bog (after Aaby, 1975, 1976).

the peak of the climatic optimum. The lowering of the tree line for pine reflects the general climatic deterioration after its optimum. Part of the climatic deterioration that caused this lowering was probably caused by the isostatic uplift (Eronen, 1979; Moe, 1979), lifting the Scandinavian mountains about 50 to 100 m higher above sea level during the last 7 ka. The effect of the uplift on the Flandrian vegetational history, particularly in the mountains, has generally been considered to have been of minor importance (Berglund, 1983). In the northern parts, such as Finnish Lapland where the lowering of the tree limit for pine was less than 200 m, the effect of the land uplift, however, was comparatively more important.

This climatic deterioration after its Flandrian optimum as witnessed by the southward withdrawal of the plants and the lowering of the tree limit of pine is also reflected at the surfaces of the mires as an increase in features related to frost action, such as the hummocks and ridges rising from the surfaces of the waterlogged northern fens, particularly conspicuous in the string bogs. Of special interest are the up to 7 m-high peat hummocks called palsas, which underneath an approximately 0.5 m thick protecting layer of *Sphagnum* peat have a 2–3 m-thick perennially frozen core of segregation ice (Lundqvist, 1962). The palsas are mainly restricted to a small area in the north (Fig. 12.2), north and northwest of the tree line of pine, to an area with a mean annual temperature of $-1°$ C (Ruuhijärvi, 1960, 1962), but with some occurrences further south along the Scandinavian mountains (Sollid & Sørbel, 1974). The palsas occur outside the area of continuous permafrost, at the border of the area of discontinuous permafrost (Lundqvist, 1969b). Their stratigraphy and development have been studied in many areas (H. Svensson, 1962; Salmi, 1968), but the time of their formation has been difficult to date. By using pollen diagrams from the palsas and from the peats in surrounding mires it has been concluded that their formation started in Subatlantic time, thus after 5 ka B.P., following the withdrawal of pine from the mountains (Ruuhijärvi, 1962). The general spread of ombrogenous peat at this time made it possible for palsas to develop, as shown in the detailed studies by K.-D. Vorren (1972) of the stratigraphy of mires with palsas in northern Norway. The radiocarbon ages of peats interpreted as dating the palsa formation showed, however, that they were formed late in Subatlantic time, especially after about 600 B.P. (K.-D. Vorren, 1979). According to these dates many palsas would thus be comparatively young periglacial features. The possible preservation of palsas from the time before the climatic optimum, as suggested by Salmi (1968) on the basis of his studies of palsas in Finnish Lapland, still needs confirmation, but it seems clear from the pollen stratigraphy and radiocarbon dates he presented that palsas were forming already in the beginning of the Subboreal. These various results indicate that there may be some regional differences as regards the beginning of the palsa formation, but that they generally can be regarded as features connected with the deterioration of climate after its Flandrian optimum.

In the Flandrian climatic changes an important part was played by changes in humidity, a result of the interaction between changes in precipitation and in temperature. The changes in humidity have been recorded as fluctuations in water level in the stratigraphy of mires and in lakes, the extremes being caused by a high precipitation and low temperature that resulted in a high water level and a low precipitation and a high temperature that resulted in a low water level (Digerfeldt, 1975b). The stratigraphical changes in the mires in Sweden and changes in lake basins were already used by Sernander (1908) to draw a curve of water-level fluctuations in two lake basins, Vänstern and Kalven, in the county of Småland, used as evidence for the division of the Flandrian into dry periods (the Boreal and Subboreal) and humid periods (the Atlantic and Subatlantic), the names of which have later been adopted in the division of the chronozones, but without the original climatic connotations. As a result of his detailed study of the Ageröd raised bog (Ageröd mosse) T. Nilsson (1964b) produced a curve of the water-level fluctuations in the basin of the mire. The curve, as shown in Fig. 12.3, was redrawn using the ages for the zones given in Table 12.1 for Scania and Blekinge. In the Boreal chronozone the water level was at its lowest, whereas it rose to its maximum position in the middle of the Subatlantic, with an earlier smaller rise in the early Atlantic. There was thus according to this curve an Early Flandrian period of low precipitation and comparatively high temperature before 8 ka B.P., whereas from about 5 ka B.P. onwards the precipitation increased and the temperature fell, culminating after about 2 ka B.P. in the Subatlantic. The curve for the water level changes from about 5 ka B.P. is a mirror image of the other curves in Fig. 12.3, all together reflecting the change to a cooler and more humid climate after the Flandrian optimum. The studies of the lake basins in southern Sweden by Digerfeldt have given additional evidence of the Flandrian water level changes. The results from the two lakes in the southern Swedish uplands near the town of Växjö, Trummen (Digerfeldt, 1972) and Växjösjön (Digerfeldt, 1975b), and from the Ranviken bay in Lake Immeln in northeastern Scania (Digerfeldt, 1974, 1975a), showed that there were two main low positions of the water level, the older one at the boundary of the Preboreal and Boreal periods, as defined for southern Sweden, and the younger one with its lowest position in the Subboreal period. The low positions in the lakes are thus slightly out of phase with those in the curve presented by T. Nilsson (1964b), but essentially reflect the same main changes in humidity. The evidence from the lakes, however, shows that the water level in lakes was low already in the Preboreal, as also found in the study of a lake, Lyngsjö, in central Halland on the Swedish west coast (Digerfeldt, 1976). Summarizing the evidence from the lakes in southern Sweden, Berglund (1983) concluded that the two main lowerings of water level culminated at about 9.3 ka to 9.5 ka B.P. and at about 2.9 ka to 3.1 ka B.P. Water-level fluctuations found in lake basins with an amplitude of several meters have been recorded throughout Sweden and Finland, and the Early Flandrian low position found in southern Sweden has also been recorded in southern Finland, as in Työtjärvi west of Lahti (Donner et al., 1978), and in the small lake Hakojärvi, also in southern Finland (Huttunen, Meriläinen, & K. Tolonen, 1978). The conclusions were, however, mainly based on the analysis of diatoms and cladoceran remains; detailed stratigraphical evidence of the lake sediments is lacking. The fluctuations in water level have not generally affected the rates of sedimentation in the lakes as determined with the help of radiocarbon dates. The rates in southern Sweden, as in Lake Trummen, and in lakes Vakojärvi and Työt-järvi in southern Finland, vary between 0.2 mm to 0.5 mm a year (Digerfeldt, 1972; Donner, 1972; Donner et al., 1978), whereas they are slightly smaller in northernmost Finland, 0.2 mm a year or slightly less (Hyvärinen, 1975a). The 2–2.5 m-thick lake sediments were on the whole regularly deposited in the deeper parts of the lakes throughout the Flandrian and possible

irregularities caused by water-level fluctuations can therefore only be traced in the shallow parts of the lakes near their shores.

In addition to the major changes of Flandrian temperature and humidity, as depicted by the curves in Fig. 12.3, climatic fluctuations of shorter duration are recorded in the ombrotrophic peats of raised bogs. The changes in climate are reflected in these bogs, which receive all their water in the form of precipitation, as changes from dark humified *Sphagnum* peat to light less humified *Sphagnum* peat. Several of these recurrence surfaces, which show shifts to increased humidity as a result of changes in temperature and/or precipitation, as mentioned earlier, have been identified throughout northwestern Europe, but a detailed chronological comparison has shown that there are age differences between them within this area. There is, however, a clear change in the mire stratigraphy at the beginning of the Subatlantic as a result of the climatic deterioration, seen in the raised bogs as the recurrence surface originally described as the "boundary horizon" (*Grenzhorizont*) in stratigraphical investigations (Godwin, 1975). In studies of raised bogs in Denmark, Aaby (1975, 1976) demonstrated that humidity changes are best recorded in the peats of the hollows, whereas smaller variations are not registered in hummocks that have remained in place for a long time, in Draved bog for more than 2.5 ka. In order to study in detail the changes in the degree of humification in Draved bog the accumulation rates of the peat in the 2.5 m-thick bog were determined by using 55 calibrated radiocarbon dates from a single section (Aaby & Tauber, 1974). The rates varied between 0.16 mm to 0.80 mm/year, being within the normal range in north European bogs as shown in the studies cited. The exact horizons at which there were shifts from dark humified peat to light peat were determined by a study of the degree of humification and an analysis of two diagnostic rhizopod genera, *Sphagnum* leaves and the pollen frequencies of some plants growing on the bog (Aaby, 1975, 1976). It was concluded from these studies that the climatic changes reflecting an increased humidity were cyclic during the past 5.5 ka, with a periodicity of about 260 years. This same cyclicity, shown in Fig. 12.3, was also detected in other Danish bogs with which the results from Draved bog were compared and is in the same order of magnitude as the cycles of 200–300 years found to characterize the Flandrian climate generally (Karlén, 1982). As the ombrotrophic peats suitable for a study of the changes are comparatively young, the study of the short-term fluctuations cannot be extended much further back in time than in the study of Draved bog. As pointed out by Aaby (1975) the periods of 260 years are the shortest that can be identified in the peats of raised bogs, whereas cyclic variations with a periodicity of 520 years can be identified in the curves of changes in sea level according to Aaby.

A periodicity in the formation of recurrence surfaces in the raised bogs, similar to that demonstrated by Aaby, has also been mentioned as a possibility in detailed studies of raised bogs in Sweden and Finland. Thus T. Nilsson (1964b) in his study of the Ageröd bog remarks that it is possible that the recurrence surfaces in this bog reflect a periodicity, but that this could not be proven with certainty due to the lack of enough material, particularly of enough radiocarbon dates. A number of recurrence surfaces in ombrotrophic peats were, however, already identified in detailed investigations before radiocarbon dating was generally applied. In the study of the bogs of southern Ostrobothnia in the coastal area of western Finland, Brandt (1949), for instance, identified seven surfaces, the oldest of them over 3 ka old, and in a detailed study of the raised bog Varrassuo in southern Finland, using rhizopod analyses, K. Tolonen (1966) identified a number of dry and wet phases in the peat growth. There is thus evidence from a large area of how humidity changes affected the growth of ombrotrophic peats, but apart from the studies in Denmark by Aaby a cyclic nature of these changes has not been demonstrated, only indicated as a likely possibility.

In addition to the search for a possible climatic cyclicity in the growth of peats in raised bogs, a combined effect of periods of different lengths has been investigated in the growth of trees, as in the study of the tree rings of living pines, standing dead trees, and stumps near the northern tree line of pine in northernmost Finland during the last 800 years (Sirén, 1961). The biological evidence from the trees in Lapland cannot, however, directly be linked with the evidence from the raised bogs further south.

The short-term climatic fluctuations described above are all in the Subboreal and Atlantic chronozones (Fig. 12.3). They can be assumed to have also taken place earlier during the Flandrian but there is no biostratigraphical evidence of them. It is particularly from the time of the climatic deterioration after the climatic optimum that suitable evidence of short-term humidity changes has been preserved in the bogs.

12.3. Tephras and impact craters

By carefully analyzing the mineralogical composition of the inorganic content of the peats in the mires of Central Sweden, Persson (1966) was able to show that they contain particles of volcanic ash and that separate ash horizons can be identified in these peats. Further studies of ash horizons in mires on the west coast of Norway (Persson, 1967b) and on the Faeroe Islands (Persson, 1968), and comparative studies in Iceland (Persson, 1967a), enabled Persson (1971) to demonstrate that several Icelandic ash falls can be traced throughout the area, investigated, and radiocarbon-dated. The Icelandic origin of the volcanic ash is supported both by the fact that the concentration of ash particles in each horizon increases toward the west and northwest and by the composition of the ash, consisting of acid tephra as shown by the studies of the refractive indices of the glass particles. As the volcanic ash particles were concentrated in distinct horizons that could be radiocarbon-dated, Persson (1971) grouped them into six volcanic ash units, tephra units, as listed in Table 12.2. No older tephras, from the period of about 4 ka to 7 ka B.P., were traced in the peats. The best-known ash fall is the youngest, that of the Icelandic volcano Askja in A.D. 1875, the ash of which is found close to the surface of the mires and of which the ash rain was followed and recorded in detail. The distribution of this ash in

Table 12.2. *Flandrian tephra units identified in Scandinavian mires and their correlation with eruptions in Iceland and impact craters in Estonia*

Tephra units with radiocarbon ages	Dated eruptions in Iceland
1. 75 B.P. (A.D. 1875)	Askja 75 B.P. (A.D. 1875)
2. 330–575 B.P. (A.D. 1375–1620)	Öraefa 588 B.P. (A.D. 1362)
3. 900–1100 B.P. (A.D. 850–1050)	Hekla I 846 B.P. (A.D. 1104), or Layer g 900–1050 B.P. (A.D. 850–900)
4. 1535–1585 B.P. (A.D. 365–415)	No equivalent tephra in Iceland
5. 2585–3050 B.P. (635–1100 B.C.)	Hekla III 2820 ± 70 B.P. (870 ± 70 B.C.)
6. 3475–3800 B.P. (1525–1850 B.C.)	Hekla IV 3830 ± 120 B.P. (1880 ± 120 B.C.)

Impact craters in Estonia	
Kaali on the island of Saaremaa	4000 B.P.
Ilumetsa southwest of Lake Peipsi	6000 B.P.
Tsöörikmäe southwest of Lake Peipsi	9500 B.P.

Source: Persson, 1971

Norway and Sweden is shown on the map in Fig. 12.4. The mires in which Persson (1971) recorded the older ash falls lies broadly within the area of the Askja ash fall; only a few mires in Sweden are south of the area. The earlier mentioned Late Weichselian Vedde Ashe Bed (Fig. 11.15), restricted to the Norwegian coast, has been recorded somewhat further south than the Flandrian tephras.

The radiocarbon ages for the tephra units in Norway and Sweden and the eruptions in Iceland were given A.D/B.C. ages by Persson (1971), but in Table 12.2 the ages are also given in converted B.P. ages. As seen from the table, five of the tephra units have been correlated with known eruptions in Iceland, the ages of the two oldest eruptions, that of Hekla III and Hekla IV, being based on radiocarbon dates. The tephra of unit 4 was recorded only in one mire in Sweden and in the Faeroes. It may not represent one single ash fall and no acid tephra of this age has been recorded in Iceland. According to the radiocarbon dates the tephra unit 3 is slightly older than the Hekla I eruption in Iceland, and it is therefore possible that the unit corresponds to a slightly older eruption in Iceland, "layer g" or "layer VII a + b" (Persson, 1971), dated at 900–1050 (A.D. 850–900), practically the same age as that for unit 3. From the distribution of the ash falls in Norway and Sweden (Fig. 12.4) it can be assumed that at least some of them spread further eastwards, even into Finland. This assumption is supported by the spread of the ash from the Hekla eruption in March 1947 into a zone from southwestern Finland northeastwards across the country. The reddish brown ash was easily noticed on the snow (Salmi, 1948).

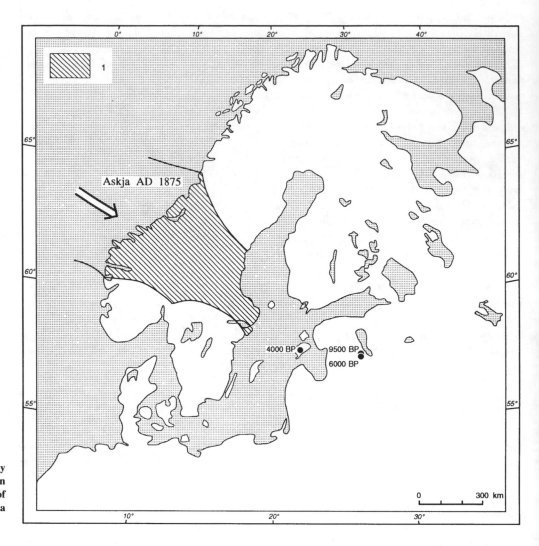

Figure 12.4. Ash fall (1) in Norway and Sweden from the Askja eruption in Iceland A.D. 1875 and locations of Flandrian impact craters in Estonia (listed in Table 12.2).

In addition to volcanic ash falls, meteorites hit the area surrounding the Baltic Sea during the Flandrian. Of the 12 impact craters identified in the rocks of the Fennoscandian Shield, most younger than 600 Ma, none is younger than 10 Ma (Pesonen, Grieve, & Leivo, 1991). But in Estonia there are three Flandrian impact craters or groups of craters, listed in Table 12.2 and shown on the map in Fig. 12.4. The group of craters at Kaali on the island of Saaremaa are in Silurian dolomite and the main crater has a diameter of 110 m and a depth of 16 m, and is filled with limnic sediments. On the basis of two radiocarbon dates from the basal sediments, the older of them being 3390 ± 35 B.P., the formation of the craters was first dated at about 3.5 ka B.P., or possibly 4 ka B.P. (Pirrus & Tiirmaa, 1990). On the basis of later detailed work it was concluded that the crater was formed no later than 4 ka B.P. (Saarse et al., 1990). At Ilumetsa southwest of Lake Peipsi in southeast Estonia there are three craters with diameters of 80 m, 50 m, and 24 m respectively, the largest being 12.5 m deep, in Devonian sandstone. The pollen diagram of the limnic sediments in the largest crater (Fig. 12.5), together with the radiocarbon age of the lowermost organic sediment, 6030 ± 100 B.P. (TA-310),

showed that the crater was formed about 6000 B.P. (Aaloe, 1979; Liiva, Kessel, & Aaloe, 1979; Pirrus & Tiirmaa, 1990). There is shattered rock mixed with boulder clay at the base of both the Kaali and Ilumetsa craters, which shows that they were formed after the deglaciation of Estonia and that they therefore cannot be older than about 13 ka B.P. The third crater, the single crater at Tsöörikmäe with a diameter of 40 m and a depth of 6.5 m situated close to the Ilumetsa craters, did not penetrate through the till into the Devonian bedrock. On the basis of the pollen diagram from the peat in the crater its formation has been dated at 9.5 ka B.P. (Pirrus & Tiirmaa, 1990). The Tsöörikmäe crater is not as distinct as the younger craters because of later denudation.

The impacts of the larger meteorites that formed the craters in Flandrian time can, as seen, be dated, whereas the ages of the finds of smaller meteorites most often cannot be determined. The number of historically recorded meteorite falls, however, gives some idea of the frequency of these falls even if the total number of falls must have been greater than the observed falls. Of the 43 meteorites recorded from Denmark, Norway, Sweden, and Finland up to January 1984 (Graham, Bevan, & Hutchinson,

Figure 12.5. Ilumetsä meteorite crater in Estonia from 6 ka B.P. (Photo J. Donner)

1985), 24 are observed falls after the beginning of the nineteenth century; in addition, one fall had been recorded in Denmark in 1654. The other recorded meteorites are finds of unknown age and a few of doubtful origin. The finds include pre-Quaternary meteorites, such as one found in Ordovician limestone in Sweden.

12.4. Glacier fluctuations

By studying the advances of the present-day glaciers in the Scandinavian mountains as witnessed by their end moraines and their withdrawals, additional evidence has been obtained of the Flandrian climatic history. The upper curve in Fig. 12.6 shows the main fluctuations of the firn line in western Norway (B. Andersen, 1965). It has been concluded that all glaciers in this area melted by about 8 ka B.P. as a result of the climatic optimum and that the present glaciers were formed later, mainly as late as during the Subatlantic, starting from 2.5 ka B.P., as noted by B. Andersen. There was a marked advance about 2.5 ka ago, but it was in A.D. 1750 that the glaciers reached their furthest and lowest positions after the climatic optimum, during the time described as "the little ice age." The lower curve in Fig. 12.6 shows the changes of the altitude of the firn line, in relation to its present altitude, of the glacier of Jostedalsbre between Nordfjord and Sognefjord in western Norway (Mangerud, 1990), the glacier being the largest in Europe, consisting of an ice cap and 24 long valley glaciers, the longest of them about 14 km long (O. Holtedahl, 1953). The broken line in the figure spans the time from 8.1 ka to 5.3 ka B.P. during which the glacier probably

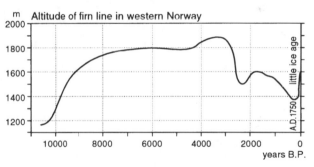

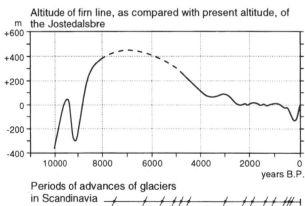

Figure 12.6. Schematic curves of changes of the firn line in western Norway, with a separate curve for the glacier of Jostedalsbre, and periods of glacier advances in Scandinavia (after Karlén, 1982).

did not exist (Mangerud, 1990), in agreement with the conclusion about the Norwegian glaciers generally. The expansion of the ice during the "little ice age" was also here clearly recorded by the marked advances of the long glaciers down the valleys to their lowest positions for the last 9 ka. Smaller fluctuations during the last 2.5 ka have also been demonstrated. In an estimation of temperature changes necessary for the fluctuations of the firn line the mean temperatures were estimated to have been about 2.5° C higher during the climatic optimum (with no ice), whereas they were just under 1° C lower during "the little ice age" (Mangerud, 1990). It is particularly the end moraines of the glaciers that show the effect of the climatic deterioration of "the little ice age," a deterioration not seen in the curves presented in Fig. 12.3 based on changes in vegetation caused by temperature changes and on groundwater fluctuations caused by humidity changes.

In a summary of all available radiocarbon ages associated with advances of glaciers in Scandinavia, Karlén (1982) was able to list a number of periods of cold/wet climate during which the glaciers advanced. This fact shows that even small changes in climate can be glaciologically recorded but not with the same accuracy as in the previously mentioned peats of the raised bogs. The record of glacier advances, however, includes evidence from northern Sweden and Nordland in Norway of advances from the time of the climatic optimum. This means that the glaciers in the northern parts of Scandinavia persisted through the warmest period of the Flandrian, whereas further south in the higher parts of western Norway they had melted in Middle Flandrian time and were later formed again, as already described.

By comparing the evidence of climatic changes presented in the curves of Fig. 12.3 and Fig. 12.6 with that of short-term fluctuations, it is evident how complex the interplay between different factors is and how difficult it is to define the parameters involved in causing the described changes. No simple curve can be drawn for the Flandrian climatic changes that would explain all changes encountered in the whole area dealt with, from Denmark in the south to northernmost Norway, Sweden, and Finland. One reason for the differences is caused by the fact that the response to a climatic shift has been different, a glacier, for instance, reacting more slowly and less sensitively to a small change of climate than trees, as witnessed by thickness variations of the tree rings.

13   Late Weichselian and Flandrian land/sea-level changes

13.1. Contemporary land uplift

The uplift of the earth's crust induced by the deloading of the thick Late Weichselian ice sheet still takes place in Scandinavia, with the center of uplift in the northern part of the Gulf of Bothnia. The oldest document in which the relative lowering of the water level along the coasts of Sweden was mentioned is from the late fifteenth century. In the seventeenth century, when this lowering was first investigated, it was explained as the result of a decrease in the amount of ocean water. The observed differences in the land/sea-level changes between various parts of the coast led, however, to the conclusion in the beginning of the nineteenth century that these changes were caused by an uplift of the earth's crust (M. Ekman, 1991). In contrast to these early observations the present knowledge of the contemporary land uplift is based on accurate instrumental measurements. The uplift has been determined by using a network of 54 mareographs that record the fluctuations of water level on the coasts of the Baltic Sea and the North Atlantic, and by repeated precise levelings in the inland areas (Kakkuri, 1985). In determining the Scandinavian uplift – based on geodetic observations – the uplift of the crust relative to mean sea level is called the apparent land uplift, Ha, which in the northern part of the Gulf of Bothnia, in the center of the area of uplift, is 9.2 mm/year (M. Ekman, 1989). As a result of a global change in climate there has been a eustatic rise of sea level, He, after the end of the last century of about 1.0 mm/year. This rise, after A.D. 550, has varied between −1 mm/year and +2 mm/year (M. Ekman, 1989, 1991). In earlier maps giving the isobases of the uplift in Scandinavia, with observations from Denmark, Norway, Sweden, and Finland, and also from the Baltic States and Russia, the value of +0.8 mm/year for He was added to the uplift values to show the uplift relative to the geoid (Kukkamäki, 1968, 1971). The value for He was subsequently determined to about 1.0 mm/year and used for the values of uplift called the leveled land uplift (M. Ekman, 1989). Later maps of the uplift, however, show the apparent land uplift, such as the map presented by Kakkuri (1985), which included recomputed uplift values for Finland by Suutarinen (1983), as

does the map in Fig. 13.1 based on a map drawn in 1989 (Kakkuri, 1992). In more detailed studies of the geodynamics of the uplift of the crust, values for the rise of the geoid have also been determined (Kakkuri, 1985; M. Ekman, 1989). The value for the geoid rise, N, reaches +0.7 mm/year or at most +0.8 mm/year in the center of the area of uplift, and about half of this in the marginal parts of the geoid rise (Ekman, 1989). If the leveled land uplift is correlated for the geoid rise the amount of uplift to the crust relative to the ellipsoid can be determined. The value obtained for this absolute land uplift, h, is thus (Ekman, 1989; see also Kakkuri, 1985): h = Ha + He + N. The map in Fig. 13.1 shows how the area covered by the thick Late Weichselian ice sheet still continues to have a domelike land uplift, with its center in the northern part of the Gulf of Bothnia, an uplift that, however, is small compared with that during and immediately after deglaciation. The geological evidence of the land uplift suggests, as will be seen later, that the pattern of uplift was essentially the same as that shown in Fig. 13.1, but the possibility of a shift in the position of the center of uplift during the Flandrian cannot be ruled out (Lundqvist, 1965; Donner, 1969b). Furthermore, there are results that suggest that the earth's crust was sensitive to local readvances of the ice during deglaciation, causing irregularities in the early land uplift. The pattern of the present uplift, as shown by its isobases (for the use of the term isobase see Flint, 1971, p. 362), cannot therefore in its details be used as such in the reconstruction of the land uplift of earlier times after deglaciation. Furthermore, this pattern changes as the accuracy of measuring the present uplift improves, as seen from the series of maps presented over the last decades. The difficulties in using the wealth of geological data in the study of the Late Weichselian and Flandrian land/sea-level changes are reflected in the numerous interpretations presented during more than a century of investigations of the land uplift.

13.2. History of the Baltic Sea

The stages in the history of the Baltic Sea, not to be confused with the chronostratigraphic warm and cold stages, were governed by the retreat of the Weichselian ice sheet and by

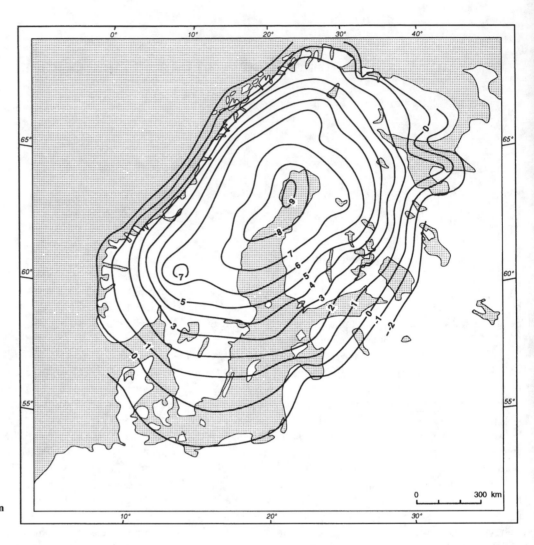

Figure 13.1. Contemporary uplift in mm/year (after Kakkuri, 1992).

the land uplift as a result of the deglaciation. The formation of the Baltic Ice Lake after the retreat of the ice margin from northern Estonia and southern Sweden has been mentioned, as have the fluctuations of the water level before and during the formation of the Fennoscandian moraines, in southern Finland, reconstructed on the basis of the marginal terraces of the two first Salpausselkäs Ss I and Ss II (Chapter 11, Section 1). It was also noted in Chapter 11 that the final drainage of the Baltic Ice Lake, resulting in a drop of the water level by 27–28 m, took place at Billingen in Sweden at about 10.6 ka B.P. according to the varve chronology; the radiocarbon dating of this event has yielded approximately the same age (see Table 11.2). Because of these changes in water level it has been possible, with the help of marginal terraces, to determine the altitude of the Baltic Ice Lake just before its drainage, along the line at which the ice margin stood at that time. The water level of the Baltic Ice Lake was then 10 m below its highest position, which in Finland coincided with the formation of the Salpausselkä I moraine or parallel moraines immediately north of it. There are thus two former positions of the water level of the Baltic Ice Lake, the BI

and B III levels noted in Chapter 11, Section 1 – with an immediate third level, B II, between them – in a 10 m vertical bracket and separated from the 27–28 m lower level of the time after the drainage of the Baltic Ice Lake. Because of this clear difference in altitude it has been possible to single out the raised beaches of the Baltic Ice Lake in areas away from the Fennoscandian moraines and to draw paleogeographic maps that show the extent of this lake just before its drainage at Billingen. Before this drainage it had had successive outlets south, through the Strait of Öresund and through valleys in southern Sweden to the ocean in the west, whereas there is no evidence of an outlet of the Baltic Ice Lake to the White Sea in the northeast, an area still covered by ice in the beginning of the Flandrian.

The extent of the Baltic Ice Lake prior to its drainage and the isobases for this ice-dammed lake, presented in Fig. 13.2, are based on a number of morphological observations of raised beaches around the coasts of the Baltic. Using the marginal deltas of the Fennoscandian moraines formed just before the drainage of the Baltic Ice Lake (see Table 11.1) as a starting point, the corresponding contemporaneous raised beaches have

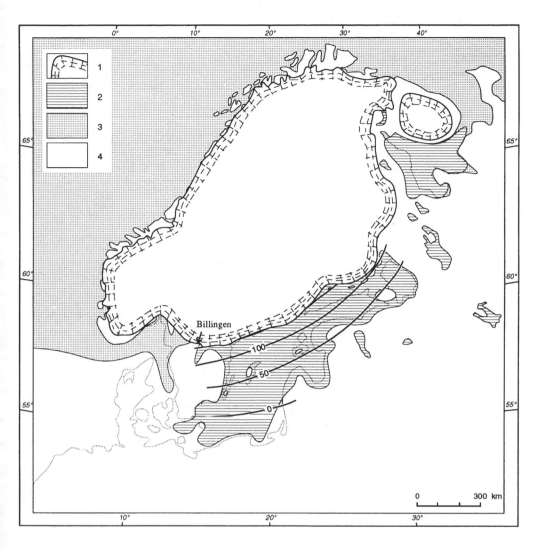

Figure 13.2. Baltic Ice Lake about 10.5 ka B.P. before its drainage at Billingen, with isobases. 1, ice; 2, lake; 3, sea; 4, dry land. (After various sources mentioned in the text)

been traced along the coasts of Lake Ladoga and the St. Petersburg area in the east, and further along the coasts of Estonia, Latvia, and Lithuania. In the southern Baltic the former coastline of the Baltic Ice Lake is submerged, but on the Swedish east coast – south of the Billingen moraine – it is again above the present sea level. The designations used in the eastern parts at various times for the shorelines of the Baltic Ice Lake equivalent to the B III level in Finland have been summarized by Donner and Raukas (1989, 1992), including details of the history of the investigations, as also discussed earlier by Hyvärinen (1973, 1975b). It can be seen from the map that the Lake Ladoga area, including most of the Karelian Isthmus between the lake and the Gulf of Finland, was part of the Baltic Ice Lake. The extent of the Baltic Ice Lake in Fig. 13.2 is essentially drawn according to previously published maps constructed by Eronen and Haila (Eronen, 1983; Alalammi, 1992), but with minor changes from some local studies. The isobases, however, are according to the map drawn by N.-O. Svensson (1989). The coastline of the sea beyond the Swedish west coast was, at the time of the Baltic Ice Lake, below the present sea level off the

coast of Jutland and further west, because at that time, at about 10.3 ka B.P., the sea level was considerably lower, still at −65 m (Jelgersma, 1979). As a result of this low sea level the White Sea basin in the northeast, at the outer limit of the area of isostatic uplift and with its threshold 25 m below present sea level, formed a large separate lake at the time of the Baltic Ice Lake (Hyvärinen, 1975b). Parts of the west coast of the White Sea were submerged but not up to Kalevala moraine, which has here been correlated with Salpausselkä II in Finland (see Table 11.1). The map in Fig. 13.2 differs for this northeastern area from these previously mentioned maps in which the ice sheet had a greater extension in the east.

The map in Fig. 13.2 shows the extent of the Baltic Ice Lake at its final stage before the drainage at Billingen. The beginning of the Baltic Ice Lake has, as mentioned before, been correlated in the Gulf of Finland with the withdrawal of the ice margin from the Pandivere moraine in northern Estonia, and in Sweden with the drainage of the Baltic through Tyringe in southernmost Sweden, the two events falling into the same age bracket of about 12.2 ka to 12 ka BP in radiocarbon years (Table 13.1). The

Table 13.1. *Stages of the Baltic Sea after the Weichselian glaciation*

Baltic stages	Changes in the Baltic Sea	Ages
Litorina Sea		
Mya Sea		
	Spread of *Mya arenaria*	0.3ka B.P.
Limnaea Sea		
	First appearance of *Lymnaea peregra* f. *ovata*	4 ka B.P.
Litorina Sea		
	First appearance of littoral "Clypeus flora"	7.5–7 ka B.P.
		8 ka in S Baltic
Mastogloia Sea		
	End of regression, equilibrium between level of the Baltic and sea level	8.5 ka B.P.
Ancylus Lake	Final closing of outlet at Degerfors, culmination of transgression and new outlet in the south through Great Belt	9 ka B.P.
	Beginning of transgression at coastal sites in the Baltic as a result of the narrowing of the outlet through Närke–Degerfors straits	9.6 ka B.P.
Yoldia Sea	Opening of Närke strait	
	Influx of brackish water in S Finland	10.3 ka B.P.
	Influx of salt water in Stockholm area	10.4 ka B.P.
	Final drainage at Billingen	10.6 ka B.P. in varve years (10.4–10.5 ka B.P. in radiocarbon years)
Baltic Ice Lake		
	Opening of connection between ice-dammed lakes in the eastern part of the Gulf of Finland with the rest of the Baltic basin as a result of the recession of the ice from the Pandivere moraine, at about the time of a drainage of the Baltic at Tyringe in southernmost Sweden c. 12.2–12 ka B.P. The Baltic has had an outflow through the Öresund Strait since c. 12.7 ka B.P.	

Baltic Ice Lake thus, according to these datings, lasted about 1,500 years or somewhat more. Cores from the southern Kattegat area, however, show that there was an outlet of the Baltic through the Öresund Strait already from 12.7 ka B.P. until its final drainage (Bergsten & Nordberg, 1992). The lake formed a cold Late Weichselian environment into which minerogenous sediments were deposited, with varved glacial clays close to the ice margin and with a low density of microfossils, such as freshwater diatoms dominated by the planktonic species *Melosira islandica* ssp *helvetica*. A biostratigraphical definition of the Baltic Ice Lake is not possible.

The final drainage of the Baltic Ice Lake at Billingen in Central Sweden marked the beginning of the Yoldia Sea stage (Table 13.1). As mentioned earlier, great amounts of water from the Baltic, during a period of rapid retreat of the ice margin, drained westwards to the coast of Skagerrak at the Swedish west coast until about 400 years after the drainage (Cato, 1982). The drainage took place at Billingen over the Lake Vänern basin, but the connection between the Baltic and the sea widened when the Närke Strait opened further north (Fig. 13.3). This resulted in an influx of salt water into the Baltic, causing a change from freshwater to brackish water in the Mälaren valley in the Stockholm area about 200 years after the drainage and in southern Finland about 300 years after the drainage (Table 13.1). Shells of the marine bivalve *Portlandia (Yoldia) arctica* were recorded in the glacial clays in the Mälaren valley, into which it penetrated from the sea about 200 years after the drainage, as a result of this influx of salt water, where it persisted for only about 100 years (Agrell, 1980; Fredén, 1980, 1981, 1988). In addition to this evidence of salt water in the Baltic, brackish water conditions prevailed for 100–150 years south along the Swedish east coast down to the area of Oskarshamn and the island of Gotland, as shown by the presence of brackish water diatoms in the Baltic sediments (N.-O. Svensson, 1989). Further east, in the sediments of the formerly submerged southern coastal areas of Finland, there are also brackish water diatoms (Alhonen, 1971), whereas the brackish water diatoms in younger Yoldia Sea sediments further north in Ostrobothnia have been interpreted as having been reworked from interglacial marine sediments (Eronen, 1983). Even if the littoral and offshore deep-water Yoldia sediments in the Baltic have slightly brackish diatom floras, in addition to freshwater forms, the Yoldia Sea stage cannot stratigraphically be defined on the basis of the diatoms (Alhonen, 1971; Fredén, 1980; Winterhalter et al., 1981; Eronen, 1983; Hyvärinen, 1988). The boundaries are related to the opening and closing of the connection with the ocean in central Sweden.

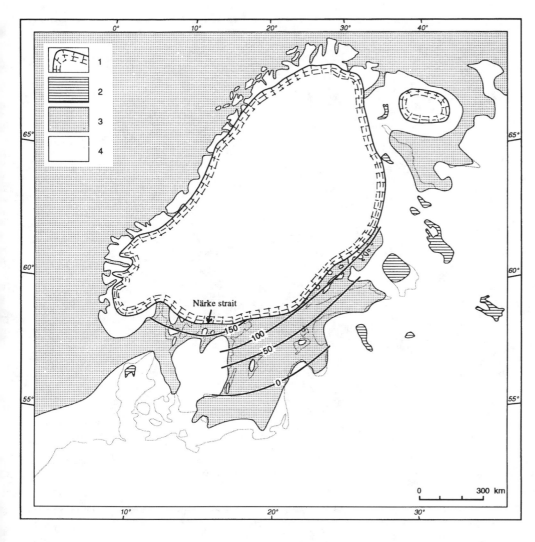

Figure 13.3. Yoldia Sea about 10.2 ka B.P. with isobases. 1, ice; 2, lake; 3, sea; 4, dry land.

The map of the Yoldia Sea (Fig. 13.3) shows its extent shortly after the drainage of the Baltic Ice Lake when the Närke Strait in central Sweden had already opened, and is essentially based on the map by Eronen and Haila published in the *Atlas of Finland* (Alalammi, 1992), but with minor revisions particularly in the eastern parts, taking into account the reconstruction of the deglaciation. An exact synchronous position of the ice margin for the whole of Scandinavia is, however, difficult to reconstruct as the ice recession at this time was comparatively rapid. The shoreline was considerably lower within the Baltic in the beginning of the Yoldia Sea stage than toward the end of the Baltic Ice Lake, mainly because of the drop of the level of the ice-dammed lake by 27–28 m, but also because of a rapid relative lowering of the water level as a result of the isostatic uplift of land in the beginning of the fast thinning and withdrawal of the ice sheet. The isobases for the highest position of the Yoldia Sea are also shown in Fig. 13.3. Lake Ladoga was at this time isolated from the Gulf of Finland and was restricted to the northern parts of the present lake. At the same time as the water level within the Baltic basin dropped, there was a considerable

eustatic rise in sea level (Jelgersma, 1979), which in the northeastern parts of Scandinavia resulted in the White Sea basin becoming connected with the ocean and in the North Sea area resulted in a submergence of formerly dry areas, such as the areas west of Jutland in Denmark.

The connection between the Baltic basin and the sea that had opened through the Närke Strait over the Lake Vänern basin to the west soon became more and more narrow and shallow as a result of the isostatic uplift. Thus the Yoldia Sea stage was followed by that of the Ancylus Lake, so called after the fresh-water gastropod *Ancylus fluviatilis* found in the littoral deposits of the island of Gotland and also on the island of Öland and in Estonia. Lake Vänern was initially part of the Ancylus Lake, which had outlets in the west at Otteid and Uddevalla and southwest through the Göta Älv valley (Fig. 13.4), but later the outflow of the lake was regulated by the Närke–Degerfors straits. The "Svea River" outlet at Degerfors, forming a threshold of the Ancylus Lake, did not, however, have a waterfall toward the west as was earlier assumed but merely acted as part of the straits where the water level on each side of

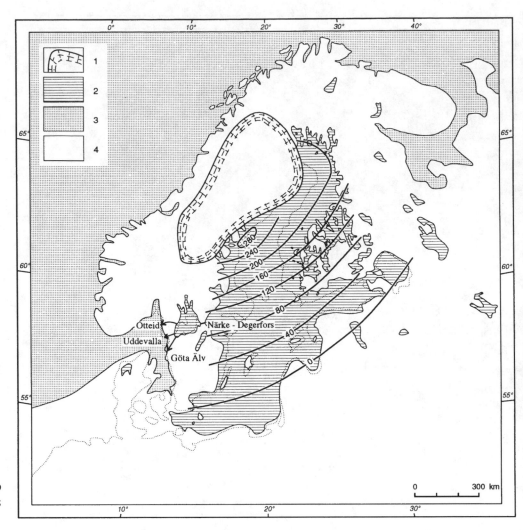

Figure 13.4. Ancylus Lake about 9 ka B.P. with isobases. 1, ice; 2, lake; 3, sea; 4, dry land.

them was in equilibrium, until the connection between the Baltic, the Lake Vänern basin, and the sea was finally closed (Fredén, 1967, 1980, 1981, 1988; Björck, 1987; N.-O. Svensson, 1989). As a result of the narrowing of the outlet of the Ancylus Lake in central Sweden the water level of the Baltic started to rise in the areas which were south and southeast of the isobase of the threshold at 113 m above present sea level (Lundqvist, 1965). The onset of this transgression at coastal sites has been used in the definition of the beginning of the Ancylus Lake stage, as mentioned in summaries of the history of the Baltic (Eronen, 1983; N.-O. Svensson, 1989). It is not an exact boundary and cannot stratigraphically be defined for the whole Baltic basin, especially as the water level started to rise in the southern part of the Baltic basin immediately after the drainage of the Baltic Ice Lake. The transgression has, however, been recorded at a number of coastal sites with organic sediments underlying beach deposits and in lake basins where the stratigraphical changes reflect a temporary rise in water level. At the latter sites detailed reconstructions of the timing of the transgression have been possible with the help of pollen analysis and a study of diatoms.

The dating at all these sites has been based on radiocarbon ages. The radiocarbon ages, however, show a great scatter, as already seen from the comparison of the dates obtained from Sweden given by N.-O. Svensson (1989) and by Haila and Raukas (in Raukas & Hyvärinen, 1992) of dates from southern Finland, Estonia, and Latvia. Haila and Raukas concluded that the transgression started at about 9.6 ka B.P., culminated at about 9 ka B.P., and came to an end about 8.5 ka B.P., a conclusion in general agreement with the recent results from Sweden (N.-O. Svensson, 1989). In contrast to the scatter of the radiocarbon dates, the biostratigraphical evidence of the Ancylus transgression clearly shows that it started in the Preboreal chronozone and ended in the Boreal chronozone, as seen in the study of a number of sites in southern Finland (Ristaniemi, 1984; Ristaniemi & Glückert, 1987). The relative rise of water level during the transgression was about 10 m in the area of the Gulf of Finland, less at the south coast of Finland, but in the southern parts of the Baltic basin it exceeded 30 m (Kolp, 1986). These stratigraphical investigations combined with a study of the beach formations formed as a result of the rise in water level have been used in reconstructing the extent of the Ancylus Lake at about 9 ka B.P. The map in Fig. 13.4 is, as is the map of the Yoldia Sea, mainly based on the map by Eronen and Haila (Alalammi, 1992). In the northern parts of the Baltic basin the position of the shoreline of the Ancylus Lake is based on a number of studies of the lowering of the water level, which because of the isostatic uplift and the positions of the outlets of the Baltic was not interrupted by a transgression. Large coastal areas were still submerged during the time of the Ancylus Lake, including central Finland, and Lake Ladoga formed a bay of the Ancylus Lake after earlier having been a separate lake (Kvasov, 1979; Delusin, 1991). In the southern parts of Lake Ladoga, as in the southern parts of the Baltic, the shoreline of the Ancylus Lake was below the present water level.

The offshore sediments of the Ancylus Lake mainly consist of a homogeneous black sulphide clay with a sharp upper boundary to an overlying gray clay. Whereas this boundary can be used as marking the end of the Ancylus Lake stage, the older boundaries between sedimentary facies are time-transgressive in the Baltic area as a result of the proximity to the retreating ice margin (Winterhalter et al., 1981). Of the diatom species recorded in the Ancylus sediments, *Melosira arenaria* is common in littoral sediments and *Melosira islandica* ssp. *helvetica* and *Stephanodiscus astraea* are frequent in the plankton assemblages of the offshore sediments. As some of the clays, now referred to as early Ancylus Lake sediments, contain brackish-water diatoms, they were interpreted to indicate that there was a separate transitional stage before the Ancylus Lake stage, called the Echeneis Sea stage after the diatom *Campylodiscus echeneis* (Sauramo, 1958). Later investigations have, however, shown that the brackish-water diatoms in the clays, as in some of the older Yoldia clays, are most likely redeposited from marine Eemian sediments (Eronen, 1974), and that there is no reason for adding an Echeneis stage between the Yoldia and Ancylus stages.

The transgression of the Ancylus Lake culminated, as mentioned above, at about 9 ka B.P., at the time of the final closing of the outlets at Degerfors and the Närke Strait (Fredén, 1981, 1988). During the Ancylus transgression the Närke–Degerfors straits had become too shallow to be able to let out the waters from the Baltic (Björck, 1987). When the outlets were finally closed in central Sweden, the Baltic found a new outlet in the south and the level of the Ancylus Lake fell (Table 13.1). During the Ancylus transgression the water level rose above the Darss threshold ("Darsser Schwelle") at −19 m between Germany and the Danish island of Falster and the waters eroded a channel to the Bay of Mecklenburg, from where they found an outlet to the sea through the Great Belt, with a drop in water level as a result (Krog, 1965, 1968, 1973; Ludwig, 1980; Kolp, 1986; Björck, 1987; N.-O. Svensson, 1989). The rise in sea level, flooding the fresh water basins in the Great Belt, eventually reached the level of the falling Ancylus Lake and an equilibrium between the levels of the Baltic basin and the ocean was reached. The rapid regression of the Ancylus Lake that thus came to an end, dated

at 8.5 ka B.P., has been used as marking the end of the Ancylus Lake stage (Table 13.1).

The decreasing isostatic uplift coupled with a relatively rapid eustatic rise of the ocean level resulted in the widening of the connection between the Baltic and the sea. Salt water could therefore penetrate into the Baltic basin through the Danish Straits and through Öresund between Denmark and Sweden, and the Baltic became brackish. Eventually, when the straits widened even more, it became marine. The changeover to brackish conditions is older in the southwestern part of the Baltic than elsewhere and thus time-transgressive. This fact is demonstrated by the dates obtained for changes in the diatom flora or for the first appearance of some mollusk species requiring brackish water. In the general division of the evolution of the Baltic, the Ancylus Lake stage was followed by the Litorina Sea stage named after the gastropod *Littorina littorea*, requiring a water salinity of at least 8.1 ‰, slightly more than the other *Littorina* species *Littorina saxatilis* var. *rudis*, which also invaded the Baltic and requires a salinity of at least 7.9‰ (Eronen, Uusinoka, & Jungner, 1979). These gastropods invaded the southern and central parts of the Baltic only after about 7 ka B.P., at a time when the salinity reached its highest values during the Flandrian. This was at a time when the Danish Straits and Öresund were broader than at present because of the highest position of the Flandrian sea level during the climatic optimum. Later the straits became more narrow as a result of continuing uplift and a slightly falling sea level. Because of the changes in the Baltic during the Litorina Sea stage Munthe (1910) divided the stage into substages, separating the Mastogloia Sea as an initial transitional substage before the Litorina Sea proper, which was succeeded by the Limnaea Sea (Fredén, 1980). A fourth substage, the Mya Sea, was later added to this division (Hessland, 1945). Here, as generally in the literature on the history of the Baltic, the original spelling of Litorina has been maintained for the Litorina Sea, even if it differs from the spelling of the name of the mollusk.

The transitional Mastogloia Sea substage was named after littoral brackish-water diatoms of the genus *Mastogloia,* of which *M. braunii* and *M. smithii* were particularly mentioned by Sauramo (1958), even if *M. elliptica* is also frequently represented in the same assemblage (Mölder & Tynni, 1973). *Mastogloia* diatoms invaded the southern parts of the Baltic already about 8.5 ka B.P. (Berglund, 1964), but spread to the Stockholm area, southern Finland, and Estonia around 8 ka B.P. (Hyvärinen et al., 1988) and to the coast of Ångermanland in northern Sweden about 7.5 ka B.P. (Fredén, 1980). Even if the appearance of *Mastogloia* diatoms in littoral sediments at coastal sites is time-transgressive, a homogeneous clay, clearly distinguishable from the sulphide clay of the Ancylus Lake, was deposited in the deeper parts of the Baltic during the Mastogloia substage, with a distinct upper boundary to the muds of the Litorina Sea proper that contain 10–15 percent organic matter. The Mastogloia Sea sediments can thus be identified in the offshore sediments as a synchronous stratigraphic unit in contrast

Figure 13.5. Litorina Sea about 7.5 ka to 7 ka B.P. with salinity values and with isobases, which outside the Baltic are those for the Tapes I shoreline. Symbols: 1, lake formed during Ancylus Lake stage before 8.5 ka B.P.; 2, lake formed after Ancylus Lake stage; 3, sea; 4, dry land. The numbers for the lakes refer to those in Table 13.2.

to the older sediments of the Baltic Ice Lake, Yoldia Sea, and Ancylus Lake (Winterhalter et al., 1981). But as there is no clear evidence of brackish water conditions in the deeper parts of the Baltic, Winterhalter et al. extended the Ancylus Lake stage to cover the Mastogloia Sea substage. They placed the upper boundary of Ancylus at the changeover to the sediments of the Litorina Sea proper.

Even if the Litorina Sea proper receives its name from the *Littorina* gastropod its identification in littoral sediments is based on diatoms. With the increase in salinity a "Clypeus flora" invaded the shallow waters – chiefly represented by *Campylodiscus clypeus*, but with several other brackish-water species. The first appearance of the Clypeus flora in shallow-water sediments is stratigraphically clearly defined and can be used as the lower boundary of the Litorina Sea proper. But even this boundary is time-transgressive, its age being 7.5 ka to 7.3 ka B.P. in southern Finland and about 7 ka B.P. in the northern part of the Gulf of Bothnia (Eronen, 1974, 1983), in agreement with ages obtained in Sweden (Fredén, 1980). In the southwestern parts of the Baltic it is about 8 ka B.P. (Hyvärinen

et al., 1988). By studying a number of lake basins it has been possible to determine the highest altitude at which sediments with a Clypeus flora occur and in this way trace the "Clypeus limit" of the Litorina Sea. This reaches an altitude of about 120 m in the area in Sweden with the highest isostatic uplift and about 100 m in parts of the Finnish coast in the northern parts of the Gulf of Bothnia, as seen on the map in Fig. 13.5 drawn after a similar map by Eronen and Haila (Alalammi, 1992). In the southern parts of the Baltic, including the southern parts of Finland, the Flandrian rise of sea level overtook the land uplift and resulted in one or many Litorina transgressions, as seen from the land/sea-level curves to be discussed in Section 4. In the area with transgressions the stratigraphical horizon of the Clypeus limit is found in sediments below the altitude to which the transgression(s) reached and thus also below the highest limit of the Litorina Sea. This makes the upper limit of the Litorina Sea even more diachronous than the ages of the Clypeus limit would suggest. The map in Fig. 13.5 thus shows the areas that were submerged by the Litorina Sea during approximately a time between 7.5 ka and 7 ka B.P. At this time, during the Flandrian climatic optimum, the Baltic Sea had its greatest salinity as judged from the distribution of the mollusks. The salinity values (after Hyvärinen et al., 1988) given in Fig. 13.5 for the Litorina Sea differ from the present ones by being about 5 ‰ higher in the inner parts of the Gulf of Bothnia and Gulf of Finland and perhaps up to 10 ‰ higher in the southern parts of the Baltic. It was both the relatively warm coastal waters and the high salinity that furthered the spread of the marine or brackish water fauna in the Litorina Sea.

With the narrowing of the Danish Straits and Öresund the salinity decreased. The immigration into the Baltic of the freshwater gastropod *Lymnaea peregra* f. *ovata,* earlier referred to as *Limnaea ovata,* about 4 ka B.P. has been used as the beginning of the Limnaea substage (Fredén, 1980). The gastropod, still present in the Baltic, has also been recorded in Ancylus sediments. The *Littorina* species, on the other hand, disappeared from the Baltic during the Limnaea substage.

The last substage in the Baltic, the Mya stage, was named after the brackish water bivalve *Mya arenaria* that spread to the European coasts about 300 years ago (Hessland, 1945; Fredén, 1980). The Mya Sea can thus be considered to represent the present Baltic Sea. Recent conventional radiocarbon datings of *Cardium edule* shells and an AMS date of *Mya arenaria* from the same faunal assemblage in northern Jutland, however, show that *Mya* was brought to Europe before the voyages of Columbus, presumably by Vikings (Petersen et al., 1992). But the exact time of its spread remains to be determined.

Table 13.1 shows the criteria on which the stages and the substages (or phases) were divided, the main stages being the Baltic Ice Lake, the Yoldia Sea, the Ancylus Lake, and the Litorina Sea, the last subdivided into four substages. The boundaries between the stages and substages are after the Yoldia Sea all more or less time-transgressive, and furthermore differently defined; a chronostratigraphic division based on radiocarbon years could perhaps be used here as in the division of the Flandrian biostratigraphical chronozones. Even if the boundaries between the stages are not always distinct each stage has its own character, even stratigraphically, as seen from the four maps in Figs 13.2 to 13.5. But the spatial differences in the Baltic during the stages are great, with variations between the offshore and the coastal areas, sometimes with a broad archipelago as in southwestern Finland at present. And there are also great differences in the land/sea-level changes between various parts of the Baltic, as will be seen from the curves of these changes presented in Section 4. Even if the broad outlines of the history of the Baltic are known there are still differences in the interpretations of the details in its development, including the interpretation of the numerous dates used. These dates, also used in dating the pollen-assemblage zones, often have a scatter that shows the magnitude of the errors inherent in the method but may also reflect real differences between the ages. The ages quoted in Table 13.1 are approximate dates largely based on averages of a number of radiocarbon dates.

13.3. History of the major lakes

The short-lived ice-dammed lakes were already mentioned in the chapter on deglaciation (Chapter 11, Section 2). In addition to these, the isostatic land uplift resulted in the formation of some major lakes or lake complexes in Sweden and particularly in Finland (Fig. 13.5). The connections of these lake basins to the sea or to the Baltic were broken at different times in the Flandrian due to their positions and altitudes. Further, similar to the Baltic basin, they were tilted because of the differential uplift, and this in many cases led to changes in their outlets. If the original outlet was situated toward the center of isostatic uplift it had, as a result, a rise in the lake level until a new outlet was formed in the part of the basin toward which the basin was tilted and the basin then was partly drained. But several lakes had a number of intermediate outlets before their present outlets were formed. The methods employed in the study of the lakes are the same as those used in the study of the Baltic. Organic sediments submerged under sediments formed during transgressions have been studied through pollen analysis and beach formations have be used to identify the positions of the shorelines. It is especially the shorelines of the upper limit of the lake levels before the drainages that have been well documented because they are morphologically distinct. The chronology has as a rule been used on radiocarbon ages.

Of the two major lakes in central Sweden the larger one, Vänern, has already been mentioned in connection with the history of the Baltic as being part of the early connection with the sea. The Vänern area with the water level well above the present lake level was for a short time part of the Ancylus Lake with outlets at Otteid and Uddevalla in the west and the Göta Älv valley in the southwest, preceded by an inland sea stage (Fredén, 1988). According to a number of radiocarbon dates the Otteid and Uddevalla straits ceased to function as outlets about 9.3 ka B.P., whereas the connection with the Baltic through the Närke Strait came to an end about 9 ka B.P., at the peak of the Ancylus transgression, as noted earlier. At the same time the Vänern basin became isolated from the sea according to the detailed investigation by Fredén (1988). As the outlet was from the beginning in the southwest through the Göta Älv valley the subsequent tilting of the lake basin resulted in a lowering of the water level throughout the lake.

The long and narrow Lake Vättern at 88.5 m a.s.l. had its first outlet in its northeastern end north-northeast of Askersund when it was isolated from the Baltic. This took place about 9.6 ka B.P. according to E. Nilsson (1968), a date close to the age of about 9.7 ka B.P. obtained earlier by Norrman (1964a) on the basis of calculations of the relationship of the gradients and ages of the shorelines of the Vättern basin. The lake was thus formed at the time of, or just before, the beginning of the Ancylus Sea stage. After the first stage in the development of Vättern named "Ancient Lake Vättern" (*Fornvättern* in Swedish), during which there was a transgression in the whole lake, a new outlet was formed in the northeastern part at Motala and the water level dropped by 2 m. This happened according to E. Nilsson (1968) at the time of the Ancylus transgression at 8.5 ka B.P., whereas the opening of the Motala outlet was dated at 7.7 ka B.P. by Norrman (1964a). There is thus some discrepancy in the dating of this event, but the general trend of the water level rise in the lake basin seems to support the younger age suggested by Norrman (1964b) and it was therefore used in Table 13.2. The *Fornvättern* stage ended with the formation of the Motala outlet and a transgression has since that time taken place in Vättern south of the outlet, as there has not been a further change in outlets. At Jönköping in the southernmost part of the lake the transgression of *Fornvättern* was about 25 m and of Vättern after the opening of the outlet at Motala about 20 m, the transgression only having been interrupted by the 2 m drop in level caused by the new outlet at Motala (Norrman, 1964b).

Among the major lakes in Finland, Inari is situated in northernmost Finland close to the Arctic Ocean. Its northeastern part was initially an ice-dammed lake outside the ice margin that had retreated from the Tromsø–Lyngen moraine on the coast, as described in Chapter 11, Section 2. According to the studies of the shorelines (Tanner, 1930; Synge, 1969) and the biostratigraphy of its sediments (Alhonen, 1969) the lake was ice free about 9 ka B.P. – the time of the Ancylus Lake in the Baltic. The ice-dammed lake drained when the ice margin had reached Virtaniemi at its eastern shore, and a short-lived connection with the ocean was opened through the valley of Paatsjoki. The meltwaters from Inari, however, prevented salt water from penetrating into the lake basin, as seen from the freshwater bottom sediments, and the lake was soon isolated from the sea. The tilting of the basin caused a continuous lowering of the water level south of the outlet at Virtaniemi on the eastern side of the lake, furthered by an erosion of the threshold by 5 m. The present lake level, however, is regulated by a dam keeping the water level about 5 m above its natural altitude at 114 m.

The numerous lakes or lake complexes in southern Finland (Fig. 13.5) each had its own often complex history, documented by morphological and biostratigraphical studies. The westernmost of these lakes, Kyrösjärvi, is only 36 km in a north–south direction and has, since it became isolated from the Baltic Mastogloia Sea at about 8 ka B.P. (Alhonen, 1967), had its outlet over the threshold at Kyröskoski to the Kokemäenjoki valley. The isolation from the Baltic is clearly documented in the bottom sediments of the lake. As a result of the tilting of the basin the water level has fallen throughout the lake since the time it became an independent lake. The larger Lake Näsijärvi northeast of Kyrösjärvi was isolated from the Baltic at about the same time, about 8 ka B.P., but it had its first outlet northwest toward the Gulf of Bothnia (Virkkala, 1962; Saarnisto, 1971a). This resulted in a transgression throughout the lake, as seen from submerged tree trunks underneath limnic sediments, until a new outlet, Tammerkoski, was eroded through the Pyynikki esker at Tampere in the southernmost part of the lake. On the basis of the shoreline gradients this is estimated to have taken place about 6.3 ka B.P. (Saarnisto, 1971b), but – according to three radiocar-bon dates from a section with a stratigraphy showing the transgression – an older age of about 7 ka B.P. was suggested (Grönlund, 1982). A fourth date lower down in the profile is, however, clearly too old, which makes the dating somewhat uncertain. A younger age between 5 ka and 6 ka B.P. for the formation of Tammerkoski was, on the other hand, suggested by Alhonen (1981) on the basis of a profile with a number of radiocarbon ages. The estimated age of 6.3 ka B.P. for the formation of the final outlet for Näsijärvi must, because of the scatter of the radiocarbon ages, still be taken as a tentative age. Näsijärvi found its final outlet into Pyhäjärvi and further into the Kokemäenjoki valley, the outlet also for Kyrösjärvi. The relatively small Lake Vanajavesi south of Näsijärvi has from the beginning had its outlet towards the northwest, with the result that there has been a transgression throughout the lake and several mires became submerged by the rising lake level. The transgression was at least 9.5 m in the southeastern part of the lake and the pollen-analytical evidence of this transgression is substantial (Auer, 1924, 1968; Simola, 1963). On the basis of the pollen stratigraphy and the gradient of the shoreline at the time of isolation Vanajavesi became a separate lake about 7.5 ka B.P., at the end of the Mastogloia Sea substage in the Baltic before the influx of salt water marking the beginning of the Litorina Sea. To lessen the effect of the transgression the threshold of Vanajavesi was lowered in 1857 form 82 m to 79.4 m.

Of the larger lakes further east Lake Päijänne and the Lake Saimaa complex, the latter consisting of a number of lakes joined together, had to begin with a common outlet northwest to the Gulf of Bothnia at Kotajärvi to the Hinkuajoki River valley. The different basins of Päijänne were isolated from the Baltic within a short period after the maximum extent of the Ancylus Lake, and the final isolation took place before the Litorina Sea stage, as shown by studies of the shoreline displacement and by the pollen stratigraphical studies combined with a study of diatoms (Tolvanen, 1922; R. Aario, 1965b; Saarnisto, 1971a, 1971b). The isolation has been dated at 8 ka to 7.5 ka B.P., which is during the Mastogloia substage. After this there was a transgression in Päijänne, stratigraphically well documented in the southern parts of the lake, where the lake level rose about 10 m. The transgression ended when a new outlet was formed in the southeastern part 6.1 ka B.P. at Heinola to the present Kymi river, which flows into the Gulf of Finland. The extent of the "Ancient Lake Päijänne" (*Muinais–Päijänne* in Finnish), the lake at its largest before it was drained, can be traced with the help of its shoreline now above the present lake level.

The development of the Lake Saimaa complex is essentially similar to that of Päijänne but had some intermediate outlets before its final outlet was formed in the southeast. Its first outlet at Selkäyslampi in the northwest, an outlet via Päijänne to the Gulf of Bothnia, was formed about 8 ka B.P. at the time of the Mastogloia Sea (Hellaakoski, 1922; Lappalainen, 1962; Saarnisto, 1970). The subsequent transgression resulted in the formation of the large independent lake of "Great Lake Saimaa" (*Suursaimaa* in Finnish), dammed by the ridge of Salpausselkä I in the southeast.

As a result of the rise of water level an outlet was formed at Matkuslampi in the southwestern part of the lake about 6 ka B.P., but it was not until the opening of another outlet further south at Kärenlampi 5.5 ka B.P. that the water level was noticeably lower. Both of these outlets led to the Kymi river, which at that time already functioned as the outlet for Päijänne, as mentioned, but the original outlet in the northwest was for a short time still in use, until 5.4 ka B.P. The water level still continued to rise in the southeasternmost part of the lake until the lake found a new and final outlet through Salpausselkä at Vuoksenniska through Imatra and the Vuoksi river to Lake Ladoga. The drop in water level was 2–2.5 m and has been dated to 5 ka B.P. (Saarnisto, 1970). The transgression can be seen and dated in submerged mires within the lake basin, but has also been recorded in the mire of Linnansuo at Imatra that was flooded when the new outlet was formed. A thin layer of silt was deposited on the surface of the bog and is now covered by peat formed after it was flooded, thus providing a stratigraphical horizon suitable for radiocarbon dating.

The present level of Lake Puulavesi is nearly 20 m higher than the levels of Päijänne and Saimaa on each side of Puulavesi, and it was therefore isolated earlier. Its history has been reconstructed by comparison of the altitudes of its former and present outlets with the general emergence of the area, as deduced from studies of the shorelines (Hellaakoski, 1928; Saarnisto, 1971a, 1971b). Its first threshold at Tammijärvi, over 100 m a.s.l., rose from the Baltic just before 9 ka B.P., from the Ancylus Lake. The outlet was in the northwest and a transgression led to the formation of new outlets, first further south at Vannijärvi and then even further south at Koskipää, resulting in a drop of water level by 3 m. This took place 6.1 ka B.P., at the same time as the final drainage south of Päijänne, the basin into which Puulavesi drained. The present level of Puulavesi is regulated by a manmade outlet in the south at Kissakoski that was dug in 1831–54 and lowered the lake by 2 m.

In addition to these lakes there are two lakes in eastern Finland that have been studied in detail. Both drain into the Saimaa lake complex. Lake Höytiäinen had its forst outlet west at Polvijärvi (Sauramo & Auer, 1928; Saarnisto, 1968). According to the pollen stratigraphical investigations the lake was formed about 9 ka B.P. and was thus isolated from the Ancylus Lake. The tilting of the Höytiäinen basin resulted in a transgression south of its outlet. In 1859, when a canal was being built from the southern part of the lake through the Jaamankangas moraine, the lake waters formed a new channel to the south at Puntarkoski and the water level dropped 9.5 m in a month. Some 170 km² became dry land along the shores and the sediments formed during the transgression became exposed, as well as the shoreline with its various beach formations formed before the drainage. The long Lake Pielinen northeast of Höytiäinen was also isolated from the Baltic about 9 ka B.P., as deduced from pollen-stratigraphical studies combined with radiocarbon dating of the shoreline displacement in the area (Hyvärinen, 1966a, 1966b). The first outlet was in the northwestern end of the lake at Kalliojärvi, but because of a transgression of a few meters within the lake a new outlet was formed in the

southernmost part at Uimaharju already 8.5 ka B.P. As the long lake, now 94 m a.s.l., has since then been tilted toward the southeast there has been a considerable regression in it northern parts, at Nurmes nearly 50 m, thus exposing large areas formerly submerged by the lake.

The history of Ladoga, Europe's largest lake with an area of about 18,000 km², has already been partly dealt with in the description of the development of the Baltic. After having been a part of the Baltic Ice Lake, Ladoga formed a small independent lake separated from the Yoldia Sea. But the Ancylus transgression again opened a connection between the Gulf of Finland and Ladoga's northwestern part, with the threshold at Vetokallio near Heinjoki east of Vyborg (Kvasov, 1979; Delusin, 1991). This northern connection functioned only for a short period during the Ancylus transgression that was tentatively interpreted as having affected the general rise of water level in the southern parts of Ladoga (Delusin, 1991). Ladoga was finally isolated from the Baltic after the Ancylus transgression about 9 ka B.P., or shortly after, with its outlet still at Vetokallio. The tilting of the lake basin resulted in a transgression submerging mires around the shores. The transgression, which reached over 10 m above the present level of Ladoga, had its peak after about 3 ka B.P. (Delusin, personal communication). It was preceded by the discharge of Saimaa through the river Vuoksi into Ladoga 5 ka BP (Saarnisto & Siiriäinen, 1970). But it was not until about 2 ka B.P. that the level of Ladoga was lowered as a result of the erosion of the new outlet through the watershed between the rivers Mga and Tosna, and the formation of the Neva river (Kvasov, 1979; Delusin, 1991). The drop in water level exposed the sedimentary sequences of the transgression at the shores of Ladoga and at the mouths of the rivers flowing into the southern part of the lake from the south and southeast.

The lakes mentioned above are all shown on the map in Fig. 13.5 and the details of their outlets listed in Table 13.2. The lakes are large enough to have been noticeably tilted as a result of the land uplift. Many had, as seen, their first outlet toward the depressed area in the center of the Scandinavian glaciation. When this area rebounded during and after the deglaciation the lake basins were tilted away from this center and new outlets were formed. This lowered the water level in most lakes; of the 12 lakes mentioned, transgression continues only in lakes Vättern and Vanajavesi. It can thus be seen that when the areas that had been submerged under the Baltic or the sea had risen above water level, most of the large lakes formed as a result of this have later diminished in area. The total effect has been an increase in the area of dry land. Added to this fact is the overgrowth of all lakes, including the numerous small lakes, reducing their sizes or resulting in their being completely covered by mires. The interplay between the previously discussed climatic changes recorded in the mires as water-level changes (see Fig. 12.3) and water-level changes caused by the tilting of the larger lake basins has not been especially studied, even if it is likely that some transgressions in the lakes were at least partly induced by the known changes in humidity.

Table 13.2. *Approximate ages for the formation of the outlets of the major lakes shown in Figure 13.5*

Lakes	Radiocarbon ages B.P.	Outlets
1. Vänern	9 ka B.P.	Göta Älv
2. Vättern	9.7–9.6 ka B.P.	Askersund
	7.7 ka B.P.	Motala
3. Inari	9 ka B.P.	Virtaniemi–Paatsjoki
4. Kyrösjärvi	8 ka B.P.	Kyröskoski–Kokemäenjoki
5. Näsijärvi	8 ka B.P.	Lapuanjoki
	6.3 ka B.P.	Tammerkoski–Kokemäenjoki
6. Vanajavesi	7.5 ka B.P.	Kuokkalankoski–Pyhäjärvi–Kokemäenjoki
7. Päijänne	7.5–8 ka B.P.	Hinkuajoki
	6.1 ka B.P.	Konnivesi–Vuolenkoski–Kymijoki
8. Saimaa	8 ka B.P.	Pielavesi–Päijänne
	6 ka B.P.	Matkuslampi–Kymijoki
	5.5 ka B.P.	Kärenlampi–Kymijoki
	5 ka B.P.	Vuoksi
9. Puulavesi	9 ka B.P.	Tammijärvi
		Vannijärvi
	6.1 ka B.P.	Koskipää
	A.D. 1831–1831	Kissakoski
10. Höytiäinen	9 ka B.P.	Polvijärvi
	A.D. 1859	Puntarkoski
11. Pielinen	9 ka B.P.	Kalliojärvi
	8.5 ka B.P.	Uimaharju–Pielisjoki
12. Ladoga	9 ka B.P.	Vetokallio
	2 ka B.P.	Neva

Swedish	älv	= river
Finnish	lampi	= pond, small lake
"	järvi	= lake
"	joki	= river
"	koski	= rapid

13.4. Shoreline diagrams and curves of land/sea-level changes

The isobases shown on the maps of the stages in the history of the Baltic Sea give an idea of the domelike uplift of the formerly glaciated area in Scandinavia, reflected also in the isobases of the contemporary uplift. The isobases of the Baltic stages cannot be extended to cover the areas bordering the ocean in the west and north for the times when the Baltic formed an independent lake and had a water level above that of the ocean, that is, for the Baltic Ice Lake and the Ancylus Lake. It is only the isobases for the Litorina Sea which can be extended to western Scandinavia where the Litorina shoreline corresponds in time with the Tapes I shoreline, which is conspicuous especially in areas where it was formed as a result of a transgression, as in many parts of Norway and in Denmark (Fig. 13.6). Some of the isobases for this approximately 7 ka-old shoreline were therefore

in Fig. 13.5 extended from the Baltic to the coasts of Denmark as presented by Hansen (1965), and to southern Norway as drawn by O. Holtedahl (1960) and B. Andersen (1965). The isobase for 20 m in Finnmark in northern Norway was drawn according to Marthinussen (in O. Holtedahl, 1960) and Sollid et al. (1973). The Tapes isobases, added to those of the Litorina Sea in the Baltic, show the pattern of land uplift, as do the isobases on more generalized maps from which the isobases were extrapolated over the whole area of uplift, including areas that were never submerged (Eronen, 1974).

In studies of former shorelines, often identified on the basis of raised beaches, diagrams have been constructed to show their tilt outwards from the center of uplift toward the marginal areas of uplift. The most commonly used diagram is a distance diagram, also called an equidistant diagram, in which the x-axis or abscissa gives the distances between the sites studied, at right angles to the isobases for a particularly well-defined shoreline or to the isobases for the recent uplift of land. The y-axis or ordinate gives the altitudes of the shorelines presented as tilted lines. In addition, another type of diagram has been used in which the x-axis gives the altitude of a particular shoreline used as a reference level in plotting the altitudes of all other shorelines. Because this diagram shows the altitudes of the shorelines in relation to one particular shoreline the diagram has been called a relation diagram or – by Tanner (1930) – an epeirogenetic spectrum. As there is not a constant relationship between the shorelines in a large area, as already pointed out by Ramsay (1926) and also by Tanner (1930), and as the relation diagram does not give the true shape of the shoreline nor direct information about their tilting, its use has been limited in Scandinavia, as mentioned by Donner (1965b) in a discussion, with examples, of the merits of the two types of diagrams. Relation diagrams were, however, commonly used in general summaries of the uplift by Sauramo (1934, 1958) and Hyyppä (1937, 1966) after the early works by Ramsay (1926) and Tanner (1930). The presence of hinge lines in the form presented by Sauramo in relation diagrams, particularly in his major work on the history of the Baltic (Sauramo, 1958), have not been demonstrated in later shoreline investigations in which the biostratigraphical evidence has been backed by radiocarbon dating in the determination of the former positions of the shorelines (Donner, 1987). But, as will be seen later, there are indications of the presence of hinge lines in the marginal parts of the area of uplift.

Thus, only distance diagrams have been used in recent shoreline investigations. But even in them it is difficult to draw lines for synchronous shorelines over great distances. This is particularly true for shorelines formed at times when there was a transgression in the marginal parts of the area of isostatic uplift as a result of a eustatic rise of ocean level, as in the case of the Litorina–Tapes shoreline, or as a result of a rise within the Baltic basin, as in that forming the Ancylus shoreline. The ages of these shorelines decrease toward the marginal parts of uplift, whereas the highest beach level is time-transgressive because of the ice

Figure 13.6. Beach ridges in Porsangerfjord, Norway. The best-developed ridge in the middle represents the Tapes shoreline. (Photo B. Andersen)

recession, the age decreasing toward the center of uplift. These nonsynchronous shorelines were originally misleadingly called metachronous (Ramsay, 1921) instead of diachronous.

In constructing distance diagrams for smaller areas the shorelines can normally be drawn as tilted straight lines, but in diagrams for the marginal parts of uplift or across the whole area of uplift this is not possible because the shorelines follow the domelike shape of the uplifted area. This is seen in the schematic diagram of a few of the best identified shorelines along a 1,090 km-long profile from Estonia in the southeast over Finland and Sweden to the Norwegian coast in the northwest, over the central area of uplift and broadly perpendicular to the isobases (Fig. 13.7; for direction of diagram see Fig. 13.8). The diagram is a corrected and simplified version of a diagram constructed earlier (Donner, 1980), which in turn was a corrected version of an even earlier diagram (Donner, 1969b) constructed on the basis of more detailed diagrams. The observations used for the profile and projected onto the baseline were in Estonia from a 130 km-wide area, in Finland and Sweden from a 90 km-wide area, and in Norway from an area 125 km wide, from a comparatively narrow strip to minimize the effect of errors caused by projecting the observations onto a single baseline. The methods used in identifying the shorelines in Fig. 13.7 vary. The older shorelines were identified with the help of marginal deltas or sandur-deltas in Finland, Sweden and Norway and by beach formations in these areas as well as in Estonia, the chronology based on the previously mentioned revised varve chronology. In Norway radiocarbon ages of marine shells were used. The altitudes of the younger shorelines in the Baltic area, especially in Finland, were determined with the help of pollen analysis, diatoms, and

radiocarbon dating of sediments in lake basins and mires at various altitudes (Donner, 1980). The diagram is thus constructed on the basis of a variety of criteria. In its southeastern part it shows the levels of the B I and B III shorelines of the Baltic Ice Lake and the subsequent Yoldia shoreline, as well as the Ancylus and Litorina shorelines, both transgressive in Estonia and southern Finland, the diachronous character of the latter shown by two lines of which the upper one is younger. The shorelines of 9 ka B.P. and 7 ka B.P. rise toward the Gulf of Bothnia. In Sweden the altitude of the older could be determined on the basis of the shoreline displacement during the final deglaciation, as dated by varved clays. The earlier-used results based on studies by Hörnsten (1964) were correlated by taking into account the revised varve chronology (Cato, 1985). At the Norwegian coast the highest shoreline in the diagram is the Main shoreline, P_{12}, formed at the time of the earlier described Tromsø–Lyngen moraine.

The general shapes of the shorelines in Fig. 13.7 reflect an asymmetric dome-shaped uplift with the summit of the dome in the eastern part of the Scandinavian mountains that had the last remnants of ice (Donner, 1969b). Further, the shorelines in Norway at the Atlantic coast have a steeper gradient than those in the marginal part of the Fennoscandian Shield in the southeast. There is in the latter area a clear difference in the gradients of the shorelines of southern Finland as compared with those in Estonia (Donner, 1970), a difference found also further east between the shorelines of Lake Saimaa and those of Lake Ladoga (Saarnisto, 1970; Saarnisto & Siiriäinen, 1970). It was therefore concluded that there is a hinge line or hinge zone in the area of uplift from the Gulf of Finland northeastwards to the

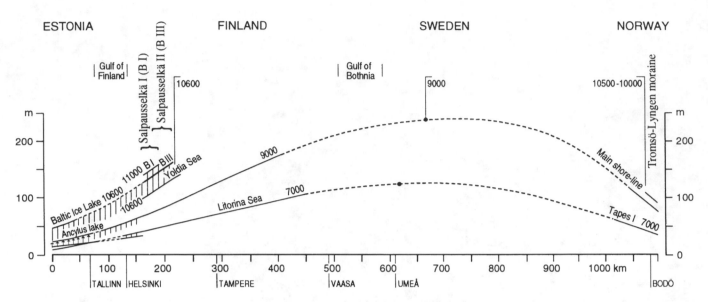

Figure 13.7. Shoreline diagram across Fennoscandia along line shown in Figure 13.8 (A–B). The positions of the shorelines of the Baltic Ice Lake and Ancylus lake above sea-level are shown with vertical lines (after Donner, 1980; Fig. 5, with corrections).

area north of Lake Ladoga, a zone at least 400 km long (Donner, 1970, 1980). The shorelines would as a result of this have been bent along this hinge. In addition, Jantunen (1990; Miidel & Jantunen in Raukas & Hyvärinen, 1992) demonstrated the presence of a fault line, visible in the bedrock, from the coast west of Helsinki to Lahti and further northeastwards, partly parallel to the hinge from the Gulf of Finland to the area north of Lake Ladoga (Fig. 13.8). The fault line also to some degree affected the land uplift and acted as a hinge, even if it cannot be seen in the general shoreline diagram in Fig. 13.7. There has thus been a revival of the previously mentioned ideas of Sauramo (1958) about hinge lines affecting the shorelines in the southeastern marginal area of uplift, but the effect of these hinges, if real and still active after deglaciation, on the shape of the shorelines does not seem to have been as great as earlier assumed by Sauramo. The hinge related to a fault line is mentioned here because it crosses the line for the shoreline diagram in Fig. 13.7; other faults formed during and after deglaciation will be noted in Section 5.

As raised beach formations occur throughout the area of uplift in the formerly submerged areas they have been extensively used for the determination of shoreline positions at the emerging coasts. The raised beaches are particularly conspicuous in northern Scandinavia where their study is furthered by the open treeless vegetation; further south in the Baltic region the forest cover as well as the generally more even landscape make the identification of former shorelines more difficult. Some of the most detailed investigations of raised beaches in Norway have been made in the northernmost parts; Marthinussen (1974) separated a number of shorelines in addition to the Tapes beach and the main shore-lines, which he here named S_O, in the Varangerfjord area. And Sollid et al. (1973) similarly produced

diagrams from three transects in Finnmarken, their L_o shoreline corresponding to the S_O line in Marthinussen's diagram, which as mentioned corresponds to the P_{12} line further south, even if this line in all investigations is called the main (shore-) line. An earlier detailed diagram for West Finnmarken was constructed by Marthinussen (O. Holtedahl, 1960), in which the main shoreline was still called P_{12}. Because it corresponds in age to the Tromsø–Lyngen moraine and is the youngest Late Weichselian shoreline it can be morphologically singled out from the Early Flandrian shorelines formed during a relatively rapid emergence. Of other studies in Norway the investigation by B. Andersen (1968) of Western Troms in North Norway of areas both inside and outside the Tromsø–Lyngen moraine may be mentioned, as well as the detailed study of Sørlandet in southernmost Norway, also by B. Andersen (1960), in which the land/sea-level changes were related to the deglaciation.

Of the studies in Sweden where raised beaches have been used in reconstructing former positions of the shoreline the study by E. Nilsson (1953) of southern Sweden, including the island of Gotland (Fig. 13.9), covers the largest area. E. Nilsson constructed a number of separate distance diagrams and a main diagram along a 480 km-long line in a south–north direction from southernmost Scania to the area north of the Middle Swedish moraines. In this diagram 77 different shorelines were separated on the basis of observations from over 100 sites. Whereas E. Nilsson used distance diagrams, the shoreline diagrams have mainly been relation diagrams in Finland, as noted before, even in studies where the shorelines have been reconstructed mainly on the basis of the distribution of raised beaches (Fig. 13.10). Examples of such diagrams are the ones of the shorelines on the slopes of Salpausselkä I in Lohja (Glückert, 1970), of southwestern Finland generally (Glückert, 1976) and of

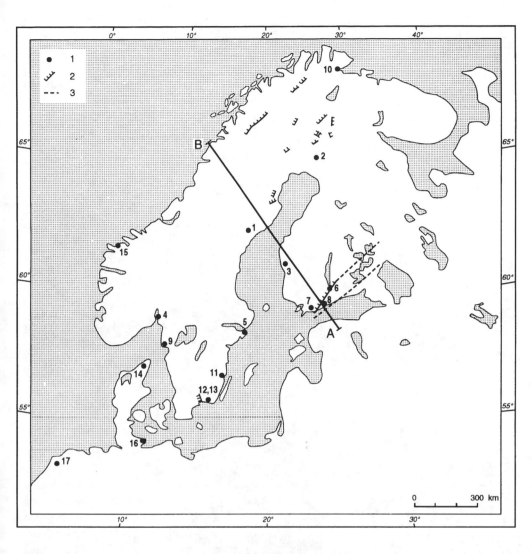

Figure 13.8. Direction of baseline (A–B) for shoreline diagram in Figure 13.7. Symbols: 1, site listed in Table 13.3 for curve of land/sea-level changes shown in Figure 13.11; 2, fault line with direction of downthrow; 3, fault zone.

Åland (Glückert, 1978). In his study of the Lake Saimaa complex, Saarnisto (1970), on the other hand, used distance diagrams, of which the longest showing the former lake levels covered a distance of 320 km. In another diagram for southern Saimaa, over a distance of 100 km, several shorelines were identified on the basis of ice-marginal formations and raised beaches, and their distribution in relation to the deglaciation.

In addition to these shoreline diagrams several others have been constructed for more limited areas in Scandinavia. They all show how the gradients of the shorelines decreased with time, which led to the use of gradient/time curves in estimating the ages of individual shorelines. In Finland this method was used by Ramsay (1926) for the shorelines of southern Finland and by Lappalainen (1962) for the shorelines of Lake Saimaa; in Sweden for the shorelines of Lake Vättern (Norrman, 1964a); and for the shorelines in Norway first by Grönlie (1941) and later by others such as Kaland (1984). A clear difference could, for instance, be demonstrated between the gradient/time curves for the shorelines in southern Finland and those in Estonia (Donner, 1970). In contrast to these curves, considered to represent

exponential curves, other gradient/time curves have been presented that show that the rate in the changes of the gradients of the oldest shorelines was initially relatively slow and then increased before a slowing down of the rate as in the other curves (N.-O. Svensson, 1989). This would indicate, according to N.-O. Svensson, that the greatest rate in shoreline tilting was delayed and took place some time after deglaciation.

From these few examples it can be seen that a detailed reconstruction of the land/sea-level changes with the help of shoreline diagrams still depends on the interpretation of the material used in these diagrams. As a result, there are a number of different diagrams, but in most of them the main outlines are the same. In addition to shoreline diagrams, which depict the warping of the earth's crust as a result of the uplift, curves have been used to show the relative land/sea-level changes in particular areas. By comparing the curves with one another, regional differences can be demonstrated. A number of curves published comparatively recently and based on radiocarbon dating, were collected from various parts of the area of the Scandinavian glaciation, including a curve from The Nether-

Figure 13.9. Sea-stacks in limestone at Litorina level in Boge near Slite on the island of Gotland, Sweden. (Photo J. Donner)

Figure 13.10. Highest beach at 208 m a.s.l., representing the Ancylus Lake about 9 ka B.P. Aavasaksa west of Rovaniemi in northern Finland. (Photo J. Donner)

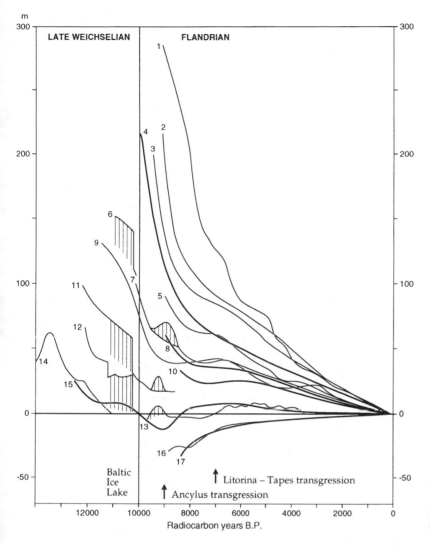

Figure 13.11. Late Weichselian and Flandrian curves of land/sea-level changes. Numbers refer to sites listed in Table 13.3 and shown in Figure 13.8. The positions of the water level above sea level in the Baltic at times when it was a lake are shown with vertical lines.

lands, as shown in Fig. 13.8, and presented in Fig. 13.11. Similar comparisons of land/sea-level curves have been made earlier by, for instance, B. Andersen (1960). The details of the areas used here are listed in Table 13.3 with references. The curve for Vendsyssel in Denmark goes further back in time than the curves from Norway, Sweden, and Finland. Northern Jutland was submerged in Late Weichselian time and the land/sea-level curve was constructed on the basis of radiocarbon dates of mollusks from the beds of marine sediments, from the Yoldia Clay and the overlying Upper Saxicava Sands, including the littoral Zirphaea Beds, fauna communities that reflect different water depths at the time of submergence (Petersen, 1984b) and not an increase in water temperature. A separate Zirphaea transgression, earlier assumed, cannot therefore be identified. The Late Weichselian transgression reached its peak of just over 60 m between 14 ka and 13 ka B.P. A Late Weichselian submergence similar to that in northern Jutland has also been demonstrated in those parts of the Norwegian coast that were well outside the Fennoscandian moraines. Thus the sea level was 30–40 m above the present

level after about 18.5 ka until about 15 ka B.P. on Andøya in northern Norway (T. O. Vorren et al., 1988). Curves 6, 11, and 12 in Fig. 13.11 show the Late Weichselian changes in the Baltic during the time of the Baltic Ice Lake. Curve 11 for Oskarshamn in Sweden shows a regression until the drainage of the lake, whereas curve 12 shows a transgression of about 5 m after an early drainage of 15 m in Allerød time (Björck, 1981). Curve 6 for the Lahti area in Finland also records the final drainage of the Baltic Ice Lake, which reached an altitude of 150 m at about 11 ka B.P., a date that, however, is slightly older according to the varve chronology (see Fig. 11.11). There is no conclusive evidence in the Salpausselkä zone of an earlier transgression up to the highest position of the Baltic Ice Lake, as noted before. Among the earlier published Late Weichselian land/sea-level curves, a curve for the St. Petersburg area (Kvasov, 1979) agrees with those shown in Fig. 13.11, whereas another curve for the same area, in addition to some curves for Estonia, Latvia, and Lithuania, shows a transgression before the high position of water level of the Baltic Ice Lake (Donner, 1982). The evidence

Table 13.3. *References to Late Weichselian and Flandrian curves of land/sea-level changes shown in Figure 13.11*

1. N Ångermanland, Sweden (U. Miller, 1986)
2. Rovaniemi–Pello area, Finland (Saarnisto, 1981)
3. S Ostrobothnia, Finland (Salomaa & Matiskainen, 1985)
4. Ski area, Norway (Sørensen, 1979)
5. Söderörn, Sweden (Risberg, U. Miller, & Brunnberg, 1991)
6. Lahti, Finland (Donner, 1982, with changes from Donner, 1992, and with corrected dates)
7. Salpausselkä I area, Finland (Ristaniemi & Glückert, 1988; Glückert, 1991)
8. Helsinki area, Finland (Hyvärinen, 1980)
9. Bohuslän, Sweden (U. Miller & Robertsson, 1988)
10. Varangerfjord, Norway (Donner, Eronen, & Jungner, 1977)
11. Oskarshamn, Sweden (N.-O. Svensson, 1989)
12. Blekinge, Sweden (Björck, 1981)
13. Blekinge, Sweden (Berglund, 1964, 1971)
14. N Vendsyssel, Denmark (Petersen, 1984b)
15. Sunnmøre, Norway (Svendsen & Mangerud, 1990)
16. Heiligenhafen, Germany (Winn et al., 1986)
17. Dutch coastal plain, The Netherlands (Jelgersma, 1979)

for a transgression in these areas is, however, not conclusive. That some areas had early land/sea-level changes different from other areas is shown by the previously mentioned study of the emergence of Hunneberg at the southern part of Lake Vänern in Sweden (Björck & Digerfeldt, 1982b). The curve for Hunneberg shows that the initial regression after 12.2 ka B.P. was relatively slow and was rapid only about 2 ka after deglaciation, thus being different from other areas (Fig. 13.11). One possible reason for the different uplift history of Hunneberg as compared with other areas is that there was extensive faulting in the area during deglaciation (Björck & Digerfeldt, 1982b). It can further be noted that observations from the coast of southwestern Norway suggest that there was a downwarping caused by a 15–20 km Younger Dryas ice advance that resulted in a transgression of 10–12 m (Anundsen & Fjeldskaar, 1983; Anundsen, 1985). This would imply that a downwarping could have taken place in a zone close to the ice margin as a result of an advance, at the same time as there was an uninterrupted uplift further away from the ice margin.

Clearly, from these examples, the land/sea-level changes during the Late Weichselian deglaciation are still inadequately known as the details of tectonic movements and the response of the earth's crust to readvances of the ice have not been fully demonstrated with stratigraphical, well-dated evidence (N.-O. Svensson, 1989, 1991). The drainage of the Baltic Ice Lake is, however, well recorded in the Baltic (Fig. 13.11), the only discrepancies being caused by the small differences in the datings of the event, as mentioned earlier (Table 13.1). In Fig. 13.11 only radiocarbon ages were used and therefore the varve date for the drainage at Billingen was excluded.

The curves showing the Flandrian land/sea-level changes reflect the rapid isostatic uplift in the center of glaciation in contrast to the marginal areas with little or no uplift but affected by the eustatic rise of sea level. This rise is shown by the curves for Heiligenhafen in the southwestern part of the Baltic and for the Dutch coastal plain, both curves rising from about −25 m at 8 ka B.P. to reach the present-day sea level only recently. The shape of the former curve was affected by the early Flandrian history of the southern Baltic before the marine incursion into the area over the Great Belt threshold (Winn et al., 1986). The difference in the smoothness of the Flandrian curves from the Baltic region and the ocean coast partly reflect differences in the interpretations of the stratigraphical evidence in tracing small fluctuations in the land/sea-level changes. In addition, some areas yield more detailed information than others. It is on the whole easier to obtain detailed evidence from the Baltic coasts, where there are practically no tides, than from the tidal coasts of the ocean. Three of the curves from the Baltic, 7, 11, and 13, show the transgression of the Ancylus Lake, with a slight but probably insignificant difference in its dating. The Litorina–Tapes transgression, caused by the interplay between the slowing down of the isostatic uplift and the eustatic rise of sea level, is reflected more or less clearly in many of the curves, but in spite of detailed stratigraphical investigations not, for instance, in the curve for the Helsinki area (8). In many detailed investigations of the Litorina Sea stage, including the Limnaea Sea, at least four transgressions, called L 1–L 4, have been identified in Sweden, as in the Stockholm area where L 1 and L 2 together have a peak at about 7 ka B.P., L 3 was dated at about 5 ka B.P., and L 4 at just under 4 ka B.P., as seen in the curve (5) for Söderörn (Risberg, U. Miller, & Brunnberg, 1991). In Bohuslän (9) there is also a clear transgression peak at about 3 ka B.P. In areas with little uplift the maximum altitude of the Flandrian transgression was reached after the peak of the main Litorina–Tapes transgression at about 7 ka B.P., as seen from the curve for Blekinge (13) in which six smaller fluctuations were recorded. It was with these six transgressions that Aaby (1975) linked the previously mentioned cyclic variations in climate with a periodicity of 520 years, identified in the peats of the Danish raised bogs. Påsse (1983), on the other hand, concluded that the small sea-level fluctuations on the Swedish west coast had a periodicity of 455–460 years. In earlier investigations on land/sea-level changes in Finland several Litorina transgressions were identified, as summarized by Donner (1965a), but these cannot directly be correlated with the transgressions in Sweden because they were described before the general use of radiocarbon dates. In later studies only two Litorina transgressions were identified that could be correlated with two transgressions in Estonia (Donner, 1966), but in more recent studies only one main transgression has been accepted, as seen from the study by Eronen (1974). A similar development can be seen in Norway. Marthinussen (1962), for instance, identified four Tapes transgressions, but in later land/sea-level curves, as in curve 15, there is only one general Tapes transgression.

In contrast to the undulating curves in which the previously noted transgressions are reflected, most curves from the area with the greatest relative uplift of land are smooth and show the slowing down of the rate of emergence. Two of the curves, 2 and 3, are for Finland, whereas curve 4 is for the Ski area near Oslo in Norway. The land/sea-level changes in the Oslo area have been studied in great detail and the well-dated curves for S. Vestfold (Henningsmoen, 1979) as well as other earlier published curves, as summarized by Hafsten (1983), are all similar in shape to curve 4.

If all curves presented in Fig. 13.11 are taken into account it can be seen that they can essentially be divided into two sets. Some of the curves, drawn with thicker lines, show the general trend of the land/sea-level changes, in the center of uplift without any signs of transgressions, but in the marginal parts with a Flandrian transgression caused by the eustatic rise of sea level reflected in the curve for the Dutch coast (17). Several curves from the Norwegian coast have a shape similar to curve 15 for Sunnmøre, as seen from the curves presented by Hafsten (1983) and by Svendsen and Mangerud (1987). In the second set of curves the influence of the Flandrian transgression is stronger, resulting in a temporary rise of sea level even in the Stockholm area (5) and affecting the curve of Ångermanland (1) in northern Sweden as well. They also show smaller fluctuations after the first Litorina–Tapes transgression, fluctuations that have been correlated with one another chiefly in the marginal areas of uplift, as noted above. There are no great differences, however, between various studies as regards the identification of the Ancylus Lake stage, even if there are some differences in its dating and in the estimates of the magnitude of the transgression, as shown by curves 7, 11, and 13.

The differences between the two sets of curves in Fig. 13.11 are not caused by the methods used in the studies of the land/sea-level changes. The chronology is based on radiocarbon dates mainly of pollen-analytically-studied organic sediments in basins with stratigraphical changes that can be linked to relative sea-level changes and also on dates of marine mollusks. There is no obvious geological reason for the differences between the two sets of curves, such as regional variations within the area they represent. It can therefore be concluded that the curves reflect the personalities at play, as also seen in attempts at presenting general eustatic curves. In some of the investigations in Scandinavia the overall trend of changes has been stressed, in others there has been a search for stratigraphical changes that can be interpreted as showing irregularities in the general trend of land/sea-level changes, particularly transgressions. It is thus a question of how good the resolutions of the methods used are considered to be in tracing small fluctuations in the relative sea-level changes. Questioning the validity of the conclusions in detailed investigations may in some cases have led to the presentation of oversimplified curves instead. It can therefore not be said that one set of curves in Fig. 13.11 is more correct than the other. They show only two different approaches, which are both subject to revisions as more detailed stratigraphical evidence is obtained.

13.5. Faults related to land uplift

The pre-Quaternary bedrock is broken up by a great number of fault lines and fault zones forming a mosaic of fault blocks. Most of the faults in the Fennoscandian Shield have a Precambrian origin, but some major faults were also formed during the Alpine revolution in the Tertiary. Some of the faults in Scandinavia have been shown to have been active during or after the last Quaternary deglaciation. Two faults in the Gulf of Finland and the southern and southeastern parts of Finland (Fig. 13.8) have been already mentioned. The hinge line or zone from the Gulf of Finland to the northern shore of Lake Ladoga, demonstrated with the help of shorelines, probably represents a fault zone, whereas the fault line northwest of this zone can clearly be seen in the bedrock topography. The downthrow of this fault is, at least in its western part, to the northwest of this reactivated Precambrian fault, but the land uplift was at one time slower on the southeastern side, even if the contemporary uplift is the same on each side (Jantunen, 1990). This indicates that the uplift of the fault blocks varied in relation to one another at different times. In Blekinge in southern Sweden there is a fault line between the Karlshamn granite and the coastal gneiss that displaced the highest shoreline in the area. The shoreline is about 5 m higher in the granite area than in that of the gneiss, the downthrow thus being to the southwest (Björkman & Trägårdh, 1982). The age of the movement was estimated at 10.9 ka B.P. on the basis of the age of the shorelines. In addition to the examples from the southern areas, several faults have been described from northern Sweden and Finland. After the description of young faults in Finnish Lapland by Kujansuu (1964) the Pärve fault in Swedish Lapland was described by Lundqvist and Lagerbäck (1976). The latter fault is 150 km long with a 10 m-high fault scarp and with the downthrow to the northwest (Fig. 13.12). It is the only one that it has been possible to date directly, because it is related to the deglaciation of the area at 8.5 ka to 9 ka B.P. By comparing the fault with sediments and striae formed during the deglaciation it was concluded that the movement along the fault was diachronous. From the summaries by Kuivamäki and Vuorela (1985) and by Lagerbäck (1990) of the faults reported from northern Sweden, Norway, and Finland, of which the major ones are shown in Fig. 13.8, it can be seen that the downthrows on the whole are away from the center of uplift, to the northwest in northern Sweden. The lengths of the faults in northern Finland shown in Fig. 13.8 vary from 4–5 km to 40 km and the heights of the fault scarps from 1 to 8 m. The faults described as postdating the last glaciation follow old fault lines or fracture zones, usually along boundaries between different rock units, but they also cut through different rock types, as observed about the Pärve fault (Lundqvist & Lagerbäck, 1976). As the fault scarps in Finland occur in glacial sediments the faulting has been dated to the period of deglaciation at about 9 ka B.P., in agreement with the dating of the Pärve fault. There are no signs of faulting after that time (Lagerbäck, 1990).

The faulting of the bedrock during the time of deglaciation in

Figure 13.12. The Pärve fault from about 8.5 ka B.P. northeast of Lake Torneträsk in Swedish Lapland, with smaller parallel faults. (Photo J. Lundqvist)

northern Sweden, Norway, and Finland, resulting in fault scarps of the magnitude described above, was evidently associated with strong earthquakes. As a result there are traces of a great number of landslides in the tills of the areas in which the fault movements took place, as seen from the maps presented of the landslide occurrences (Kuivamäki & Vuorela, 1985; Lagerbäck, 1990). The observations of fault scarps and associated landslides are mainly restricted to the northern parts of Scandinavia (Fig. 13.8). This is at least partly because the features can here – in an area with an open, sparsely forested vegetation – more easily be identified on aerial photographs than further south where the forest cover is dense. It is likely that there were a number of active faults throughout Scandinavia at the time of deglaciation, as a result of the comparatively rapid uplift of land at that time. Later during the Flandrian the seismic activity decreased and it has not been possible to connect younger earthquakes with movements of particular faults in the bedrock. The anomalies found in the contemporary uplift have, however, been taken to indicate that there is a tectonic component in the vertical movement, but that it does not exceed 10–20 percent of the total uplift rate (Kakkuri, 1987). Further, it has been concluded that the seismic events recorded during the past 500 years were at least partly caused by tectonic stresses in the Fennoscandian Shield.

13.6. Marine and freshwater fauna

It can be seen from the earlier chapters that the foraminifera have been successfully used in the study of marine sediments in Denmark and Norway, as well as of sediments from the North Sea area and the northern Atlantic, to identify warm and cold stages as well as substages in the latter. The Flandrian sediments, although relatively thin, can also be identified and clearly have a different foraminiferal fauna than the Late Weichselian sediments, as seen in studies of northern Jutland (Fredericia & Knudsen, 1990) and the North Sea (Feyling-Hanssen, 1982). But foraminifera have also been used in the study of sediments that can be related to land/sea-level changes, in combination with other methods. Thus Feyling-Hanssen (1964; Feyling-Hanssen & Knudsen, 1979) identified one Late Weichselian (A) and six Flandrian (B–G) foraminiferal assemblage zones in the marine clays of the Oslofjord area below the marine limit at 221 m. In these sediments the variations in the abundance of some foraminifera were interpreted as reflecting the changes in water temperature during the Late Weichselian Older Dryas, Allerød, and Younger Dryas chronozones. The changes in the younger Flandrian assemblages were largely influenced by the shallowing of the waters as a result of the land uplift. In Scandinavian studies of foraminifera the species have often been grouped according to the zones in the zoogeographical division shown in Fig. 13.13, applied both to foraminifera and mollusks (Knudsen, 1986b). In general studies of deep-sea sediments, pelagic species of the surface waters have been separated from the benthic species of deeper waters. The planktonic species of the former are those that best reflect the temperature changes of the surface waters, which can be correlated with the terrestrial evidence of climatic changes.

Whereas foraminifera have a limited use in the study of land/sea-level changes, marine mollusks have been used more extensively, with renewed interest after the introduction of

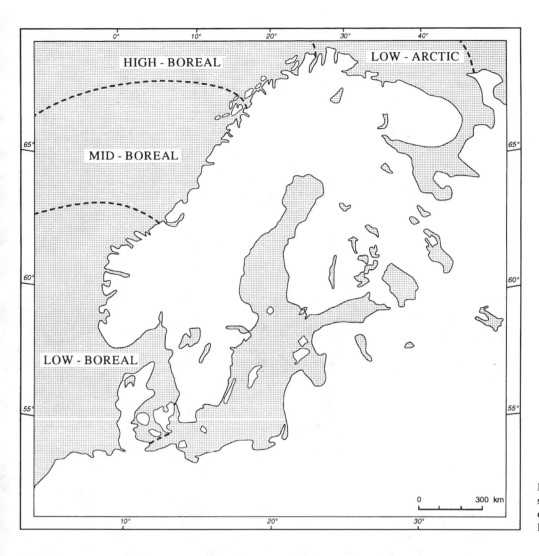

Figure 13.13. Zoogeographical division of the marine fauna of the coasts of northwestern Europe (after Feyling-Hanssen, 1955).

radiocarbon dating. But even earlier, changes in the mollusk fauna were linked with the shoreline displacement. The investigation of the Oslofjord area is perhaps the best example of this. It is an area where, as seen before, the land/sea-level changes have been reconstructed by pollen-analytical studies combined with radiocarbon dating, and where foraminifera have been used in the study of the clays. In the inner Oslofjord area several stratigraphical units of beach sediments below the marine limit at 221 m were separated by Øyen and Brøgger, and by the latter also of clays deposited in deeper water, on the basis of changes in the mollusk fauna (O. Holtedahl, 1960; B. Andersen, 1965). Brøgger's division of beach deposits and clays is shown in Table 13.4 with the taxonomy he used, even if later revisions of the older units have been presented. The essential features of the faunal changes are, however, reflected in the division by Brøgger. In the long list of mollusks in the deposits (O. Holtedahl, 1960) many species have at present a wide depth range as well as a wide range over the climatic zones (Fig. 13.13). The fauna of the Yoldia clay deposited at the time of the Ra moraine has, however, a fauna with *Portlandia (Yoldia) arctica*, a

species at present found only in the high arctic zone, in addition to the arctic species of *Yoldia hyperborea* and *Sipho togatus*. In the younger Arca–Portlandia clay *Arca glacialis* is an arctic–high-boreal species. In the fauna of the Flandrian climatic optimum there are only two species, *Tapes decussatus* and *Tellina crassa*, which are Lusitanian species. On the whole the fauna shows that there was a gradual rise of the water temperature until a rather late maximum in Flandrian time in the Isocardia clay, at the time when the Upper Tapes beds were formed (O. Holtedahl, 1960). Thus the changes in the Oslofjord area of the coastal waters were, according to the mollusks, from arctic conditions at the time of deglaciation to a Lusitanian environment at the time of the climatic optimum, followed by a slight late Flandrian cooling. The correlation of the Yoldia clay with the Ra moraine was confirmed by the radiocarbon dating of *Portlandia arctica* shells from the Yoldia clay (B. Andersen, 1965). In other cases the radiocarbon ages of shells have led to reinterpretations of earlier conclusions based entirely on changes in the marine fauna. The investigation of the already mentioned Late Weichselian sediments in northern Jutland is an example of this. Here the Yoldia Sea in which Yoldia

Table 13.4. *Division by Brøgger of beach deposits and clays in the Oslofjord area based on mollusk assemblages (after B. Anderssen, 1965)*

Moraines	Beach deposits	Clay zones
		Mya arenaria clay
		Scrobicularia clay
	Lower Tapes beds	
	Upper Tapes beds	Isocardia clay
	Ostrea beds	Upper Ostrea clay
	Cardium beds	
		Cardium clay
	Lower Mya beds	
	Upper Mya beds	
		Mytilus & Cyprina clay
	Mytilus gravel	
		Young Arca–Portlandia clay
Aker		Arca clay
Ås–Ski		
		Yoldia clay
Ra		

clay was deposited with, among others, *Portlandica arctica, Hiatella arctica,* and *Mya truncata,* reached its highest position 14 ka to 13 ka B.P. The younger, shallow-water sediments with a temperate fauna, the Zirphaea beds, were deposited at the same time as the Upper Saxicava sands and do not represent a separate transgression (Petersen, 1984b). The importance of these results lies in the realization that different faunal communities existed at different depths at the same time, with the arctic communities in the clays of deeper water. In the light of these conclusions the detailed chronological scheme for the Oslofjord area based on mollusks is likely to be revised. It is particularly the use of molluskan assemblages in shell-bearing beach deposits in studies of land/sea-level changes that have to be questioned, as a result of detailed studies of the deposition of these deposits and the assemblages in them.

In his detailed study of the shell-bearing sediments in Bohuslän on the Swedish west coast Hessland (1943) demonstrated that they are essentially submarine aggradation deposits and cannot therefore directly be linked with raised beaches. This was also taken into account by Fredén (1988) in his study of the paleoenvironments of the marine faunas at the Swedish west coast and further inland in which he, with the help of radiocarbon dating, compared the faunal changes with the deglaciation history. Fredén demonstrated that clays have shell layers of which some are autochthonous with complete bivalves and others allochthonous in which the shells, often broken, have been transported

down into the clays from nearby habitats, but that most shells in shell banks are accumulations in which the shells are not in situ. The shells in these sediments represent mixed death assemblages. The arctic–boreal mollusk assemblages recorded from southwestern Sweden existed according to the radiocarbon ages along the west coast 13 ka to 10.2 ka B.P., in the southern part of the Vänern basin 11.8 ka to 10 ka B.P., and in the western part of the basin 9.9 ka to 9.6 B.P., showing the migration northeastwards after the retreating ice margin. In the study of the shoreline displacement in Finnmark in northernmost Norway (curve 10 in Fig. 13.11) the younger Flandrian regression after 5.5 ka B.P. was dated with the help of radiocarbon ages of shells in beach deposits. By dating different shell species it was demonstrated that the beaches had mixed death assemblages in which only ages of *Mytilus edulis* and *Modiolus modiolus* could be used for dating the regression, whereas shells of other species – such as of *Mya truncata* and *Arctica islandica* – that had been mixed into the beach deposits, gave up to about 2.5 ka older ages (Donner, Eronen, & Jungner, 1977). These results were thus in keeping with the results obtained by Fredén (1988) about shell banks, and with the earlier conclusions by Hessland (1943). In Finnmark, however, the shells dated were from beach ridges formed near sea level, in contrast to the submarine shell banks or aggradation deposits at the Swedish west coast. These examples illustrate how radiocarbon dating has provided an opportunity to study in detail the origin of the mollusk assemblages found in sediments related to land/sea-level changes.

The stages and substages in the Flandrian history of the Baltic were, as previously mentioned (Table 13.1), originally named after mollusks, with the exception of the Mastogloia Sea named after a diatom genus. But in the identification of these stages the mollusks have been of limited use because of their scarcity in littoral sediments. The shells of *Portlandia (Yoldia) arctica* were, as noted, recorded in clays from a limited area in Sweden. The shells of *Ancylus fluviatilis* could, however, after first having been found in beach sediments on the Estonian island of Saaremaa (Ösel) in 1877 by Schmidt (Ramsay, 1900), be used in the identification of raised beaches of the freshwater Ancylus Lake on the Swedish islands of Gotland and Öland, the first shells of *Ancylus* having been recorded on Gotland by Munthe in 1887 (Fredén, 1980). The mollusk fauna in the lower beach sediments of the Litorina Sea were clearly different from that of the Ancylus Lake. The main changes in the Flandrian history of the Baltic can in a few areas be separated solely on the basis of the mollusk faunas in the beach deposits. The emerged coast of Estonia is such an area, where the sediments of the Ancylus Lake could be separated from those of the Litorina Sea and the subsequent Limnaea Sea (Kessel & Raukas, 1980). In this threefold division the freshwater mollusk assemblage of the Ancylus Lake is dominated by *Lymnaea peregra* (65 percent) but with a comparatively strong representation of *Ancylus fluviatilis* (19 percent). The Litorina sediments contain mainly shells of *Hydrobia ventrosa* (31 percent) and *Cerastoderma (Cardium) edule* (28 percent), both still common in the Baltic, but also with a fair proportion of *Littorina* species, which are absent in Estonia in the younger sediments of the Limnaea Sea. The change during the latter stage was gradual, as was the decrease in salinity, and the changeover from the Litorina Sea is only recorded by a marked decrease in the porportion of the *Littorina* species. At the time of the recent invasion of *Mya arenaria* into the Baltic, some other species also spread into the area or were recorded for the first time, such as *Unio timidus*.

The beach sediments in Estonia described above have been related to the land/sea-level changes in that area, but similar stratigraphical schemes based on mollusk faunas have not been established for the mainlands of Sweden and Finland. On the contrary, in the extensive investigation of 159 shell beds in southwestern Finland, Segerstråle (1927) showed that the shells of mollusks that lived in shallow water were later washed down and accumulated in deeper water on the lee side of hills and ridges, and in some places transported further away from their original habitats. These conclusions agree with those made by Hessland (1943) in his study of the Swedish west coast, and also with those made by Fredén (1988). The thickest accumulation of shell beds described by Segerstråle from Finland was 6.5 m thick, but most of them were less than 1 m thick. Some of the shell accumulations had a $CaCO_3$ content as high as 70–80 percent. All shell beds were below the shoreline of the Litorina Sea and contained shells of the bivalves *Mytilus edulis*, giving the beds a violet color, *Macoma (Tellina) baltica*, and *Cerastoderma edule*, in addition to some gastropods, among them the *Littorina* species. These and *Mytilus edulis* could during the time of the Litorina Sea penetrate far into the Gulf of Bothnia and the Gulf of Finland because of the higher salinity at that time, as noted earlier, and were as a result also larger in size than at present. The *Littorina* species have since then withdrawn to the southwestern parts of the Baltic.

From the nature of the shell accumulations in Finland it can be seen that they cannot here, nor in Sweden, be directly used in studies of the shoreline displacement, but that they give additional information about changes in salinity of the coastal waters. As the present salinity requirements for the most common bivalves in the Baltic are known, detailed reconstructions of the salinity changes throughout its Flandrian history are possible, as are also more detailed studies of the shell species involved. Thus the species generally referred to as *Cardium edule*, common at present along the Baltic coasts, in fact represents *Cerastoderma glaucum* with a salinity amplitude of 4–15 ‰, whereas *Cerastoderma (Cardium) edule* is a marine species requiring a higher salinity.

The changes in the Baltic governed by the relative land uplift affected not only the mollusk fauna but also the rest of the fauna. As the freshwater stage of the Ancylus Lake interrupted the initial marine influence started at the time of the Yoldia Sea, a freshwater fauna could invade the Baltic, and this fauna was further trapped in the lake basins formed at the emerging coasts. This process continued throughout the Flandrian, but as seen

from Table 13.2 many of the major lakes became independent lakes at a time when the Baltic was a freshwater basin, before the salt water stage of the Litorina Sea. This scenario led Segerstråle (1956, 1957, 1976, 1982) to conclude that what he called the glacial relics of mainly crustaceans, which had survived the last glaciation in ice-dammed lakes further east, invaded the Baltic and the Scandinavian lakes over Lake Onega and the White Sea, a lake during the deglaciation, and also over ice-dammed lakes south of the Baltic. Thus the crustaceans followed the retreating ice margin, entered into the Baltic when it was a freshwater lake and eventually into the lakes formed as the submerged areas rose from the Baltic. Segerstråle concluded that the crustaceans *Gammaracanthus lacustris, Limnocalanus grimaldii, L. macrurus, Mesidotea entomon, Mysis relicta, Pallasea quadrispinosa,* and *Pontoporeia affinis* at the time of the higher salinity in the Baltic in Litorina time could only have survived in the innermost parts of the coastal waters of the Gulf of Finland and the Gulf of Bothnia, from where they again spread when the salinity decreased. *Gammaracanthus*, however, became extinct in the Baltic, whereas *Mesidotea* and *Pontoporeia* again thrived in the Baltic. The fish *Myoxocephalus (Cottus) quadricornis* was also included in the group of relics. This explanation for the present distribution of the relict species in the Baltic and the lakes is only a very schematic presentation. It is to be expected that the faunal elements during the Flandrian were mobile and that there was a continuous interchange between the lakes through the numerous waterways; very few lake basins are effectively isolated from others.

In the discussion of the glacial relics in the fauna the present distribution of the ringed seal *Phoca hispida* has been included. On the basis of biometrical comparisons of the skulls of the four subspecies of this seal, Müller-Wille (1969) could draw conclusions about its presumed migration from the east into the Baltic area. The subspecies *Phoca hispida pomorum*, now living in the White Sea and off the coasts of the Kola Peninsula and Novaya Zemlya, differs most from the other subspecies because it has been isolated from them since the Late Weichselian deglaciation over 10 ka ago. After that there were no connections over ice-dammed lakes into the Baltic. Forstén and Alhonen (1975), however, concluded that *P. hispida* came into the Baltic through the Närke Strait at the time of the Yoldia Sea. The subspecies *P. h. ladogensis* in Lake Ladoga differs more from the subspecies *P. h. botnica* in the Baltic than the subspecies *P. h. saimensi* in Lake Saimaa, which is in keeping with the history of these lakes (Table 13.2). Lake Ladoga became an independent lake earlier than Lake Saimaa, although there was a connection between Lake Ladoga and the Gulf of Finland later. The ringed seal is an example of the way that, using statistical analysis, it can be shown how populations isolated for 8–10 ka have differentiated, but, as pointed out by Müller-Wille (1969), the environmental influences in the lake basins of Saimaa and Ladoga have probably been more important than the time element. As seals generally were extensively hunted it can be assumed that they were more widespread in the major lakes earlier in Flandrian time than they are at present. They are still seriously threatened, as seen from the observed reduced number of seals in Lake Saimaa. Remains of seals, occasionally complete skeletons, have been found both in Flandrian clays and at many archaeological sites where there are a number of seal bones. Thus, for instance, it can be shown that the harp seal *Phoca groenlandica* was common in the Litorina Sea of the Flandrian climatic optimum. It is no longer a native species in the Baltic. In addition to subfossil finds of the two mentioned seal species, bones of the gray seal *Halicoerus grypus,* still present in the Baltic, have been recorded from a number of cultural refuse sites, showing that also this seal was hunted by early settlers of the Baltic coasts

(Forstén & Alhonen, 1975). Many of the sites in which seal bones or complete skeletons have been found in Baltic sediments have been studied in detail by pollen and diatom analysis combined with radiocarbon dating. These studies show that most finds are from Ancylus and Litorina sediments. The finds of harp seal, however, are restricted to the sediments of the Litorina Sea and of the early Limnaea Sea. Its disappearance is possibly due to both climatic and physical factors (Forstén & Alhonen, 1975). There are thus in the distribution and history of the seals in the Baltic and the major lakes trends directly related to the changes caused by the geological history of the area, with its land uplift, in addition to changes caused by man.

14   Land mammals

The evolution of the Quaternary mammalian fauna in Europe has been correlated with the stages and substages divided on criteria outlined in the previous chapters. In the area of the Scandinavian glaciations dealt with here the mammalian remains are restricted to a few sites with bones in situ, a number of single finds of bones incorporated into glacial sediments, particularly of the last glaciation, and to undisturbed Late Weichselian and Flandrian (Holocene) sites. In contrast to this area are the marginal areas of glaciation and the nonglaciated areas further south with their sites with rich faunal remains, mostly open-air sites but also caves (Kurtén, 1986). These have provided a detailed basis for an informal classification of the mammalian faunas, a division into ages that reflect the major evolutionary changes in the European land mammals. Thus the Villafranchian mammalian age between about 3 Ma and 1 Ma covers a period from the late Pliocene to the early Pleistocene above the Tertiary–Quaternary boundary. The younger land mammal ages in Europe, Galerian and Steinheimian, representing the last 1 Ma and thus the palaeomagnetic Jamarillo event and the Brunhes normal epoch (see Table 1.2), have, however, been correlated with the warm and cold stages listed in Table 3.2, the Early Steinheimian correlated with the Holsteinian (Kurtén, 1986). Detailed lists and accounts of the Villafranchian and subsequent faunas have been given in general accounts of the Quaternary faunas (Kurtén, 1968; Flint, 1971; Kahlke, 1981; T. Nilsson, 1983).

The changes in the Villafranchian fauna at the time of a general change in the vegetation from a steppe to a forest environment in a warm to temperate climate included the appearance of some new mammal groups, among them many ancestral species. Thus, for instance, the Middle to Late Pleistocene woolly mammoth (*Mammuthus primigenius*) evolved from the Villafranchian ancestral mammoth species *Mammuthus meriodinalis*. Some Villafranchian species survived to the present day, as the beaver (*Castor fiber*) and hippopotamus (*Hippopotamus amphibius*), and some bats (Kurtén, 1986). At the transition from the Villafranchian mammal age to the Galerian in the Middle Pleistocene there were great changes in the fauna. Whereas the Early Pleistocene forest fauna was rather uniform,

the later Middle Pleistocene cooling of the climate led to a differentiation into "warm" and "cold" faunas, with the appearance of arctic mammals of the climax community, the biome, of that climate at the time of the extensive glaciations of Europe. In addition to the mammals that can be used as climatic indicators, there are a number of climatically indifferent species (Koenigswald, 1988). Of the 119 living forms of mammals of which there is a fossil record, 51 made their appearance in the early Middle Pleistocene, 27 in the late Middle Pleistocene, and 28 in the Late Pleistocene (Kurtén, 1968). The remaining forms appeared earlier, some over 3 Ma ago. The general Pleistocene cooling of the climate also led to the extinction of taxa, as seen in the faunas of the warm stages, as well as in the withdrawal of some taxa from northern Europe. The Eemian fauna in this part still included southern elements, such as the hippopotamus (*Hippopotamus amphibius*) as far as 54½° N, elephant (*Elephas antiquus*), and the rhinos *Dicerorhinus kirchbergensis* and *D. hemitoechus* (Kurtén, 1986). The general cooling climate that led to a change to a more open vegetation is also reflected in the fauna, for instance, by the presence of *Equus*, bison, and two forms of elephant (Flint, 1971; West, 1977; Kahlke, 1981). The grazing of large herbivores probably also maintained and extended the areas with an open vegetation (Stuart, 1991). In addition to these changes in the population of large mammals, the fossil remains of rodents show that they evolved more rapidly and that their stratigraphical value therefore is of great importance, as seen from the changes summarized by Kahlke (1981). Of all the changes in the mammalian fauna the global extinction of nearly all large land mammals at the end of the Late Pleistocene was unique, as pointed out by Stuart (1991), but such extinctions did not reduce the fauna equally everywhere. In Europe, however, a change at the end of the Pleistocene is clearly documented. These changes are closely linked with the environmental changes during the Weichselian cold stage and therefore also stratigraphically important.

Stuart (1991) has given the most comprehensive account of the late Weichselian extinctions of mammals of both northern Eurasia and North America, including lists of the available radiocarbon dates of the finds and a critical discussion of the reliability of these dates. The best known of all extinct large mammals is the woolly mammoth (*Mammuthus primigenius*). Its Late Pleistocene distribution stretched from western Europe, from Britain down to Spain, in a broad belt through Asia and over Alaska to North America. The oldest finds are from the Saalian stage, but it may have been present earlier, though the majority of the finds represent the last cold stage. This means that woolly mammoth survived through the last (Eemian) warm stage, as some finds in Europe show (Stuart, 1991). Most finds are skeletal remains, bones, teeth, or tusks; but in the areas with permafrost, frozen carcasses have also been found, such as the Berezovka mammoth excavated in 1901, and Dina, the baby mammoth found in 1977, both from northeastern Asia (Guthrie, 1990). The mammoth remains in the area of the Scandinavian glaciations have been found in a secondary position and have been transported by ice and/or meltwater streams. The finds in the four Nordic and in the three Baltic countries are shown in Fig. 14.1, based on earlier separate maps for Denmark, Norway, and Finland by Berglund, Håkansson, and Lagerlund (1976) and Lundqvist and Pleijel (1976); for Finland by Donner, Jungner, and Kurtén (1979); for Denmark by Aaris-Sørensen, Petersen, and Tauber (1990); and for Norway, Sweden, Finland, and the Baltic states by Lepiksaar (1992). Details of the Norwegian finds, including a list of the radiocarbon dates, were also published by N. Heintz, Garnes, and Nydal (1979). Taking into account the possible errors of the radiocarbon dates Stuart (1991) concluded that woolly mammoth became extinct in Europe 12 ka B.P. and in Siberia 10.2 ka B.P., thus in Late Weichselian time. Recent finds from Wrangel Island in the Arctic Ocean of a dwarfed-mammoth population, dated between 7 ka and 4 ka, show that mammoth could survive in some areas until the Holocene (Vartanyan, Garutt, & Sher, 1993). The finite radiocarbon ages for northern and central Europe have a gap between about 20 ka and 15 ka B.P., corresponding to the time of the extensive Late Weichselian glaciation (see Fig. 10.1), whereas there is no corresponding gap in the ages for Siberia (Stuart, 1991). Most of the finds in the area of the Scandinavian glaciation represent either the Early Weichselian or Middle Weichselian interstadials and only a few the Late Weichselian time after the maximum glaciation (Lepiksaar, 1992). The finds and the finite radiocarbon ages are not generally at variance with the previously noted results about the interstadials, as presented in Fig. 10.1, but some are problematic and difficult to fit into the general pattern of nonglacial intervals, as was earlier noted about radiocarbon dates used in dating the Middle Weichselian interstadials (see Fig. 9.3). The broad outlines of the age distribution of the mammoth finds can, however, be reconstructed. The finds in Norway and Sweden inside the Fennoscandian moraines represent in Norway the Brumunddal–Gudbrandsdalen Interstadial (N. Heintz, Garnes, & Nydal, 1979) and in Sweden the Jämtland Interstadial (Lundqvist & Pleijel, 1976), both Early Weichselian, the finds grouping themselves to the areas in which the sediments of these interstadials occur (Fig. 14.1). Both of these areas were probably, as concluded earlier, glaciated in Middle Weichselian

time. The majority of the finds in southern Sweden, in Denmark, and probably in the Baltic countries, on the other hand, represent the comparatively long nonglacial Middle Weichselian interval at about 45 ka to 20 ka B.P. before the Late Weichselian glaciation maximum. One finite age of mammoth from the west coast of Finland at about 25 ka B.P., in addition to a date of a tundra reindeer antler from northern Finland at about 34 ka B.P. (Donner, Jungner, & Kurtén, 1979), would indicate that the finds are Middle Weichselian in age. But as there is no stratigraphical evidence of an ice-free interval at that time and as the older, finite, radiocarbon dates can in many cases be considered erroneous, as has been mentioned, it would be premature to use these dates alone as proof of a Middle Weichselian nonglacial interval in Finland. Furthermore, a date of about 15 ka B.P. for a mammoth find from near Helsinki is at variance with the Late Weichselian deglaciation history of that region.

This temporal and spatial distribution of the mammoth finds agrees with the results of the environmental changes during the Weichselian. There is, according to Lepiksaar (1992), no evidence of long-distance transport of mammoth finds by ice. During the Early Weichselian interstadials mammoth could thrive in the ice-free areas of central Scandinavia, which partly at least had a tundra vegetation, whereas the suitable habitats during the Middle Weichselian interstadials were further away from the Scandinavian mountains, in Denmark, the Baltic countries, and further south in Europe. After the Late Weichselian glaciation mammoth could again return and live in the cold environment close to the retreating ice margin, but finds from this time are few. This fact is shown by the finds of a tusk from Rosmos in Denmark dated at 13,240 +760 −690 B.P. (K-3697, Aaris-Sørensen, Petersen, & Tauber, 1990) and a tusk from Lockarp in southern Sweden for which there are three dates between 13,090 ±120 B.P. and 13,360 ±95 B.P. (Lu-796:2, Lu-796; Berglund, Håkansson, & Lagerlund, 1976). In explaining the Weichselian distribution of land mammals Guthrie (1990) assumed that the environment they lived in was a "mammoth steppe," a grassland with no analogs, unlike the present treeless tundra that could not have maintained such a fauna during the Weichselian. By measuring the bones of dated skeletal remains of mammoth in Siberia, A. E. Heintz and Garutt (1965) showed that it was 20–25 percent taller at about 37–34.6 ka B.P. than before at 47.5–40.5 ka B.P. or later at 11.7–11.2 ka B.P. To relate these changes in size to temperature (Kurtén, 1968) is, however, difficult in view of the results about the climatic fluctuations during the Weichselian.

Apart from these mammoth finds there are very few records within Scandinavia of the large land mammals that became extinct at the end of the Pleistocene. There are, however, some dated finds of giant deer (Irish elk), *Megaloceros giganteus*, from southern Sweden and Jutland in Denmark. The four dates for four different finds are all between 11,330 ±110 B.P. and 11,630 ±120 B.P., thus from Allerød time (Liljegren, 1977; Stuart, 1991). Mammoth and giant deer belong to the group of mammals that became extinct in Europe before the end of the Late Weich-

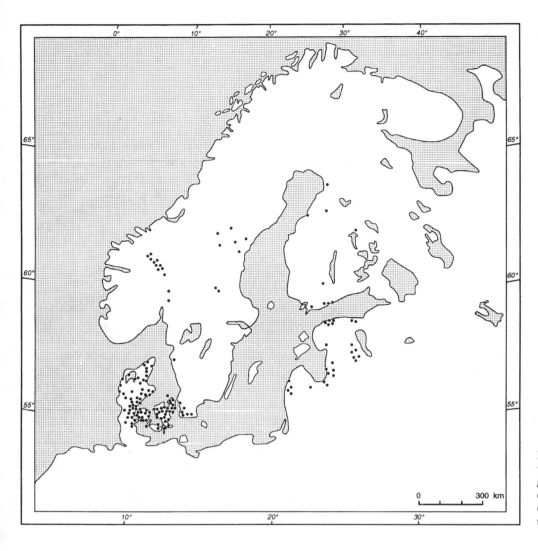

Figure 14.1. Finds of remains of woolly mammoth (*Mammuthus primigenius*) in Denmark, Norway, Sweden, Finland, and the Baltic states (based on sources mentioned in the text).

0   300 km

selian, at a time of great environmental changes as is seen from the vegetational history. From the evidence from Ireland it seems that giant deer became extinct when the grasslands of Allerød time changed to a less favorable vegetation for grazing in Younger Dryas time (Stuart, 1991). But the general extinctions of large mammals have been linked with an overkill by modern man with new, efficient hunting tools. However, the overkill happened at a time when the populations of large mammals had already been greatly reduced as a result of the environmental changes at the end of the Late Weichselian (Stuart, 1991).

The periglacial environment at the time of the Late Weichselian deglaciation was, according to the finds in southern Scandinavia, suitable both for mammal species now restricted to the treeless tundras and for species of the present mixed boreal forest (Aaris-Sørensen & Petersen, 1984). Polar bear, *Ursus maritimus*, belong to the former group. There are several finds from southern Sweden with radiocarbon ages between 12.7 ka and 10.4 ka B.P. and a find from northern Jutland in Denmark dated at about 11 ka B.P. (Aaris-Sørensen & Petersen, 1984; Berglund, Håkansson, & Lepiksaar, 1992). A complete skeleton of a polar bear was found

on the coast of southwestern Norway and dated at about 10.6 ka B.P. (Blystad et al., 1983). A find from southern Finland of arctic fox, *Alopex lagopus*, was – on the basis of the pollen analysis of the material found in the cavities of the specimen – dated as being from the Younger Dryas time (Kurtén, 1966). Furthermore, there are several Late Weichselian finds of reindeer, *Rangifer tarandus*, from Denmark (Iversen, 1973) and from southern Sweden (Liljegren, 1975). But in addition there are from the same periods finds of wild horse, *Equus przewalskii*, and wisent, *Bison bonasus*, from Denmark – steppe animals both. The finds of wisent from southern Sweden are younger, from the Early Flandrian Preboreal time (Liljegren, 1975). There are, however, some taxonomic problems in the rather complicated relationships between identified species. The European bison (wisent) *Bison bonasus* probably developed from *Bison priscus* common during the Eemian and Weichselian, but there are some difficulties in distinguishing between finds of the aurochs *Bos primigenius* from *Bison* spp. (Stuart, 1991). The oldest find in Scandinavia of the aurochs, which during the Weichselian lived further south in Europe in an area with open vegetation, is from Denmark and dated to the end of the Late Weichselian. It was, however, still present in the forests during the Flandrian, later than the wisent; the last aurochs died in Poland in A.D. 1627 (Liljegren & Lagerås, 1993). Of the few finds of musk-ox, *Ovibos moschatus*, in Denmark, Sweden, and Norway (Borgen, 1979) none can with certainty be dated, but they are probably older than the Late Weichselian glaciation, even if musk-ox was still present in Europe at that time before it retreated to northern Siberia, where it survived until late in the Holocene (Stuart, 1991).

In addition to the just mentioned species present in the periglacial zone south of the ice margin in southern Scandinavia there were also mammals such as elk, *Alces alces,* and brown bear, *Ursus arctos*. There are several Flandrian finds of brown bear, but it has also been recorded in Denmark from the end of the Late Weichselian, with one find a mandible of Allerød time from Jutland (Iregren, Ringberg, & Robertsson, 1990). There was thus in an environment with a vegetation without modern analogs – the previously mentioned mammoth steppe – a "disharmonious" faunal assemblage (Aaris-Sørensen, Petersen, & Tauber, 1990). It was this environment that the first humans, who were hunters, invaded from the south over Denmark into Scandinavia at the end of the Weichselian, an area north of Early or Middle Paleolithic sites in Europe (Guthrie, 1990) and therefore colonized for the first time. The change in climate, with the general spread of forest at the beginning of the Flandrian, together with the human colonization, resulted in a change of the fauna – if not an extinction, a withdrawal north and northeast, of several mammal species.

The Flandrian mammalian fauna was impoverished compared with earlier warm-stage faunas, but "the human invasions of Europe in postglacial time brought with them many smaller mammals such as rats," as so aptly put by Flint (1971, p. 756).

The history of the mammalian fauna in Scandinavia and the Baltic countries, based on numerous finds, has been dealt with in detail by Lepiksaar (1986); the finds from Finland have been discussed by Kurtén (1988) and the finds from Scania in southern Sweden by Liljegren (1975). Lepiksaar also plotted the ways of immigration from the south and southeast, as well as the climatic barriers restricting the distribution of some southern species, the most important barrier being *Limes norrlandicus* in central Sweden and southern Norway. In the beginning of the Flandrian the distribution of the mammals was still affected by the melting of the last remnants of the ice sheet, but from the Atlantic time onwards the warm stage latitudinal and altitudinal life zones were established (Lepiksaar, 1986). In addition to the wisent and aurochs, some other mammals disappeared during the Flandrian, such as wild boar (*Sus scrofa*) from Sweden and Norway, beaver (*Castor fiber*) from Finland, and wild cat (*Felis silvestris*). According to Lepiksaar there had still been a balance between the Mesolithic hunters and the environment, whereas with the introduction of farming the natural landscape was reduced. The domestic animals introduced from the south started to compete with the wild animals, for example, when livestock was freely grazing in the forest. But the smaller climatic fluctuations at the end of the Flandrian also affected the mammalian fauna. During the little ice age there was a noticeable reduction in the elk population, and it was at that time that the wild boar and the beaver suffered heavily and disappeared from many areas. But modern agriculture, forestry, and industrialization have had a much greater effect on the destruction of the indigenous mammalian fauna, a destruction furthered by the spread or introduction of new species into the area, such as the raccoon dog and muskrat of the Far East, and the American mink escaped from fur farms (Lepiksaar, 1986).

15   Quaternary chronology in Scandinavia

15.1. Present basis for chronology

The general division of the Quaternary in Scandinavia has been based on the stratigraphical evidence of warm and cold stages in which the boundaries have been defined by pollen analysis. It is essentially a division based on climatic changes as reflected in the vegetational history. Similarly the substages have been identified on the same basis. The stratigraphical division of the Quaternary of Scandinavia is one based on terrestrial sediments in central and northwestern Europe, established in the marginal zone or just outside the areas glaciated during the cold stages. The stratigraphy from within the area glaciated in Scandinavia has been placed in this frame, assuming that the stratigraphical evidence of any stage or substage is always more completely represented in the terrestrial successions outside the central area of glaciations, even if they also are known to be incomplete.

The chronology of the Quaternary stages (see Table 3.2) is, as shown in Cahpter 3, Section 1, based on various methods and the ages quoted in Tables 1.1, 1.2, and 3.3 are based on results from outside Scandinavia. The age for the Eemina is based on the correlation with the deep-sea chronology, on the age for stage 5e as obtained by correlating it with a particular reef terrace of Barbados, which was dated by ^{230}Th/^{234}U measurements. Similarly the dates used in the subdivision of the Weichselian (see Fig. 10.1) are based on ages estimated for the boundaries of the stages in the deep-sea stratigraphy, as mentioned in Chapter 7. The correlation of the Holsteinian with the deep-sea chronology is, on the other hand, already more problematic, as seen from Table 3.3. Some control of the chronology has been obtained from the dating of the boundaries between the palaeomagnetic epochs and events listed in Table 1.2. None of these methods has directly dated the Quaternary sequence in Scandinavia.

The subdivision of the Early, Middle, and Late Pleistocene, as far as to the upper part of the Middle Weichselian, is solely based on stratigraphical evidence. The three younger Middle Weichselian interstadials have been bracketed by radiocarbon ages (see Table 7.1), whereas for the Late Weichselian (see Table 11.2) and the Flandrian (see Table 12.1) a chronostratigraphic division

based on conventional radiocarbon ages has been applied. It has been assumed that the radiocarbon method is exact enough to provide a basis for subdivision, that radiocarbon ages rather than stratigraphical boundaries can be used in defining the Late Weichselian and Flandrian succession. However, near the limit of the radiocarbon dating method its use has been restricted, because the sediments dated have been poor in organic matter. As a result the dating of the Middle Weichselian interstadials has been problematic, as shown in Fig. 9.3, as have the interpretations of some of the stray finds of mammals with Middle Weichselian ages. But even the Late Weichselian sediments have in many cases, when poor in organic matter, been difficult to date. It is, however, in dating the organic sediments of the Flandrian that the radiocarbon method has been most effective. It has provided an opportunity to reconstruct time-transgressive changes in the vegetational history of the Flandrian, changes that cannot be reconstructed with the same accuracy for the older warm stages. But the radiocarbon ages, even if the errors such as those caused by the hard-water effect in limnic sediments are taken into account, do not provide a chronology in sidereal years. The Flandrian conventional radiocarbon ages, apart from the oldest ones, can be converted to approximate sidereal years by using the calibration curves provided by dated tree-ring chronologies, but this is not yet possible for the Late Weichselian radiocarbon ages. The detailed varve chronology used for dating the Late Weichselian and Early Flandrian deglaciation in Sweden and Finland (see Fig. 11.11) cannot therefore be directly linked with the chronology based on radiocarbon years. There are thus two independent chronologies for this time, the radiocarbon method dating the organic sediments and the varved clays dating the recession of the ice margin, including the moraines. Of these two chronologies that based on radiocarbon years has, as demonstrated, been used for the chronostratigraphic division.

Thus, accuracy by which the Quaternary history can be dated depends on the dating methods available for this period of time, and – in a restricted area such as that of the Scandinavian glaciations – on the available datable sediments. The radiocarbon method is therefore the only radiometric method that it has been possible to use in Scandinavia, and therefore it has not been possible to date the sediments in age beyond the range of this method. With the introduction of new methods, however, it may be possible to date older Quaternary sediments and sediments that cannot be dated by the radiocarbon method, and in this way to test the conclusions about the established stratigraphical sequence, which in Scandinavia is mostly restricted to the Late Pleistocene. Some of these more recent methods and their applications will be mentioned in the following section.

15.2. Additional chronological methods

Of methods applied to Quaternary sediments thermoluminescence (TL) dating of quartz and potassium feldspar grains in minerogenous sediments has a range that covers at least the last Weichselian cold stage and can possibly also be used for older sediments, such as those of Eemian and pre-Eemian age. Its use in dating dune sands and other aeolian sediments that have been exposed to sunlight for some time has been clearly demonstrated, whereas sediments formed in ice-marginal environments have been problematic to date because of the incomplete zeroing of their minerals (Wintle, 1991). The measurement of the optically stimulated thermoluminescence (OSL), particularly using infrared wavelengths, is a refinement of the technique that is expected to provide more exact dates than conventional TL dating.

The Weichselian minerogenic sediments in the Scandinavian area of glaciation that can be dated with the TL method are mainly sediments deposited close to the ice margin and therefore, as mentioned, difficult to date. Further, whereas radiocarbon ages are obtained for organic sediments and therefore date the nonglacial intervals during which they were formed, the TL ages from a glaciated area primarily date the onset or end of a glaciation, when minerogenic sediments were deposited near the ice margin or aeolian sediments formed in the periglacial zone, also close to the ice margin. It is only at sites where the sediments are partly or totally minerogenous that the middle parts of the nonglacial intervals may directly be TL dated. And as it has a greater range than radiocarbon dating it is, if successful, of particular importance in Scandinavia in separating the Early Weichselian interstadials from the Eemian, and in demonstrating the presence of a possible Middle Weichselian nonglacial interval. The extent of the ice-free area at this time is, as seen from previous chapters, still not known, nor is the extent of the early Middle Weichselian glaciation.

Many of the TL ages calculated on the basis of the early measurements have such a wide range that they are of limited use, apart from demonstrating the applicability of the method. A number of TL dates from the Eemian and Early Weichselian sediments at Fjøsanger in Norway had a spread from 27.9 ka to 145 ka, although the averages of the ages from the different beds of silts, sands, and gravels became older with increasing depth (Hütt, Punning, & Mangerud, 1983). The TL dates did not, however, either support or change the conclusions about the age of the sediments based on the biostratigraphical studies referred

to earlier. Similarly, TL dates of sediments between tills in northern Norway were of limited use in dating the nonglacial intervals (Olsen, 1988), as were the earlier TL dates for till-covered sediments in northern Finland (Hütt, Punning, & Raukas, 1983). In contrast to the early attempts at TL dating that paved the way for later investigations, there are some later results that are significant for the Weichselian chronology. These TL ages, given in Fig. 15.1, may not be conclusive but are clearly suggestive and should be viewed against the results presented in Fig. 10.1 showing schematic glaciation curves for the Weichselian. The TL dating carried out by Hütt on sediments between tills in Estonia (Liivrand, 1991) show that the ages from 90 ka to 108 ka for sediments – stratigraphically referred to as Early Weichselian – from the three sites of Peedu, Valga, and Röngu agree with the age of the Early Weichselian. These sediments are overlain by a till and the two TL ages of 62.4 ka and 66.5 ka of the glaciolacustrine sediments above this till suggest that Estonia was glaciated in late Early Weichselian or early Middle Weichselian time, no later than about 70 ka. Taking into account an error of the TL dates of up to approximately 10 percent these ages are in keeping with the general chronology presented in Figure 10.1. The younger tills in Estonia have been placed in the Late Weichselian (Liivrand, 1991). An early Middle Weichselian glaciation, as in Estonia, is also suggested by the TL age of 67 ka for melt-water sand in Denmark, whereas two ages from northern Jutland at 20 ka and 22 ka are for till-covered sediments (Petersen, 1984a)

A number of coversand samples have been dated from western Jutland in Denmark and all are related to the Late Weichselian maximum glaciation and to the periglacial conditions persisting in Denmark to the end of the Late Weichselian, and possibly at the beginning of the Flandrian (Kolstrup et al., 1990). The coversands could be correlated with the established chronozones (see Table 11.2), with a general separation of an older coversand from a younger coversand by the Allerød Interstadial. From the TL dates presented in Fig. 15.1 it can be seen that most of them are related to the Late Weichselian glaciation and to the periglacial period after it, and that TL dating provides a means of correlating the aeolian coversands with the Late Weichselian substages. The range of TL dates for eolian sediments in central Sweden shows that the sediments were formed at the time of deglaciation and that the dates are broadly compatible with the deglaciation chronology based on varved clays and radiocarbon measurements (Lundqvist & Mejdahl, 1987). The TL dating of about 40 ka of aeolian deposits in the Gudbrandsdal valley in Norway is, however, difficult to interpret (B. Andersen & Mangerud, 1990) because if correct it would show that the deglaciation during the Middle Weichselian was more extensive than suggested in Fig. 10.1. To base such a conclusion only on TL dates seems, however, uncertain as shown by results from Sweden. A TL date of 60 ka was there obtained for a sample from Pilgrimstad, for a sediment which on biostratigraphical evidence (see Table 8.1) is Early Weichselian in age (Garcia Ambrosiani, 1990). Similarly, two TL dates of 43 ka and 69 ka of

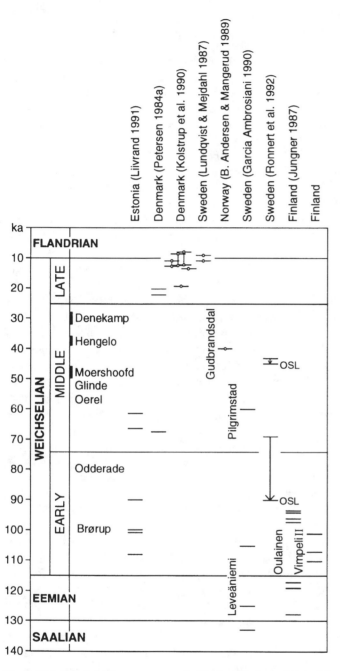

Figure 15.1. Selected TL ages compared with chronological scheme of the Eemian and Weichselian, as presented in Figure 10.1. Dates with circles represent aeolian sediments.

Gudbrandsdal valley in Norway cannot, however, be conclusive evidence for an extensive Middle Weichselian deglaciation, but a deglaciation of central Sweden at that time cannot be ruled out against the evidence from Estonia and taking into account the general pattern of Early Weichselian and early Middle Weichselian glaciations as presented in Fig. 8.4.

There are, in addition to the Middle Weichselian TL dates, a number of dates from Sweden and Finland for sediments that on stratigraphical grounds have been referred to as being either Eemian or Early Weichselian. Three datings of the waterlain silty sands biostratigraphically placed in the Brørup Interstadial gave TL ages of 105 ka, 124 ka, and 133 ka (Garcia Ambrosiani, 1990). As the ages have a wide range, from late Saalian over the Eemian to the Early Weichselian, they are of limited use in dating the sediments at this site. A better age bracket was obtained from Oulainen in Ostrobothnia in Finland, where till-covered organic sediments, also correlated with the Brørup Interstadial (see Table 8.1), overlie sands with TL ages of 117 ka, 119 ka, and 128 ka and are covered by sands which gave TL ages between 94 ka and 97 ka directly from above the organic sediments, and between 88 ka and 102 ka away from the section but from the same sand (Jungner, 1983, 1987). The TL dates from the Vimpeli II section in Ostrobothnia are for the till-covered sands with 10–20 cm-thick compressed peat layers, on the basis of the macroscopic plant remains referred to as Eemian (Aalto et al., 1989), with a pollen composition unlike that of the Vimpeli I section, which has been dated as being either Eemian (Aalto et al., 1983) or Early Weichselian (Donner, 1988). The TL dates of 99 ka, 107 ka, and 115 ka, taken as being the most reliable of seven samples (Jungner, 1987) and dating the nonglacial interval of the Vimpeli II section, correspond to the age of the interval at Oulainen between the dates for the sands below and above the organic sediments (Fig. 15.1). If all these TL dates are correct they would show that both the nonglacial intervals at Oulainen and Vimpeli are Early Weichselian and represent the Brørup Interstadial. But whereas the Oulainen TL dates have a laboratory error of only 9–13 percent the accuracy of the Vimpeli dates is less as a result of poor bleaching of the sediments used for dating. Further, four samples from Vimpeli, in addition to the three just mentioned samples, gave ages of 166 ka, 184 ka, and 225 ka, much in excess of ages even for the Eemian. As the results from Vimpeli cannot be considered conclusive the sediments at Oulainen and Vimpeli cannot on the basis of their TL ages be taken with certainty to represent the same nonglacial interval (Jungner, 1987).

The separation of Eemian and Early Weichselian interstadial sediments in Ostrobothnia is, as seen earlier, stratigraphically difficult as there is no till bed between them. There was thus a long ice-free period from the early Eemian into the Early Weichselian, an interval into which most TL dates fall. At another Early Weichselian site, Marjamurto, an interstadial correlated with Brørup was identified, with a preliminary TL date of 197 ka (Peltoniemi et al., 1989). At another site, Risåsen, which is a till-covered esker, the TL age based on four samples was 130 ka, an

a reddish-brown clay from the area between lakes Vänern and Vättern in Sweden show that the clay, deformed by an overriding ice, was deposited before the Late Weichselian glaciation but they do not give an answer as to its true age. An OSL date of 91 ka was obtained for the sample that gave a TL date of 69 ka (Ronnert, Svedhage, & Wedel, 1992). The OSL date for the sample with a TL date of 43 ka was, however, 45 ka. These three TL dates from Sweden together with the date from the

age in keeping with the conclusion that the esker was formed during the Saalian deglaciation and later covered by a Weichselian till (Donner, 1988). But then there are also TL dates that are at variance with the general stratigraphy of the area (Niemelä & Jungner, 1991), in addition to the four dates from Vimpeli mentioned.

As seen from the discussion above, and from Fig. 15.1, the TL dating of sediments related to ice-marginal environments give, in spite of their often incomplete zeroing, ages that are promising for a future TL chronology to be established that may help especially in the identification of the Early and Middle Weichselian interstadials and in separating them from the Eemian. But so far the TL dates as such cannot be used for a reinterpretation of the stratigraphy and chronology established with other methods (see Fig. 10.1); they can chiefly be used in testing the reliability of the method.

Whereas the TL dating method and the OSL dating developed from it have been used for minerogenic sediments, electron spin resonance (ESR) dating, which is also based on the received cumulative dose, has been tested in dating speleothem calcite, mollusks, and corals (Smart, 1991). In Estonia mollusk shells were ESR dated from beach deposits that have been related to the Flandrian stages in the history of the Baltic and for which their age therefore has geologically been estimated (Hütt et al., 1985). In two cases the ESR dates corresponded to the estimated geological ages, but for the four remaining samples there was either no result or the dates were too old, one being about 3 ka too old. In spite of the inaccuracies in these results the potential use of the method has been demonstrated. And it can also be used in dating older material, as shown by the results of ESR dating of marine shells from the White Sea area. A series of 10 samples from a section in Zaton northeast of Arkhangelsk gave ESR dates between 82 ka and 120 ka for shells representing the time of the Eemian transgression (Molodkov & Raukas, 1988). If these dates are correct they would imply that some of the shells are Early Weichselian. But as the ESR dates, similarly to the TL dates, are still fraught with comparatively large errors, in spite of a good laboratory precision, the ages cited must still be regarded as tentative.

In contrast to the methods described above, which belong to the group of dating methods based on the effects of the natural radiation on the material to be dated (Roth & Poty, 1985), amino acid dating has been used in the study of marine shells, although it does not directly date the shells. In dating the shells the ratio of the D/L amino acids is determined. After the death of a living organism, in this case of the mollusk, the amino acid L-isoleucine in the proteins epimerizes with D-alloisoleucine, with the D/L ratio increasing with age. The epimerization, however, is nonlinear as its rate depends on temperature – faster in warm temperatures than in cool, and also faster in the beginning of the epimerization than later (G. H. Miller & Mangerud, 1985; Sykes, 1991). The method was tested by analyzing a great number of samples of marine shells of different mollusk taxa from northwestern Europe, from a number of known pre-Elsterian, Holsteinian, Eemian, and Early, Middle, and Late Weichselian sites, as well as

from sites from the shores of the Arctic ocean (S. H. Miller & Mangerud, 1985). When the D/L ratios were plotted in diagrams against the current temperatures at the sites used it could be demonstrated that the ratios increased with the ages of the deposits that the shells represented, and that the age relationships on the whole agreed with the earlier conclusions about the sites. The D/L ratios from Eemian deposits showed that they represent one warm stage, the only uncertainty being the interpretation of the ratios for the sites in western Norway. There, the scatter of the ratios is so great that there is some uncertainty as to whether the warm stages of Fjøsanger and Bø (see Table 6.1) both belong to the Eemian or if the former is older. And the correlation of the D/L ratios from Holsteinian sites suggests that the marine sediments from the area of Esbjerg and Rögle in Denmark may be pre-Elsterian instead of Holsteinian, which, if correct, would imply that there is a pre-Elsterian till in the area. The amino acid stratigraphy, on the other hand, also suggests that the Holsteinian corresponds to the isotope stage 7, but less likely to stage 9, a conclusion that seems to be at variance with the general stratigraphical division and its comparison with isotope stages as discussed in Chapter 3, Section 1. The uncertainty about the age of the Holsteinian is thus also reflected in the interpretation of the amino acid dating. The temperature differences between the sites from the arctic areas and those in northwestern Europe make a correlation between these areas uncertain, as pointed out by G. H. Miller and Mangerud (1985), but the shells from cold waters give additional evidence of the influence of the water temperature on the epimerization. The amino acids have also been studied in foraminifera from cores from the Arctic ocean and conclusions drawn about the rate of sedimentation (Sejrup et al., 1984).

It can be seen from these results that amino acid dating on the whole supports the established Pleistocene stratigraphy of marine sediments in northwestern Europe but that some of the correlations may be questioned. Attention should therefore be paid to these discrepancies in interpretations. The results do not, however, necessitate a major revision of the stratigraphical division. Amino acid dating is a most useful additional method in the study of the marine sequences but cannot alone provide a chronological frame for the Pleistocene, as it is dependent on a calibration by other dating methods. The scatter of the D/L ratios that were referred to certain stages or substages is rather great, as for TL dates, as seen from Fig. 15.2, in which one of two alternatives of grouping the ratios for northwestern Europe is schematically given according to G. H. Miller and Mangerud (1985). Samples referred to as being Cromerian were not included because they were not from the main area dealt with here.

In the discussion of the division and dating of the Quaternary, the paleomagnetic epochs and events were mentioned (see Table 1.2), a widely used time scale that has enabled correlations to be made over wide areas. In addition reversal magnetostratigraphy curves of secular paleomagnetic fluctuations of relative declination and relative inclination have been used in correlating Late Weichselian and Flandrian cores of lake sediments with one

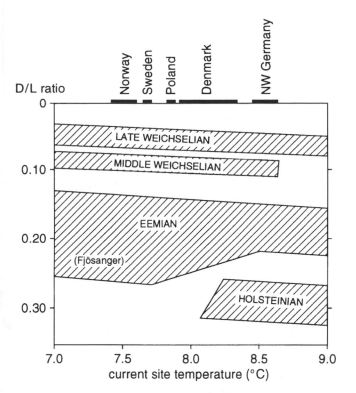

Figure 15.2. An alternative interpretation of the grouping of D/L ratios of samples from northwestern Europe (after Miller & Mangerud, 1985, Fig. 9).

from place to place, is a potential method for correlations between varved clays and cores of lake sediments dated with radiocarbon and studied pollen-analytically. It was in this way that the differences between radiocarbon years and varve years in the earlier study of Late Weichselian sediments from southern Sweden could be estimated (Björck, Sandgren, & Holmquist, 1987). Paleomagnetic curves were also obtained for the Late Weichselian varved clays in the area north of Lake Ladoga (Bakhmutov, Zagnii, & Tretyak, 1985), but here the varve chronology was not tied to the chronology in the region of the Baltic Sea further west. Paleomagnetic curves for correlations were also used for Flandrian sediments, as in an early study of the sediments of Lake Lojärvi in southern Finland, where the fluctuations in declination during the last 4 ka were compared with those in northwestern England, using radiocarbon ages for the chronology (K. Tolonen, Siiriäinen, & Thompson, 1975). These examples already show that the potential of the method and recent improvements in both coring and in the paleomagnetic measurements of the sediments have increased the possibility of useful correlations of cores from lake basins.

From the preceding discussion it can be seen how new dating methods or methods of correlation are being developed and refined, and how they are being tested on a variety of Quaternary sediments. The methods are useful only as long as the stratigraphical context of the dated sediment or find is known. A chronology not firmly linked to a stratigraphical frame is of little use. And the difficulty in converting the years obtained from various chronological methods to sidereal years is always present, even when using methods like the study of varved clays, which by their nature should give a chronology in sidereal years. No single chronological method can be used for dating the whole Quaternary history, and for the Late Pleistocene and Holocene many methods have been used side by side, even though the suggested subdivision of the Late Weichselian and Flandrian is based on radiocarbon years.

another (Thompson, 1991). The method, however, has its drawbacks as, in addition to not dealing with global features, it is sensitive to disturbances of sediments and to coring disturbances, as shown by sediment cores from Iceland in which known tephra horizons could be used for correlation (Thompson, 1991). The use of curves of paleomagentic fluctuations, even if they vary

References

Aaby, B. (1975). Cykliske klimatvariationer de siste 7500 år påvist ved undersøgelser af højmoser og marine transgressionsfaser. *Danmarks Geologiske Undersøgelse, Årbog* 1974, 91–104.

Aaby, B. (1976). Cyclic climatic variations in climate over the past 5,500 yr reflected in raised bogs. *Nature* 263, 281–84.

Aaby, B., & Tauber, H. (1974). Rates of peat formation in relation to degree of humification and local environment, as shown by studies of a raised bog in Denmark. *Boreas* 4, 1–17.

Aaloe, A. (1979). Meteoriidikraatrid Ilumetsas. *Eesti Loodus* 22, 756–61.

Aalto M., Donner, J., Hirvas, H., & Niemelä, J. (1989). An interglacial beaver dam deposit at Vimpeli, Ostrobothnia, Finland. *Geological Survey of Finland, Bulletin* 348.

Aalto, M., Donner J., Niemelä, J., & Tynni, R. (1983). An eroded interglacial deposit at Vimpeli, South Bothnia, Finland. *Geological Survey of Finland, Bulletin* 324.

Aalto, M., Eriksson, B. & Hirvas, H. (1992). Naakenavaara Interglacial – a till-covered peat deposit in western Finnish Lapland. *Bulletin of the Geological Society of Finland* 64 (2), 169–81.

Aario, L. (1940). Waldgrenzen und subrezenten Pollenspektren in Petsamo Lappland. *Annales Academiae Scientiarum Fennicae* 54(8).

Aario, R. (1965a). Die Fichtenverhäufigung im Lichte von C¹⁴ – Bestimmungen und die Altersverhältnisse der finnischen Pollenzonen. *Comptes Rendus de la Société géologique de Finland* 37, 215–31.

Aario, R. (1965b). Development of ancient Lake Päijänne and the history of the surrounding forests. *Annales Academiae Scientiarum Fennicae* A III 81.

Aario, R. (1990). Moraine landforms in Northern Finland. In *Glacial heritage of Northern Finland, Excursion Guide, International Drumlin Symposium, Oulu 1990 Finland*, ed. R. Aario, pp.13–27. Nordia Tiedonantoja, A I 1990, Oulu: Pohjois-Suomen Maantieteellinen Seura.

Aario, R., & Forsström, L. (1979). Glacial stratigraphy of Koillismaa and North Kainuu, Finland. *Fennia* 157(2).

Aaris-Sørensen, K., & Petersen, K. S. (1984). A Late Weichselian find of polar bear (*Ursus maritimus* Phipps) from Denmark and reflections on the paleoenvironment. *Boreas* 13, 29–33.

Aaris-Sørensen, K., Petersen, K., & Tauber, H. (1990). Danish finds of mammoth (*Mammuthus primigenius* (Blumenbach)): Stratigraphical position, dating and evidence of Late Pleistocene environment. *Danmarks Geologiske Undersøgelse* B 14.

Aarseth, I., & Mangerud, J. (1974). Younger Dryas end moraines between Hardangerfjorden and Sognefjorden, Western Norway. *Boreas* 3, 3–22.

Aartolahti, T. (1966). Über die Einwanderung und die Verhäufigung der Fichte in Finnland. *Annales Botanici Fennici* 3(3), 368–379.

Aartolahti, T. (1970). Fossil ice-wedges, tundra polygons and recent frost cracks in southern Finland. *Annales Academiae Scientiarum Fennicae* A III 107.

Aartolahti, T. (1972). On deglaciation in southern and western Finland. *Fennia* 114.

Aartolahti, T. (1977). Lentohiekka Suomessa. *Suomalainen Tiedeakatemia, Esitelmät ja pöytäkirjat* 1976, 83–95.

Agrell, H. (1980). The Quaternary history of the Baltic, Sweden. In *The Quaternary history of the Baltic* ed. V. Gudelis & L.-K. Königsson. pp. 219–39. Acta Universitatis Upsaliensis, Symposia Universitatis Upsaliensis Annum Quingentesimum Celebrantis, 1, Uppsala.

Ahti, T., Hämet-Ahti, L., & Jalas, J. (1968). Vegetation zones and their sections in northwestern Europe. *Annales Botanici Fennici* 5(3), 169–211.

Alalammi, P., ed (1988). *Atlas of Finland, Folio 141–143. Vegetation and flora; Fauna; Nature conservation*. Helsinki: National Board of Survey, Geographical Society of Finland.

Alalammi, P., ed. (1992). *Atlas of Finland, Folio 123–126. Geology*. Helsinki: National Board of Survey, Geographical Society of Finland.

Alhonen, P. (1964). Radiocarbon age of waternut (Trapa natans L.) in the sediments of Lake Karhejärvi, SW-Finland. *Memoranda Societatis pro Fauna et Flora Fennica* 40, 192–97.

Alhonen, P. (1967). Palaeolimnological investigations of three inland lakes in southwestern Finland. *Acta Botanica Fennica* 76.

Alhonen, P. (1969). The developmental history of Lake Inari, Finnish Lapland. *Annales Academiae Scientiarum Fennicae* A III 98.

Alhonen, P. (1971). The stages of the Baltic Sea as indicated by the diatom stratigraphy. *Acta Botanica Fennica* 92.

Alhonen, P. (1981). Stratigraphical studies of Lake Iidesjärvi sediments. Part I: Environmental changes and palaeolimnological development. *Bulletin of the Geological Society of Finland* 53(2), 97–107.

Andersen, B. G. (1960). Sørlandet i sen- og postglacial tid. *Norges Geologiske Undersökelse* 210.

Andersen, B. G. (1965). The Quaternary of Norway. In *The Quaternary*, vol. 1, ed. K. Rankama, pp. 91–138. New York: Wiley.

Andersen, B. G. (1968). Glacial geology of western Troms, North Norway. *Norges Geologiske Undersökelse* 256.

Andersen, B. G. (1975). Glacial geology of Northern Nordland, North Norway. *Norges Geologiske Undersøkelse* 320.

Andersen, B. G. (1979). The deglaciation of Norway 15,000–10,000 B.P. *Boreas* 8, 79–87.

Andersen, B. G. (1981). Late Weichselian ice sheets in Eurasia and Greenland. In *The Last Great Ice Sheets*, ed. G. H. Denton & T. J. Hughes, pp. 3–35. New York: Wiley.

Andersen, B. G., & Mangerud, J. (1990). The last interglacial–glacial cycle in Fennoscandia. *Quaternary International* 3/4, 21–9.

Andersen, B. G., & Nesje, A. (1992). Quantification of Late Cenozoic glacial erosion in a fjord landscape. *Sveriges Geologiska Undersökning* Ca 81, 15–20.

Andersen, B. G., Nydal, R., Wangen, O. P., & Østmo, S. R. (1981). Weichselian before 15,000 years B.P. at Jaeren–Karmøy in southwestern Norway. *Boreas* 10, 297–314.

Andersen, B. G., Sejrup, H. P., & Kirkhus, Ø. (1983). Eemian and Weichselian deposits at Bø on Karmøy, SW Norway: A preliminary report. *Norges geologiske undersøkelse* 380, 189–201.

Andersen, B. G., Wangen, O. P., & Østmo, S. R. (1987). Quaternary geology of Jaeren and adjacent areas, south-western Norway. *Norges geologiske undersøkning* 411.

Andersen, S. T. (1961). Vegetation and its environment in Denmark in the Early Weichselian Glacial (Last Glacial). *Danmarks Geologiske Undersøgelse* II 75.

Andersen, S. T. (1965). Interglacialer og interstadialer i Danmarks Kvartaer. *Meddelelser fra Dansk Geologisk Forening* 15(4), 486–506.

Andersen, S. T. (1970). The relative productivity and pollen representation of north European trees, and correction factors for tree pollen spectra. *Danmarks Geologiske Undersøgelse* II 96.

Andersen, S. T. (1973). The differential pollen productivity of trees and its significance for the interpretation of a pollen diagram from a forested region. In *Quaternary Plant Ecology*, ed. H. J. B. Birks & R. G. West, pp. 109–115. Oxford: Blackwell.

Andersen, S. T. (1979). The subdivision of the Quaternary of Norden: a comment. *Boreas* 8, 74.

Andersen, S. T. (1980). Early and Late Weichselian chronology and birch assemblages in Denmark. *Boreas* 9, 53–69.

Andersen, S. T., de Vries, H., & Zagwijn, W. H. (1960). Climatic change and radiocarbon dating in the Weichselian Glacial of Denmark and the Netherlands. *Geologie en Mijnbouw* 39, 38–42.

Andrén, T. (1990). Till stratigraphy and ice recession in the Bothnian Bay. *University of Stockholm, Department of Quaternary Research, Report* 18.

Anundsen, K. (1985). Changes in shore-level and ice-front position in Late Weichsel and Holocene, southern Norway. *Norsk geografisk Tidsskrift* 39, 205–25.

Anundsen, K. (1990). Evidence of ice movement over southwest Norway indicating an ice dome over the coastal district of West Norway. *Quaternary Science Reviews* 9, 99–116.

Anundsen, K., & Fjeldskaar, W. (1983). Observed and theoretical Late Weichselian shorelevel changes related to glacier oscillations at Yrkje, South-West Norway. In *Late- and Postglacial oscillations of glaciers: Glacial and periglacial forms*, ed. H. Schroeder-Lanz, pp. 133–70. Rotterdam: Balkema.

Auer, V. (1924). Die postglaziale Geschichte des Vanajavesisees. *Bulletin de la Commission géologique de Finlande* 69.

Auer, V. (1968). Die Isobasenrichtung in der Gegend des Sees Vanajavesi. *Annales Academiae Scientiarum Fennicae* A III 94.

Averdieck, F.-R. (1967). Die Vegetationsentwicklung de Eem-Interglazials und der Frühwürm-Interstadiale von Odderade / Schleswig-Holstein. *Fundamenta*, R.B., 2, 101–35.

Averdieck, F.-R., Friedrichsen, O., Ullrich, W., & Vogel, H. (1976). Geobotanische und -zoologische Untersuchungen an Eeminterglazialen in SW-Holstein. *Meyniana* 28, 1–18.

Backman, A. L. (1941). Najas marina in Finnland während der Postglazialzeit. *Acta Botanica Fennica* 30.

Backman, A. L. (1943). Ceratophyllum submersum in Nordeuropa während der Postglazialzeit. *Acta Botanica Fennica* 31.

Backman, A. L. (1948). Najas flexilis in Europa während der Quartärzeit. *Acta Botanica Fennica* 43.

Backman, A. L. (1950). Najas tenuissima (A.Br.) Magnus einst und jetzt. *Societas Scientiarum Fennica, Commentationes Biologicae* 10(19).

Backman, A. L. (1951). Najas minor All. in Europa einst und jetzt. *Acta Botanica Fennica* 48.

Bahnson, H., Petersen, K. S., Konradi, P. B., & Knudsen, K. L. (1974). Stratigraphy of Quaternary deposits in the Skaerumhede II boring: Lithology, molluscs and foraminifera. *Danmarks Geologiske Undersøgelse, Årbog* 1973, 27–62.

Bakhmutov, V., Ekman, I., & Zagnyi, G. (1987). Stratigraphic subdivision and correlation of varved clays in Lake Ladoga area based on geological, palynological and palaeomagnetic studies. In *Palaeohydrology of the temperate zone, I – Rivers and lakes*, ed. A. Raukas & L. Saarse, pp. 191–203. Tallinn: Valgus.

Bakhmutov, V. G., Zagnii, G. F., & Tretyak, A. N. (1985). Paleomagnetism of varved sediments in Karelia. *Geophysical Journal* 6(5), 730–42.

Behre, K. E. (1989). Biostratigraphy of the last glacial period in Europe. *Quaternary Science Reviews* 8, 25–44.

Behre, K. E., & Lade, U. (1986). Eine Folge von Eem und 4 Weichsel-Interstadialen in Oerel / Niedersachsen und ihr Vegetationsablauf. *Eiszeitalter und Gegenwart* 36, 11–36.

Belderson, R. H., & Wilson, J. B. (1973). Iceberg plough marks in the vicinity of the Norwegian Trough. *Nordsk Geologisk Tidsskrift* 53, 323–28.

Bergersen, O. F., & Garnes, K. (1971). Evidence of subtill sediments from a Weichselian Interstadial in the Gudbrandsdalen Valley, Central East Norway. *Norsk geografisk Tidsskrift* 25, 99–108.

Bergersen, O. F., & Garnes, K. (1981). Weichselian in central South Norway: The Gudbrandsdal Interstadial and the following glaciation. *Boreas* 10, 315–22.

Berglund, B. E. (1964). The Post-glacial shore displacement in eastern Blekinge, southeastern Sweden. *Sveriges Geologiska Undersökning*, C 599.

Berglund, B. E. (1966a). Late-Quaternary vegetation in eastern Blekinge, southeastern Sweden: A pollen-analytical study. I. Late-glacial time. *Opera Botanica* 12(1).

Berglund, B. E. (1966b). Late-Quaternary vegetation in eastern Blekinge, southeastern Sweden: A pollen-analytical study. II. Post-glacial time. *Opera Botanica*, 12(2).

Berglund, B. E. (1971). Littorina transgressions in Blekinge, South Sweden: A preliminary survey. *Geologiska Föreningen i Stockholms Förhandlingar* 93, 625–52.

Berglund, B. E. (1979). The deglaciation of southern Sweden 13,500–10,000 B.P. *Boreas* 8, 89–118.

Berglund, B. E. (1983). Palaeoclimatic changes in Scandinavia and on Greenland – a tentative correlation based on lake and bog stratigraphical studies. *Polish Academy of Sciences, Quaternary Studies in Poland* 4, 27–43.

Berglund, B. E., Håkansson, S., & Lagerlund, E. (1976). Radiocarbon-dated mammoth (*Mammuthus primigenius* Blumenbach) finds in South Sweden. *Boreas* 5, 177–91.

Berglund, B. E., Håkansson, S., & Lepiksaar, J. (1992). Late Weichselian polar bear (*Ursus maritimus* Phipps) in southern Sweden. *Sveriges Geologiska Undersökning* Ca 81, 31–42.

Berglund, B. E., & Lagerlund, E. (1981). Eemian and Weichselian stratigraphy in South Sweden. *Boreas* 10, 323–62.

Berglund, B. E., & Rapp, A. (1988). Geomorphology, climate and vegetation in north-west Scania, Sweden, during the Late Weichselian. *Geographia Polonica* 55, 13–35.

Bergman, L. (1982). Clastic dykes in the Åland Islands, SW Finland and their origin. *Geological Survey of Finland Bulletin* 317, 7–32.

Bergsten, H., & Nordberg, K. (1992). Late Weichselian marine stratigraphy of the southern Kattegat, Scandinavia: Evidence for drainage of the Baltic Ice Lake between 12,700 and 10,300 years B.P. *Boreas* 21, 223–52.

Bergström, E. (1959). Problemet om övervintring. Utgjorde Lofoten och Vesterålen ett refugium under sista istiden. *Svensk Naturvetenskap* 1959, 116–122.

Berthelsen, A. (1978). The methodology of kineto-stratigraphy as applied to glacial geology. *Bulletin of the Geological Society of Denmark* 27, Special Issue, 25–38.

Berthelsen, A. (1979). Contrasting views on the Weichselian glaciation and deglaciation of Denmark. *Boreas* 8, 125–132.

Berthelsen, A., Konradi, P., & Petersen, K. S. (1977). Kvartaere lagfølger i Vestmøns klinter. *Dansk Geologisk Forening, Årsskrift* 1976, 93–9.

Birks, H. J. B., & Berglund, B. E. (1979). Holocene pollen stratigraphy of southern Sweden: A reappraisal using numerical methods. *Boreas* 8, 257–79.

Birks, H. J. B., & Birks, H. H. (1980). *Quaternary Palaeoecology*. London: Edward Arnold.

Björck, S. (1979). Late Weichselian stratigraphy of Blekinge, SE Sweden, and water level changes in the Baltic Ice Lake. *University of Lund, Department of Quaternary Geology, Thesis* 7.

Björck, S. (1981). A stratigraphic study of Late Weichselian deglaciation, shore displacement and vegetation history in south-eastern Sweden. *Fossils and Strata* 14.

Björck, S (1984). Bio- and chronostratigraphic significance of the Older Dryas Chronozone – on the basis of new radiocarbon dates. *Geologiska Föreningens i Stockholm Förhandlingar* 106, 81–91.

Björck, S. (1987). An answer to the Ancylus enigma? Presentation of a working hypothesis. *Geologiska Föreningens i Stockholm Förhandlingar* 109, 171–76.

Björck, S., Berglund, B. E., & Digerfeldt, G. (1988). New aspects on the deglaciation chronology of South Sweden. *Geographia Polonica* 55, 37–49.

Björck, S., & Digerfeldt, G. (1982a). New ^{14}C dates from Hunneberg supporting the revised deglaciation chronology of the Middle Swedish end moraine zone. *Geologiska Föreningens i Stockholm Förhandlingar* 103, 395–404.

Björck, S., & Digerfeldt, G. (1982b). Late Weichselian shore displacement at Hunneberg, southern Sweden, indicating complex uplift. *Geologiska Föreningens i Stockholm Förhandlingar* 104, 132–55.

Björck, S., & Digerfeldt, G. (1984). Climatic changes at Pleistocene/Holocene boundary in the Middle Swedish endmoraine zone, mainly inferred from stratigraphic indications. In *Climatic Changes on a Yearly to Millennial Basis*, ed. N.-A. Mörner & W. Karlén, pp. 37–56. Dordrecht: Reidel.

Björck, S., & Digerfeldt, G. (1986). Late Weichselian–Early Holocene shore displacement west of Mt. Billingen, within the Middle Swedish endmoraine zone. *Boreas* 15, 1–18.

Björck, S., & Håkansson, S. (1982). Radiocarbon dates from Late Weichselian lake sediments in South Sweden as a basis for chronostratigraphic subdivision. *Boreas* 11, 141–50.

Björck, S., & Möller, P. (1987). Late Weichselian environmental history of southeastern Sweden during the deglaciation of the Scandinavian ice sheet. *Quaternary Research* 28, 1–37.

Björck, S., Sandegren, P., & Holmquist, B. (1987). A magnetostratigraphic comparison between ^{14}C years and varve years during the Late Weichselian, indicating significant differences between the time-scales. *Journal of Quaternary Science* 2, 133–40.

Björklund, K. R., Bjørnstad, H., Dale, B., Erlenkeuser, H., Henningsmoen, K. E., Høeg, H. I., Johnsen, K., Manum, S. B., Mikkelsen, N., Nagy, J., Pederstad, K., Quale, G., Rosenqvist, I. T., Salbu, B., Schoenharting, G., Stabell, B., Thiele, J., Throndsen, I., Wassman, P., & Werner, F. (1985). Evolution of the Upper Quaternary depositional environment in the Skagerrak: A synthesis. *Norsk Geologisk Tidsskrift* 65, 139–49.

Björkman, H., & Trägårdh, J. (1982). Differential uplift in Blekinge indicating late-glacial neotectonics. *Geologiska Föreningens i Stockholm Förhandlingar* 104, 75–9.

Björnbom, S. (1979). Clayey basal till in central and northern Sweden: A deposit from an old phase of the Würm glaciation. *Sveriges Geologiska Undersökning* C 753.

Blystad, P., Thomsen, H., Simonsen, A., & Lie, R. W. (1983). Find of a nearly complete Late Weichselian polar bear skeleton, *Ursus maritimus* Phipps, at Finnøy, southwestern Norway: A preliminary report. *Norsk Geologisk Tidsskrift* 63, 193–7.

Bogaard, P. van der, & Schmincke, H.-U. (1985). Laacher See Tephra: A widespread isochronous late Quaternary tephra layer in central and northern Europe. *Geological Society of America Bulletin* 96, 1554–71.

Borell, R., & Offerberg, J. (1955). Geokronologiska undersökningar inom Indalsälvens dalgång mellan Bergeforsen och Ragunda. *Sveriges Geologiska Undersökning* Ca 31.

Borgen, U. (1979). Ett fynd av fossil myskoxe i Jämtland och något om myskoxarnas biologi och historia. *Fauna och Flora* 74(1).

Bouchard, M. A., Gibbard, P., & Salonen, V.-P. (1990). Lithostratotypes for Weichselian and pre-Weichselian sediments in southern and western Finland. *Bulletin of the Geological Society of Finland* 62(1), 79–95.

Bouchard, M. A., & Salonen, V.-P. (1990). Boulder transport in shield areas. In *Glacier indicator tracing*, ed. R. Kujansuu & M. Saarnisto, pp. 87–107. Rotterdam: Balkema.

Boulton, G. S., Smith, G. D., Jones, A. S., & Newsome, J.

(1985). Glacial geology and glaciology of the last mid-latitude ice sheet. *Journal of the Geological Society of London* 142, 447–74.

Bowen, D. Q. (1978). *Quaternary geology,* Oxford: Pergamon.

Bowen, D. Q., Rose, J., McCabe, A. M., & Sutherland, D. G. (1986). Correlation of Quaternary glaciations in England, Ireland, Scotland and Wales. *Quaternary Science Reviews* 5, 299–340.

Brander, G. (1934). Suomen Geologinen Yleiskartta, Lehti C 3, Kuopio. *Maalajikartan selitys, Suomen Geologinen Toimikunta – Geologiska Kommissionen i Finland.*

Brander, G. (1937). Ein Interglazialfund bei Rouhiala in Südostfinnland. *Bulletin de la Commission géologique de Finlande* 118.

Brander, G. (1943). Neue Beiträge zur Kenntnis der Interglazialen Bildungen in Finnland. *Comptes Rendus de la Société géologique de Finlande* 15, 87–137.

Brandt, A. (1949). Über die Entwicklung der Moore im Küstengebiet von Süd-Pohjanmaa am Bottnischen Meerbuse. *Annales Botanici Societatis Zoologicae Botanicae Fennicae Vanamo* 23(4).

Brotzen, F. (1961). An Interstadial (Radiocarbon dated) and the substages of the last glaciation in Sweden. *Geologiska Föreningens i Stockholm Förhandlingar* 83, 144–50.

Cato. I. (1982). The landslide at Tuve 1977 and the complex origin of the clays in south-western Sweden. In *Landslides and Mudflows,* pp. 297–283. Reports of Alma-Ata International Seminar, October 1981, UNESCO, Center of International Projects, GKNT, Moscow.

Cato, I. (1985). The definitive connection of the Swedish geochronological time scale with the present, and the new date of the zero year in Döviken, northern Sweden. *Boreas* 14, 117–22.

Cato, I. (1987). On the definitive connection of the Swedish time scale with the present. *Sveriges Geologiska Undersökning* Ca 68.

Cepek, A. G. (1986). Quaternary stratigraphy of the German Democratic Republic. *Quaternary Science Reviews* 5, 359–64.

Cepek, A. G. (1990). On the correlation of Scandinavian glaciations within and before the Cromerian Complex. *The Cromer Symposium, Norwich 1990,* SEQS, INQUA Subcommission on European Stratigraphy, Abstracts, 9.

Chanda, S. (1965). The history of vegetation of Brøndmyra, a Late-glacial and early Post-glacial deposit in Jaeren South Norway. *Årbok for Universitetet i Bergen. Matematisk–Naturvitenskapelig Serie* 1965(1).

Chebotareva, N. S., & Makarycheva, I. A. (1982). Paleogeography of the treeless stage of the Valdai Russian Plain. *Quaternary Studies in Poland* 3, 15–20.

Christensen, L. (1973a). Geologisk tolkning af afgrødemønstre i landbrugsjorder. *Det nye Dansk landbrug* 6, 13–18.

Christensen, L. (1973b). Fossile polygonmønstre i jyske landbrugsjorder. *Ugeskrift for agronomer og hortonomer* 1973(2), 102–7.

Christensen, L. (1974). Crop-marks revealing large-scale patterned ground structures in cultivated areas, southwestern Jutland, Denmark. *Boreas* 3, 153–80.

Dahl, E. (1947). Litt om forholdena under og etter siste istid i Norge. *Naturen* 1947(7-8), 232–252.

Dahl, E. (1955). Biogeographic and geologic indications of unglaciated areas in Scandinavia during the glacial ages. *Bulletin of the Geological Society of America* 66, 1499–1519.

Dahl. R. (1965). Plastically sculptured detail forms on rock surfaces in northern Nordland, Norway. *Geografiska Annaler* 47 A, 83–140.

Dahl, R. (1966). Block fields, weathering pits and torlike forms in the Narvik mountains, Nordland, Norway. *Geografiska Annaler* 48 A, 55–85.

Dahl, R. (1967). Post-glacial micro-weathering of bedrock surfaces in the Narvik district of Norway. *Geografiska Annaler* 49 A, 155–66.

Dahl, R. (1972). The question of glacial survival in western Scandinavia in relation to the modern view of the Late Quaternary climate history. *Ambio Special Report* 1972(2), 45–9.

Daly, R. A. (1934). *The changing world of the ice age.* New Haven: Yale University Press.

Danielsen, A. (1970). Pollen-analytical Late Quaternary studies in the Ra district of Østfold, southeast Norway. *Årbok for Universitetet i Bergen. Matematisk-Naturvetenskapelig Serie* 1969(14).

Danielsen, A. (1971). Skandinaviens fjellflora i lys av senkvartaer vegetationshistorie. *Blyttia* 29(4), 183–209.

Dansgaard, W., White, J. W. C., & Johnsen, S. J. (1989). The abrupt termination of the Younger Dryas climate event. *Nature* 339, 532–4.

Davis, W. M. (1899). The peneplain. *American Geologist* 23, 207–39.

De Geer, G. (1896). *Om Skandinaviens geografiska utveckling efter istiden.* Stockholm: Norstedt & Söners.

De Geer, G. (1940). Geochronologia Suecica. Principles. *Kungliga Svenska Vetenskapsakademiens Handlingar* Serie 3, 18(6).

Délibrias, G. (1985). Le Carbone 14. In *Méthodes de datation par les phénomènes naturels. Applications,* ed. E. Roth & B. Poty, pp. 421–458. Collection CEA, Series Scientifique, Paris: Masson.

Delusin, I. (1991). The Holocene pollen stratigraphy of Lake Ladoga and the vegetational history of its surroundings. *Annales Academiae Scientiarum Fennicae* A III 153.

Digerfeldt, G. (1972). The Post-glacial development of Lake Trummen. *Folia Limnologica Scandinavica* 16.

Digerfeldt, G. (1974). The Post-glacial development of the Ramviken bay in Lake Immeln. I: The history of the regional vegetation; II: The water-level changes. *Geologiska Föreningens i Stockholm Förhandlingar* 96, 3–32.

Digerfeldt, G. (1975a). The Post-glacial development of Ramviken bay in Lake Immeln. III. Palaeolimnology. *Geologiska Föreningens i Stockholm Förhandlingar* 97, 13–28.

Digerfeldt, G. (1975b). Post-glacial water-level changes in Lake Växjösjön, central southern Sweden. *Geologiska Föreningens i Stockholm Förhandlingar* 97, 167–73.

Digerfeldt, G. (1976). A Pre-Boreal water-level change in Lake Lyngsjö, central Halland. *Geologiska Föreningens i Stockholm Förhandlingar* 98, 329–39.

Digerfledt, G., Battarbee, R. W., & Bengtsson, L. (1975). Report on annually laminated sediment in lake Järlasjön, Nacka, Stockholm. *Geologiska Föreningens i Stockholm Förhandlingar* 97, 29–40.

Ditlefsen, C., & Knudsen, K. L. (1990). Marine kvartaere aflejringer ved Skaershøj på Thyholm, Nordjylland. *Dansk Geologisk Forening, Årsskrift* 1987–89, 71–5.

Donner, J. (1951). Pollen-analytical studies of late-glacial deposits in Finland. *Comptes Rendus de la Société géologique de Finlande* 24, 1–92.

Donner, J. (1963). The zoning of the Post-glacial diagrams in

Finland and the main changes in the forest composition. *Acta Botanica Fennica* 65.

Donner, J. (1965a). The Quaternary of Finland. In *The geologic systems: the Quaternary, Volume 1*, ed. K. Rankama, pp. 199–272. New York: Interscience Publishers.

Donner, J. (1965b). Shore-line diagrams in Finnish Quaternary research. *Baltica* 2, 11–20.

Donner, J. (1966). A comparison between the Late-glacial and Post-glacial shore-lines in Estonia and south-western Finland. *Societas Scientiarum Fennica, Commentationes Physico-Mathematicae* 31(11).

Donner, J. (1969a). Land/sea level changes in southern Finland during the formation of the Salpausseklä endmoraines. *Bulletin of the Geological Society of Finland* 41, 135–50.

Donner, J. (1969b). A profile across Fennoscandia of Late Weichselian and Flandrian shore-lines. *Societas Scientiarum Fennica, Commentationes Physico-Mathematicae* 36(1).

Donner, J. (1970). Deformed Late Weichselian and Flandrian shore-lines in south-eastern Fennoscandia. *Societas Scientiarum Fennica, Commentationes Physico-Mathematiceae* 40, 191–8.

Donner, J. (1971). Towards a stratigraphical division of the Finnish Quaternary. *Societas Sceintiarum Fennica, Commentationes Physico-Mathematicae* 41, 281–305.

Donner, J. (1972). Pollen frequencies in the Flandrian sediments of Lake Vakojärvi, south Finland. *Societas Scientiarum Fennica, Commentationes Biologicae* 53.

Donner, J. (1978). The dating of the levels of the Baltic Ice Lake and the Salpausselkä moraines in south Finland. *Societas Scientiarum Fennica, Commentationes Physico-Mathematicae* 48, 11–38.

Donner, J. (1979). Dateringen av den geologiska utvecklingen i Finland efter istiden. *Societas Scientiarum Fennica, Årsbok-Vuosikirja* 56(3).

Donner, J. (1980). The determination and dating of synchronous Late Quaternary shorelines in Fennoscandia. In *Earth rheology, isostasy and eustasy,* ed. N. Mörner, pp. 285–93. Chichester: Wiley.

Donner, J. (1982). Fluctuations in water level of the Baltic Ice Lake. *Annales Academiae Scientiarum Fennicae* A III 134, 13–28.

Donner, J. (1983). The identification of Eemian interglacial and Weichselian interstadial deposits in Finland. *Annales Academiae Scientiarum Fennicae* A III 136,

Donner, J. (1986). Weichselian indicator erratics in the Hyvinkää area, southern Finland. *Annales Academiae Scientiarum Fennicae* A III 140.

Donner, J. (1987). Some aspects of the nature of the Late Weichselian and Holocene uplift in Finland. *Geological Survey of Finland, Speical Paper* 2, 9–12.

Donner, J. (1988). The Eemian site of Norinkylä compared with other interglacial and interstadial sites in Ostrobothnia, western Finland. *Annales Academiae Scientiarum Fennicae* A III 149,

Donner, J. (1989). Transport distances of Finnish crystalline erratics during the Weichselian glaciation. *Geological Survey of Finland, Special Paper* 7, 7–13.

Donner, J. (1991). The pollen analytical identification and definition of Quaternary interglacials and interstadials in Fennoscandia, with special reference to Finland. In *Current Perspectives in Palynological Research,* ed. S. Chanda, pp. 139–50. *Journal of Palynology* 1990–91.

Donner, J. (1992). Is there evidencce in the zone of the first

Salpausselkä moraine in Finland of a transgression of the Baltic Ice Lake before its drainage? *Sveriges Geologiska Undersökning* Ca 81, 87–90.

Donner, J., Alhonen, P., Eronen, M., Jungner, H., & Vuorela, I. (1978). Biostratigraphy and radiocarbon dating of the Holocene lake sediments of Työtjärvi and the peats in the adjoining bog Varrassuo west of Lahti in southern Finland. *Annales Botanici Fennici* 15, 258–80.

Donner, J., Eronen, M., & Jungner, H. (1977). The dating of the Holocene relative sea-level changes in Finnmark, North Norway. *Norsk geografisk Tidsskrift* 31, 103–28.

Donner, J., & Gardemeister, R. (1971). Redeposited Eemian marine clay in Somero, south-western Finland. Appendix by R. Tynni. *Bulletin of the Geological Society of Finland* 43, 73–88.

Donner, J., & Jungner, H. (1973). The effect of re-deposited organic material on radiocarbon measurements of clay samples from Somero, south-western Finland. *Geologiska Föreningens i Stockholm Förhandlingar* 95, 267–8.

Donner, J., & Jungner, H. (1974). Errors in the radiocarbon dating of deposits in Finland from the time of deglaciation. *Bulletin of the Geological Society of Finland* 46, 139–44.

Donner, J., Jungner, H., & Kurtén, B. (1979). Radiocarbon dates of mammoth finds in Finland compared with radiocarbon dates of Weichselian and Eemian deposits. *Bulletin of the Geological Society of Finland* 51, 45–54.

Donner, J., Korpela, K., & Tynni, R. (1986). Veiksel-jääkauden alajaotus Suomessa. *Terra* 98(3), 240–7.

Donner, J., Lappalainen, V., & West, R. G. (1968). Ice wedges in south-eastern Finland. *Geologiska Föreningens i Stockholm Förhandlingar* 90, 112–16.

Donner, J., & Raukas, A. (1989). On the geological history of the Baltic Ice Lake. *Proceedings of the Academy of Sciences of the Estonian SSR, Geology* 38(4), 128–37.

Donner, J., & Raukas, A. (1992). Baltic Ice Lake. In *Geology of the Gulf of Finland,* ed. A. Raukas & H. Hyvärinen, pp. 262–76. Tallinn: Estonian Academy of Sciences.

Dreimanis, A., & Lundqvist, J. (1984). What should be called till? *Striae* Uppsala, 20, 5–10.

Ehlers, J. (1983). Different till types in North Germany and their origin. In *Tills and Related Deposits,* ed. E. B. Evenson, C. Schlüchter, & J. Rabassa, pp. 61–79. Rotterdam: Balkema.

Ehlers, J. (1990a). Reconstructing the dynamics of the North-west European Pleistocene ice sheets. *Quaternary Science Reviews* 9, 71–83.

Ehlers, J. (1990b). Untersuchungen zur Morphodynamik der Vereisungen Norddeutschlands unter Berücksichtigung benachbarten Gebiete. *Bremer Beträge zur Geographie und Raumplannung* 19.

Ehlers, J., & Gibbard, P. (1991). Anglian glacial deposits in Britain and the adjoining offshore regions. In *Glacial deposits in Great Britain and Ireland,* ed. J. Ehlers, P. L. Gibbard, & J. Rose, pp. 17–24. Rotterdam: Balkema.

Ehlers, J., Gibbard, P., & Rose, J. (1991). Glacial deposits in Britain and Europe. General overview. In *Glacial deposits in Great Britain and Ireland,* ed. J. Ehlers, P. L. Gibbard, & J. Rose, pp. 493–501. Rotterdam: Balkema.

Ehlers, J., & Linke, G. (1989). The origin of deep buried channels of Elsterian age in Northwest Germany. *Journal of Quaternary Science* 4(3), 255–65.

Ehlers, J., Meyer, K.-D., & Stephan, H.-J. (1984). Pre-Weichselian glaciations of North-West Europe. *Quaternary Science Reviews* 3, 1–40.

Ehlers, J., & Wingfield, R. (1991). The extension of the Late Weichselian / Late Devensian ice sheets in the North Sea Basin. *Journal of Quaternary Science* 6(4), 313–26.

Eissmann, L. (1990). Das mitteleuropäische Umfeld der Eemvorkommen des Saale-Elbe-Gebietes und Schlussfolgerungen zur Stratigrafie des jüngeren Quartärs. *Altenburger Naturwissenschaftliche Forschungen* 5, 11–48.

Ekman, I. (1987). Minor lakes in Soviet Karelia. In *Palaeohydrology of the Temperate Zone. II: Lakes*, ed. A. Raukas & L. Saarse, pp. 43–63. Tallinn: Valgus.

Ekman, I., & Iljin, V. (1991). Deglaciation, the Younger Dryas end moraines and their correlation in the Karelian S.S.S.R. and adjacent areas. In *Eastern Fennoscandian Younger Dryas end moraines, Excursion Guide*, ed. H. Rainio & M. Saarnisto, pp. 73–101. Geological Survey of Finland Guide 32.

Ekman, M. (1989). Impacts of geodynamic phenomena on systems for height and gravity. *Bulletin Géodésique* 63, 281–96.

Ekman, M. (1991). A concise history of postglacial uplift research (from its beginning to 1950). *Terra Nova* 3, 358–65.

Elina, G., & Filimonova, L. (1987). Late-glacial vegetation on the territory of Karelia. In *Palaeohydrology of the Temperate Zone. III: Mires and lakes*, ed. A. Raukas & L. Saarse, pp. 53–68. Tallinn: Valgus.

Erd, K. (1970). Pollen-analytical classification of the Middle Pleistocene in the German Democratic Republic. *Palaeogeography, Palaeoclimatology, Palaeoecology* 8, 129–45.

Erikson, B. (1912). En submorän fossilförande aflagring vid Bollnäs i Hälsingland. *Geologiska Föreningens i Stockholm Förhandlingar* 34, 500–41.

Eriksson, B., Grönlund, T., & Kujansuu, R. (1980). Interglasiaalikerrostuma Evijärvellä, Pohjanmaalla. *Geologi, Suomen Geologinen Seura – Geologiska Sällskapet i Finland* 1980(6), 65–71.

Eriksson, B., Aalto, M., & Kankainen, T. (1992). Hazelnuts from peat deposits at Evijärvi, western Finland. *Bulletin of the Geological Society of Finland* 63(2), 141–8.

Eriksson, K. G. (1960). Studier över Stockholmsåsen vid Halmsjön. *Geologiska Föreningens i Stockholm Förhandlingar* 82, 43–125.

Eronen, M. (1974). The history of the Litorina Sea and Associated Holocene events. *Societas Scientiarum Fennica, Commentationes Physico-Mathematiceae* 44(4), 79–195.

Eronen, M. (1979). The retreat of pine forest in Finnish Lapland since the Holocene climatic optimum: a general discussion with radiocarbon evidence from subfossil pines. *Fennia* 157(2), 93–114.

Eronen, M. (1983). Late Weichselian and Holocene shore displacement in Finland. In *Shorelines and Isostasy*, ed. D. E. Smith & A. G. Dawson, pp. 183–207. Institute of British Geographers, Special Publication, 16. London: Academic Press.

Eronen, M. & Haila, H. (1981). The highest shore-line in the Baltic in Finland. *Striae* (Uppsala) 14, 157–8.

Eronen, M., & Huttunen, P. (1987). Radiocarbon-dated subfossil pines from Finnish Lapland. *Geografiska Annaler* 69 A, 297–303.

Eronen, M., Uusinoka, R., & Jungner, H. (1979). ^{14}C-ajoitettu sinisimpukkalöytö Seinäjoelta ja tietoa muista kuoriesiintymistä Itämeren piirissä. *Terra* 91(4), 209–18.

Eronen, M., & Vesajoki, H. (1988). Deglaciation pattern indicated by the ice-margin formations in Northern Karelia, eastern Finland. *Boreas* 17, 317–27.

Eskola, P. (1933). Tausend Geschiebe aus Lettland. *Annales Academiae Scientiarum Fennicae* A 39(5).

Faegri, K. (1940). Quartärgeologische Untersuchungen im westlichen Norwegen. II. Zur spätquartären Geschichte Jaerens. *Bergens Museums Årbok 1939–40, Naturvidenskapelig rekke* 7.

Faegri, K., & Iversen, J. (1975). *Textbook of pollen analysis*, 3d ed. Copenhagen: Munksgaard.

Faure, H. (1980). Base de reflexion pour la stratigraphie du Quaternaire. *Supplément au Bulletin de l'Association Française pour l'Etude du Quaternaire* N.S. 1, 343–4.

Faustova, M. A. (1984). Late Pleistocene Glaciation of European USSR. In *Late Quaternary environments of the Soviet Union*, ed. A. A. Velichko, pp. 3–12. London: Longman.

Feyling-Hanssen, R. W. (1955). Stratigraphy of the marine late-Pleistocene of Billefjorden, Vestspitsbergen. *Norsk Polarinstitutt Skrifter* 107.

Feyling-Hanssen, R. W. (1964). A Late Quaternary correlation chart for Norway. *Norges Geologiske Undersökelse* 223, 67–91.

Feyling-Hanssen, R. W. (1971). Weichselian interstadial foraminifera from the Sandnes-Jaeren area. *Bulletin of the Geological Society of Denmark* 21, 72–116.

Feyling-Hanssen, R. W. (1974). The Weichselian section of Voss-Eigeland, south-western Norway. *Geologiska Föreningens i Stockholm Förhandlingar* 96, 341–53.

Feyling-Hanssen, R. W. (1981). Foraminiferal indication of Eemian interglacial in the northern North Sea. *Bulletin of the Geological Society of Denmark* 29, 175–89.

Feyling-Hanssen, R. W. (1982). Foraminiferal zonation of a boring in Quaternary deposits of the northern North Sea. *Bulletin of the Geological Society of Denmark* 31, 29–47.

Feyling-Hanssen, R. W., & Knudsen, K. L. (1979). Foraminiferer og deres betydning i skandinavisk kvartaer geologi. *Dansk Natur, Dansk Skole, Årsskrift* 1979, 3–37.

Firbas, F. (1949). *Spät- und nacheiszeitliche Waldgeschichte Mitteleuropas nördlich der Alpen*, vol. 1. Jena: Gustav Fisher.

Flint, R. F. (1971). *Glacial and Quaternary Geology*. New York: Wiley.

Florin, M.-B. (1979). The Younger Dryas vegetation at Kolmården in southern Central Sweden. *Boreas* 8, 145–52.

Fogelberg, P. (1986). Berggrundens relief i Fennoskandien – gamla och nya åsikter om utvecklingen. *Societas Scientiarum Fennica, Sphinx, Årsbok-Vuosikirja* B 61-62, 57–69.

Forsström, L. (1982). The Oulainen Interglacial in Ostrobothnia, western Finland. *Acta Universitatis Ouluensis* A 136, *Geologica* 4.

Forsström, L., Aalto, M., Eronen, M., & Grönlund, T. (1988). Stratigraphic evidence for Eemian crustal movements and relative sea-level changes in eastern Fennoscandia. *Palaeogeography, Palaeoclimatology, Palaeoecology* 68, 317–35.

Forsström, L., & Eronen, M. (1985). Flandrian and Eemian shore levels in Finland and adjacent areas – a discussion. *Eiszeitalter und Gegenwart* 35, 135–45.

Forsström, L., Eronen, M., & Grönlund, T. (1987). On marine phases and shore levels of the Eemian Interglacial and Weichselian interstadials on the coast of Ostrobothnia, Finland. *Geological Survey of Finland, Special Paper* 2, 37–42.

Forstén, A., & Alhonen, P. (1975). The subfossil seals of Finland and their relation to the history of the Baltic Sea. *Boreas* 4, 143–55.

Fredén, C. (1967). A historical review of the Ancylus Lake and the Svea River. *Geologiska Föreningens i Stockholm Förhandlingar* 89, 239–67.

Fredén, C. (1980). The Quaternary history of the Baltic. The western part. In *The Quaternary history of the Baltic*, ed. V. Gudelis & L.-K. Königsson, pp. 59–74. Acta Universitatis Upsaliensis, Symposia Universitatis Upsaliensis Annum Quingentesimum Celebrantis, 1, Uppsala.

Fredén, C. (1981). Late Quaternary events at Degerfors, Sweden. *Striae* (Uppsala), 14, 159–62.

Fredén, C. (1988). Marine life and deglaciation chronology of the Vänern basin, southwestern Sweden. *Sveriges Geologiska Undersökning* Ca 71.

Fredericia, J., & Knudsen, K. L. (1990). Geological framework in the Skagen area. *Special Issue No. 9. Journal of Coastal Research, Proceedings Skagen Symposium Sept. 2–5 1990*:2, 647–59.

Fredskild, B. (1975). A late-glacial and early post-glacial pollen-concentration diagram from Langeland, Denmark. *Geologiska Föreningens i Stockholm Förhandlingar* 97, 151–61.

Frenzel, B. (1973). On the Pleistocene vegetation history. *Eiszeitalter und Gengenwart* 23/24, 321–32.

Frenzel, B. (1989). Theoretische Grundprobleme der botanischen Biostratigraphie des Eiszeitalters. In *Quaternary type sections: imagination or reality*, ed. J. Rose & C. Schlüchter, pp. 33–9. Rotterdam: Balkema.

Fries, M. (1951). Pollenanalytiska vittnesbörd om senkvartär vegetationsutveckling, särskilt skogshistoria, i nordvästra Götaland. *Acta Phytogeographica Suecica* 29.

Fries, M. (1965). Outlines of the Late-Glacial and Postglacial vegetational and climatic history of Sweden, illustrated by three generalized pollen diagrams. *The Geological Society of America, Special Paper* 84, 55–64.

Fromm, E. (1970). An estimation of errors in the Swedish varve chronology. In *Radiocarbon variations and absolute chronology*, ed. I. U. Olsson, pp. 163–72. Stockholm: Almqvist & Wiksell.

Fyfe, G. J. (1990). The effect of water depth on ice-proximal glaciolacustrine sedimentation: Salpausselkä I, southern Finland. *Boreas* 19, 147–64.

Gams, H. (1950). Die Alleröd-Schwankung im Spätglazial. *Zeitschrift für Gletscherkunde und Glazialgeologie* I(2), 162–71.

Garcia Ambrosiani, K. (1990). Pleistocene stratigraphy in central and northern Sweden – a reinvestigation of some classical sites. *University of Stockholm, Department of Quaternary Geology, Report* 16.

Garcia Ambrosiani, K. (1991). Interstadial minerogenic sediments at the Leveäniemi mine, Svappavaara, Swedish Lapland. *Geologiska Föreningens i Stockholm Förhandlingar* 113, 273–87.

Garcia Ambrosiani, K., & Robertsson, A.-M. (1992). Early Weichselian interstadial sediments at Härnösand, Sweden. *Boreas* 21, 305–17.

Garnes, K., & Bergersen, O. F. (1977). Distribution and genesis of tills in central South Norway. *Boreas* 6, 135–47.

Garnes, K., & Bergersen, O. F. (1980). Wastage features of the inland ice sheet in central South Norway. *Boreas* 9, 251–69.

Gerasimov, I. P., Serebryanny, L. R., & Cebotareva, N. S. (1965). Stratigraphische Gliederung des Pleistozäns im nördlichen Mittel- und Osteuropa und ihre Korrelation. *Berichte der geologische Gesellschaft DDR* 10(1), 89–94.

Gibbard, P. L. (1979). Late Pleistocene stratigraphy of the area around Muhos, North Finland. *Annales Academiae Scientiarum Fennicae* A III 129.

Gibbard, P., Forman, S., Salomaa, R., Alhonen P., Jungner, H., Peglar, S., Suksi, J., & Vuorinen, A. (1989). Late Pleistocene stratigraphy at Harrinkangas, Kauhajoki, Western Finland. *Annales Academiae Scientiarum Fennicae* A III 150.

Gibbard, P., & Saarnisto, M. (1977). Periglacial phenomena at Tohmajärvi, eastern Finland. *Geologiska Föreningens i Stockholm Förhandlingar* 99, 295–8.

Gillberg, G. (1961). The middle-Swedish moraines in the province of Dalsland, W Sweden. *Geologiska Föreningens i Stockholm Förhandlingar* 83, 335–69.

Gillberg, G. (1965). Till distribution and ice movements on the northern slopes of the south Swedish highlands. *Geologiska Föreningens i Stockholm Förhandlingar* 86, 433–84.

Gillberg, G. (1967). Further discussion of the lithological homogeneity of till. *Geologiska Föreningens i Stockholm Förhandlingar* 89, 29–49.

Gillberg, G. (1968). Lithological distribution and homogeneity of glaciofluvial material. *Geologiska Föreningens i Stockholm Förhandlingar* 90, 189–204.

Gjessing, J. (1967). Norway's Paleic surface. *Norsk geografisk Tidsskrift* 21, 69–132.

Glückert, G. (1970). Vorzeitliche Uferentwicklung am ersten Salpausselkä in Lohja, Südfinnland. *Annales Universitas Turkuensis* A II 45.

Glückert, G. (1973). Two large drumlin fields in Central Finland. *Fennia* 120.

Glückert, G. (1974). Map of glacial striation of the Scandinavian ice sheet during the last (Weichsel) glaciation in northern Europe. *Bulletin of the Geological Society of Finland* 46, 1–8.

Glückert, G. (1976). Post-glacial shore-level displacement of the Baltic in SW Finland. *Annales Academiae Scientiarum Fennicae* A III 118.

Glückert, G. (1978). Östersjöns postglaciala strandförskjutning och skogens historia på Åland. *Publications of the Department of Quaternary Geology, University of Turku* 34.

Glückert, G. (1979). Itämeren ja metsien historia Salpausselkävyöhykkeessä Uudenmaan länsiosassa. *Publications of the Department of Quaternary Geology, University of Turku* 39,

Glückert, G. (1987). Zur letzten Eiszeit im alpinen und nordäuropäischen Raum. *Geographica Helvetica* 1987(2), 93–8.

Glückert, G. (1991). The Ancylus and Litorina transgression of the Baltic in southwest Finland. *Quaternary International* 9, 27–32.

Godwin, H. (1962). Half-life of radiocarbon. *Nature* 195, 984.

Godwin, H. (1975). *The history of the British flora*, 2d ed. Cambridge University Press.

Graham, A. L., Bevan, A. W. R., & Hutchinson, R. (1985). *Catalogue of meteorites*. London: British Museum, Natural History.

Grichuk, V. P. (1961). Fossil floras as a paleontological basis for the stratigraphy of Quaternary deposits in the northwestern part of the Russian Plain (in Russian). *Relyef i stratigrafiya chetvertlichnykh olozheni Severo-Zapada Russkoi ravniny*, pp. 25–71. Moscow: Nauka.

Grönlie, A. (1941). Contribution to the Quaternary chronology. *Det Kongelige Norske Videnskabers Selskab Forhandlinger* 14(12), 43–6.

Grönlund, T. (1977). Tertiäärisiä piileväesiintymiä Lapista.

Geological Survey of Finland, Report of Investigations 17, 19–30.

Grönlund, T. (1982). Tammerkosken iästä. *Geologi, Suomen Geologinen Seura – Geologiska Sällskapet i Finland* 1982 (6), 115–17.

Grönlund, T. (1991). The diatom stratigraphy of the Eemian Baltic Sea on the basis of sediment discoveries in Ostrobothnia, Finland. *Geological Survey of Finland, Report of Investigations* 102.

Gross, H. (1954). Das Alleröd-Interstadial als Leithorizont der letzten Vereisung in Europa und Amerika. *Eiszeitalter und Gegenwart* 4/5, 189–209.

Gross, H. (1967). Geochronologie des letzten Interglazials im nördlichen Europa mit besondered Berücksichtung der UdSSR. *Schriften des naturwissenschaftlichen Vereins für Schleswig-Holstein* 37, 11–125.

Grube, F., Christensen, S., & Vollmer, T. (1986). Glaciations in North West Germany. *Quaternary Science Reviews* 5, 347–358.

Gustafson, G. (1976). Fynd av subfossila växtdelar i Dalarna. *XII Nordiska Geologvintermötet, Göteborg, Abstracts*, 20.

Guthrie, R. D. (1990). *Frozen fauna of the Mammoth Steppe*. Chicago and London: University of Chicago Press.

Hafsten, U. (1956). Pollen-analytic investigations on the late Quaternary development in the inner Oslofjord area. *Årbok for Universitet i Bergen. Naturvitenskabelig rekke* 8.

Hafsten, U. (1957). Om mistelteinens og bergflettens historie i Norge. *Blyttia* 15, 43–60.

Hafsten, U. (1963). A late-glacial pollen profile from Lista, South Norway. *Grana Palynologica* 4(2), 326–37.

Hafsten, U. (1983). Shore-level changes in South Norway during the last 13,000 years, traced by biostratigraphical methods and radiocarbon datings. *Norsk geografisk Tidsskrift* 37, 63–79.

Hafsten, U. (1985). The immigration and spread of spruce forest in Norway, traced by biostratigraphical studies and radiocarbon datings. A preliminary report. *Norsk geografisk Tidsskrift* 39, 99–108.

Hafsten, U., Henningsmoen, K. E., & Høeg, H. I. (1979). Invandringen av gran til Norge. In *Fortiden i søkelyset*, ed. R. Nydal, S. Westin, U. Hafsten, & S. Gulliksen, pp. 171–198. Trondheim: Laboratoriet for Radiologisk Datering.

Halden, B. (1948). Nya data rörande det interglaciala Bollnäsfyndet. *Sveriges Geologiska Undersökning* C 495, 24–37.

Hamberg, M. J., & Harland, W. B., ed. (1981). *Earth's pre-Pleistocene glacial record*. Cambridge University Press.

Hammen, T. van der, Maarleweld, G. C., Vogel, J. C., & Zagwijn, W. H. (1967). Stratigraphy, climatic succession and radiocarbon dating of the last glacial in the Netherlands. *Geologie en Mijnbouw* 46(3), 79–95.

Hammen, T. van der, Wijmstra, T. A., & Zagwijn, W. H. (1971). The floral record of the late Cenozoic of Europe. In *The Late Cenozoic glacial ages*, ed. K. K. Turekian, pp. 391–424. New Haven: Yale University Press.

Hansen, S. (1940). Varvighet i danske og skaanske senglaciale aflejringer. *Danmarks Geologiske Undersøgelse* II 63.

Hansen, S. (1965). The Quaternary of Denmark. In *The geologic systems; the Quaternary, Volume 1*, ed. K. Rankama, pp. 1–90. New York: Interscience Publishers.

Härme, M. (1949). On a pre-glacial weathering in Tyrvää, southwestern Finland. *Comptes Rendus de la Société géologique de Finlande* 22, 87–9.

Hart, J. K., & Boulton, G. S. (1991). The interrelation of glaciotectonic and glaciodepositional processes within the glacial environment. *Quaternary Science Reviews* 10, 335–50.

Hausen, H. (1912). Undersökning af porfyrblock från sydvästra Finlands glaciala aflagringar. *Fennia* 32(2).

Hedberg, H. D., ed. (1976). *International stratigraphic guide.* New York: Wiley.

Heikkinen, A. (1975). Radiohiilimenetelmän käytöstä moreenin sisältämän vanhan hiiliaineksen ajoituksessa. *Geologi, Suomen Geologinen Seura – Geologiska Sällskapet i Finland* 27(4), 51–3.

Heikkinen, A., & Äikää, O. (1977). Geological Survey of Finland radiocarbon measurements VII. *Radiocarbon* 19(2), 263–79.

Heinonen, L. (1957). Studies on the microfossils in the tills of the North European glaciation. *Annales Academiae Scientiarum Fennicae* A III 52.

Heintz, A. E., & Garutt, V. E. (1965). Determination of the absolute age of the fossil remains of Mammoth and Woolly Rhinoceros from the permafrost in Siberia by the help of radiocarbon (C 14). *Norsk Geologisk Tidsskrift* 45, 73–9.

Heintz, N., Garnes, K., & Nydal, R. (1979). Norske og sovjetiske mammutfunn i kvartaergeologisk perspektiv. In *Fortiden i søkelyset*, ed. R. Nydal, S. Westin, U. Hafsten, & S. Gulliksen, pp. 209–25. Trondheim: Laboratoriet for Radiologisk Datering.

Hellaakoski, A. (1922). Suursaimaa. *Fennia* 43(4).

Hellaakoski, A. (1928). Puulan järviryhmän kehityshistoriasta. *Fennia* 51(2).

Hellaakoski, A. (1930). On the transportation of materials in the esker of Laitila. *Fennia* 52(7).

Helle, M., Sønstergaard, E., Coope, G. R., & Rye, N. (1981). Early Weichselian peat at Brumunddal, southeastern Norway. *Boreas* 10, 369–79.

Henningsmoen, K. (1979). En karbon-datert strandforskyvningskurve fra søndre Vestfold. In *Fortiden i Søkelyset*, ed. R. Nydal, S. Westin, U. Hafsten, & S. Gulliksen, pp. 239–47. Trondheim: Laboratoriet for Radiologisk Datering.

Hessland, I. (1943). Marine Schalenablagerungen Nord-Bohusläns. *Bulletin of the Geological Institutions of the University of Upsala* 31.

Hessland, I. (1945). On the Quaternary Mya period in Europe. *Arkiv för Zoologi, Stockholm* 37 A 8.

Hillefors, Å. (1966). Iskilar i Norra Halland. *Svensk Geografisk Årsbok* 42, 134–44.

Hillefors, Å. (1969). Västsveriges glaciala historia och morfologi. *Meddelanden från Lunds Universitets Geografiska Institution, Avhandlingar* 60.

Hillefors, Å. (1974). The stratigraphy and genesis of the Dösebacka and Ellesbo drumlins. A contribution to the knowledge of the Weichsel-glacial history in western Sweden. *Geologiska Föreningens i Stockholm Förhandlingar* 96, 335–74.

Hillefors, Å. (1979). Deglaciation models from the Swedish west coast. *Boreas* 8, 153–69.

Hillefors, Å. (1983). The Dösebacka and Ellesbo drumlins – morphology and stratigraphy. In *Glacial deposits in North-West Europe*, ed. J. Ehlers, pp. 141–50. Rotterdam: Balkema.

Hillefors, Å. (1985). Deep-weathered rock in western Sweden. *Fennia* 163(2), 293–301.

Hintikka, V. (1963). Über das Grossklima einiger Pflanzenareale in zwei Klimakoordinatensystemen dargestelt. *Annales Botanici Societatis Zoologicae Botanicae Fennicae Vanamo* 34(5).

Hirvas, H. (1983). Correlation problems of interglacial deposits in Finnish Lapland. *IGCP Project 73/1/24, Quaternary glaciations in the northern hemisphere. Report (Paris)* 9, 129–39.

Hirvas, H. (1991). Pleistocene stratigraphy of Finnish Lapland. *Geological Survey of Finland Bulletin* 354.

Hirvas, H., Alfthan, A., Pulkkinen, E., Puranen, R., & Tynni, R. (1977). Raportti malminetsintää palvelevasta maaperä-tutkimuksesta Pohjois-Suomessa vuosina 1972–1976. *Geological Survey of Finland, Report of Investigation* 19.

Hirvas, H., & Nenonen, K. (1987). The till stratigraphy of Finland. In *INQUA Till Symposium, Finland 1985*, ed. R. Kujansuu & M. Saarnisto, pp. 49–63. *Geological Survey of Finland, Special Paper* 3.

Hirvas, H., & Niemelä, J. (1986). Ryytimaa, Vimpeli. In *17e Nordiska geologmötet 1986. Excursion Guide, excursion C 2*, ed. M. Haavisto-Hyvärinen, pp. 47–50. *Geological Survey of Finland Guide* 15.

Hirvas, H., & Tynni, R. (1976). Tertiääristä savea Savukoskella sekä havaintoja tertiäärisistä mikrofossiileista. *Geologi, Suomen Geologinen Seura – Geologiska Sällskapet i Finland* 1976(3), 33–40.

Hjort, C., & Sundquist, B. (1979). *Geologi och miljö*. Stockholm: Natur och Kultur.

Holtedahl, H. (1955). On the Norwegian continental terrace, primarily outside Møre–Romsdal: its geomorphology and sediments. *Årbok for Universitet i Bergen, Naturvidenskapelig rekke* 14.

Holtedahl, H. (1958). Some remarks on geomorphology of continental shelves off Norway, Labrador, and Southeast Alaska. *Journal of Geology* 66(4), 461–71.

Holtedahl, H. (1964). An Allerød fauna at Os, near Bergen, Norway. *Norsk Geologisk Tidsskrift* 44(3), 315–22.

Holtedahl, H. (1975). The geology of the Hardangerfjord, West Norway. *Norges Geologiske Undersøkelse* 323.

Holtedahl, O. (1953). Norges Geologi. *Norges Geologiske Undersøkelse* 164, I & II.

Holtedahl, O. (1956). Av den norske landoverflates historie. *Det Kongelige Norske Videnskabers Selskabs Forhandlinger* 28, 45–66.

Holtedahl, O. (1960). Geology of Norway. *Norges Geologiske Undersøkelse* 208.

Hoppe, G. (1952). Hummocky moraine regions with special reference to the interior of Norrbotten. *Geografiska Annaler* 1952(1–2),

Hoppe, G. (1959). Några kritiska kommentarer till diskussionen om isfria refugier. *Svensk Naturvetenskap* 1959, 123–34.

Hoppe, G., & Liljequist, G. H. (1956). Det sista nedisnings-förloppet i Nordeuropa och dess meteorologiska bakgrund. *Ymer* 1956(1), 43–74.

Hörnsten, Å. (1964). Ångermanlands kustland under isavsmält-ningsskedet. Preliminärt meddelande. *Geologiska Föreningens i Stockholm Förhandlingar* 86, 181–205.

Hörnsten, Å., & Olsson, I. U. (1964). En C¹⁴-datering av glaciallera från Lugnvik, Ångermanland. *Geologiska Föreningens i Stockholm Förhandlingar* 86, 206–10.

Houmark-Nielsen, M. (1988). Glaciotectonic unconformities in Pleistocene stratigraphy as evidence for the behavior of former Scandinavian icesheets. In *Glaciotectonics: forms and processes*, ed. D. G. Croot, pp. 91–9. Rotterdam: Balkema.

Houmark-Nielsen, M. (1989). Danmark i istiden – en tegne-serie. *Varv, Copenhagen* 1989(2), 43–72.

Houmark-Nielsen, M. (1990). The last interglacial–glacial cycle in Denmark. *Quaternary International* 3/4, 31–9.

Houmark-Nielsen, M., & Berthelsen, A. (1981). Kineto-stratigraphic evaluation and presentation of glacial-stratigraphic data, with examples from northern Samsø, Denmark. *Boreas* 10, 411–22.

Houmark-Nielsen, M., & Kolstrup, E. (1981). A radiocarbon dated Weichselian sequence from Sejerø, Denmark. *Geologiska Föreningens i Stockholm Förhandlingar* 103, 73–8.

Hultén, E. (1950). *Atlas of the distribution of vascular plants in NW Europe* Stockholm: Esselte.

Huntley, B. (1990). European post-glacial forests: compositional changes in response to climatic change. *Journal of Vegetation Science* 1, 507–18.

Huntley, B., & Prentice, I. C. (1988). July temperatures in Europe from pollen data, 6000 years before present. *Science* 241, 687–90.

Hütt, G., Molodkov, A., Kessel, H. N., & Raukas, A. (1985). ESR dating of subfossil Holocene shells in Estonia. *Nuclear Tracks*, 10 891–8.

Hütt, G., Punning, J.-M., & Mangerud, J. (1983). Thermoluminescence dating of the Eemian–Early Weichselian sequence at Fjøsanger, western Norway. *Boreas* 12, 227–31.

Hütt, G., Punning, J.-M., & Raukas, A. (1983). Application of the TL method to elaborate the geochronological scale of the Late Pleistocene for the East European Plain. In *Correlation of Quaternary chronologies*, ed. W. C. Mahaney, pp. 47–55. Toronto: Geo Books.

Huttunen, P., Meriläinen, J., & Tolonen, K. (1978). The history of a small dystrophied forest lake, southern Finland. *Polskie Archivum Hydrobiologii* 25, 189–202.

Hyvärinen, H. (1966a). Studies on the late-Quaternary history of Pielis-Karelia, eastern Finland. *Societas Scientiarum Fennica, Commentationes Biologicae* 29(4).

Hyvärinen, H. (1966b). A shore-line diagram for the easternmost section of the Salpausselkäs. *Societas Scientiarum Fennica, Commentationes Physico-Mathematicae* 33(4).

Hyvärinen, H. (1971a). Two Late Weichselian stratigraphical sites from the foreland of the Salpausselkäs in Finland. *Societas Scientiarum Fennica, Commentationes Biologicae* 40.

Hyvärinen, H. (1971b). Ilomantsi Ice Lake: a contribution to the Late Weichselian history of eastern Finland. *Societas Scientiarum Fennica, Commentationes Physico-Mathematicae* 41, 171–8.

Hyvärinen, H. (1972). Flandrian regional pollen assemblage zones in eastern Finland. *Societas Scientiarum Fennica, Commentationes Biologicae* 59.

Hyvärinen, H. (1973). The deglaciation history of eastern Fennoscandia – recent data from Finland. *Boreas* 2, 85–102.

Hyvärinen, H. (1975a). Absolute and relative pollen diagrams from northernmost Fennoscandia. *Fennia* 142.

Hyvärinen, H. (1975b). Myöhäisjääkauden Fennoskandia – käsityksiä ennen ja nyt. *Terra* 87(3), 155–66.

Hyvärinen, H. (1976). Flandrian pollen deposition rates and tree-line history in northern Fennoscandia. *Boreas* 5, 163–75.

Hyvärinen, H. (1980). Relative sea-level changes near Helsinki, southern Finland, during early Litorina times. *Bulletin of the Geological Society of Finland* 52(2), 207–19.

Hyvärinen, H. (1988). Definition of the Baltic stages. *Annales Academiae Scientiarum Fennicae* A III 148, 7–11.

Hyvärinen, H., Donner, J., Kessel, H., & Raukas, A. (1988). The Litorina Sea and Limnaea Sea in the northern and central Baltic. *Annales Academiae Scientiarum Fennicae* A III 148, 25–35.

Hyyppä, E. (1936). Über die spätquartäre Entwicklung Nordfinnlands mit Ergänzungen zur Kenntnis des spätglazialen Klimas. *Comptes Rendus de la Société géologique de Finlande* 9, 401–65.

Hyyppä, E. (1937). Post-glacial changes of shore-line in South Finland. *Bulletin de la Commission géologique de Finlande* 120.

Hyyppä, E. (1950). Helsingin ymäristö. Maaperäkartan selitys. *Geologinen Tutkimuslaitos (Geological Survey of Finland)* 11–53.

Hyyppä, E. (1960). Quaternary geology of eastern and northern Finland. *Guide to excursion No. C 35 – International Geological Congress, XXI Session, Norden.* Helsinki: Finnish Organizing Committee for the International Geological Congress 1960.

Hyyppä, E. (1966). The Late-Quaternary land uplift in the Baltic sphere and the relation diagram of the raised and tilted shore levels. *Annales Academiae Scientiarum Fennicae* A III 90, 153–68.

Ilvessalo, Y. (1956). Suomen metsät vuosista 1921–24 vuosiin 1951–53. *Communicationes Instituti Forestalis Fennicae* 47(1).

Iregren, E., Ringberg, B., & Robertsson, A.-M. (1990). The brown bear (*Ursus arctos* L) find from Ugglarp, southernmost Sweden. *Sveriges Geologiska Undersökning* C 824.

Iversen, J. (1942). En pollenanalytisk tidsfaestelse af ferskvandslagene ved Nørre Lyngby. *Meddelelser fra Dansk Geologisk Forening* 10(2), 130–51.

Iversen, J. (1944). Viscum, Hedera and Ilex as climatic indicators. *Geologiska Föreningens i Stockholm Förhandlingar* 66, 463–83.

Iversen, J. (1947). Plantevaekst, dyreliv og klima i det senglaciale Danmark. *Geologiska Föreningens i Stockholm Förhandlingar* 69, 67–78.

Iversen, J. (1953). Radiocarbon dating of the Alleröd period. *Science* 118 (3053), 6–11.

Iversen, J. (1954). The Late-Glacial flora of Denmark and its relation to climate and soil. *Danmarks Geologiske Undersøgelse* II 80, 87–119.

Iversen, J. (1958). The bearing of glacial and interglacial epochs on the formation and extinction of plant taxa. *Uppsala Universitets Årsskrift* 1958(6), 210–15.

Iversen, J. (1964). Plant indicators of climate, soil, and other factors during the Quaternary. *Report of the VIth International Congress on Quaternary, Warsaw 1961. Vol. II: Palaeobotanical Section* (Lodz), 421–28.

Iversen, J. (1973). The development of Denmark's nature since the last glacial. *Danmarks Geologiske Undersøgelse* V 7-C.

Jansen, J. H., van Weering, T. C. E., & Eisma, D. (1979). Late Quaternary sedimentation in the North Sea. In *The Quaternary history of the North Sea*, ed. E. Oele, R. T. E. Schüttenhelm, & A. J. Wiggers, pp. 175–187. Acta Universitatis Upsaliensis, Symposia Universitatis Upsaliensis Annum Quingentesimum Celebrantis, 2, Uppsala.

Jansen, E., Sejrup, H. P., Fjaeran, T., Hald, M., Holtedahl, H., & Skarbø, O. (1983). Late Weichselian paleoceanography of the southeastern Norwegian Sea. *Norsk Geologisk Tidsskrift* 63, 117–46.

Janssen, C. R. (1974). *Verkenningen in de palynologie.* Utrecht: Oostoek, Scheltema, & Holkema.

Jantunen, T. (1990). Maankohoamisen saranalinjoista Suomen kallioperässä: tutkimuskohteena Pikkalan–Lahden murroslaakso. *Terra* 102(3), 158–63.

Järnefors, B. (1963). Lervarvskronologien och isrecessionen i östra Mellansverige. *Sveriges Geologiska Undersökning, C* 594.

Jelgersma, S. (1979). Sea-level changes in the North Sea basin. In *The Quaternary history of the North Sea*, ed. E. Oele, R. T. E. Schüttenhelm, & A. J. Wiggers, pp. 233–248. Acta Universitatis Upsaliensis, Symposia Universitatis Upsaliensis Annum Quingentesimum Celebrantis, 2, Uppsala.

Jenkins, D. G. (1987). Was the Pliocene-Pleistocene boundary placed at the wrong stratigraphic level? *Quaternary Science Reviews* 6, 41–2.

Jensen, F. B., & Knudsen, K. L. (1984). Kvartaerstratigrafiska undersøgelser ved Gyldendal og Kås Hoved i det vestlige Limfjordområde. *Dansk Geologisk Forening, Årsskrift* 1983, 35–54.

Jensen, K. A., & Knudsen, K. L. (1988). Quaternary foraminiferal stratigraphy in boring 81/29 from central North Sea. *Boreas* 17, 273–87.

Jerz, H., & Linke, G. (1987). Arbeitsergebnisse der Subkommission für Europäische Quartärstratigraphie: Typusregion des Holstein-Interglazials. *Eiszeitalter und Gegenwart* 37, 145–8.

Jessen, K. (1935). Archaeological dating in the history of north Jutland's vegetation. *Acta Archaeologica* 5(3), 185–214.

Jessen, K., & Milthers, V. (1928). Stratigraphical and paleontological studies of interglacial fresh-water deposits in Jutland and Northwest Germany. *Danmarks Geologiske Undersøgelse* II 48.

Johansen, O.-I., Henningsmoen, K. E., & Sollid, J. L. (1985). Deglasiasjonsforløpet på Tingvollhalvøya og tilgrensende områder, Nordvestlandet, i lys av vegetasjonsutviklingen. *Norsk geografisk Tidsskrift* 39, 155–74.

Johnsson, G. (1956). Glacialmorfologiska studier i södra Sverige. *Meddelanden från Lunds Universitets Geografiska Institution, Avhandlingar* 31.

Johnsson, G. (1981). Fossil patterned ground in southern Sweden. *Geologiska Föreningens i Stockholm Förhandlingar* 103, 79–89.

Johnsson, G. (1986). Different types of fossil frost fissures in south Sweden. *Geologiska Föreningens i Stockholm Förhandlingar* 108, 167–75.

Jungner, H. (1983). Preliminary investigations on TL dating of geological sediments from Finland. *Council of Europe, Strasbourg, PACT* 9, 565–72.

Jungner, H. (1987). Thermoluminescence dating of sediments from Oulainen and Vimpeli, Ostrobothnia, Finland. *Boreas* 16, 231–5.

Kahlke, H. D. (1981). *Das Eiszeitalter.* Cologne: Aulis Verlag Deubner.

Kaitanen, V. (1969). A geographical study of the morphogenesis of northern Lapland. *Fennia* 99(5).

Kaitanen, V. (1985). Problems concerning the origin of inselbergs in Finnish Lapland. *Fennia* 163(2), 359–64.

Kakkuri, J. (1985). Die Landhebung in Fennoskandien im Lichte der heutigen Wissenschaft. *Zeitschrift für Vermessungswesen* 110(2), 51–9.

Kakkuri, J. (1987). Character of the Fennoscandian land uplift in the 20th century. *Geological Survey of Finland, Special Paper* 2, 15–20.

Kakkuri, J. (1992). Fennoscandian maankohoaminen, uplift of

Fennoscandia. In *Atlas of Finland, Geology Folio 123–126*, ed. P. Alalammi, pp. 35–6. Helsinki: Maanmittaushallitus.

Kaland, P. E. (1984). Holocene shore displacement and shorelines in Hordaland, western Norway. *Boreas* 13, 203–42.

Karlén, W. (1982). Holocene glacier fluctuations in Scandinavia. In *Holocene glaciers*, ed. W. Karlén, pp. 26–34. *Striae* (Uppsala), 18.

Katz, N. J., Katz, S. V., & Kipiani, M. G. (1965). *Atlas and keys of fruits and seeds occurring in the Quaternary deposits of the USSR*. Academy of Sciences of the USSR, Commission for investigations of the Quaternary Period, Moscow: Nauka.

Kejonen, A. (1985). Weathering in the Wyborg rapakivi area, southeastern Finland. *Fennia* 163(2), 309–13.

Kessel, H., & Raukas, A. (1980). The Quaternary history of the Baltic. Estonia. In *The Quaternary history of the Baltic*, ed. V. Gudelis & L.-K. Königsson, 127–46. Acta Universitatis Upsaliensis, Symposia Universitatis Upsaliensis Annum Quingentesimum Celebrantis, 1, Uppsala.

Kilpi, S. (1937). Das Sotkamo-Gebiet in spätglazialer Zeit. *Bulletin de la Commission géologique de Finlande* 117, 1–118.

Klingberg, F. (1989). A revaluation of the Göta Älv interstadial to Late Weichselian. *Geologiska Föreningens i Stockholm Förhandlingar* 111, 51–2.

Knudsen, K. L. (1978). Middle and Late Weichselian marine deposits at Nørre Lungby, northern Jutland, Denmark, and their foraminiferal faunas. *Danmarks Geologiske Undersøgelse* II 112.

Knudsen, K. L. (1984). Foraminiferal stratigraphy in a marine Eemian-Weichselian sequence at Apholm, North Jutland. *Bulletin of the Geological Society of Denmark* 32, 169–80.

Knudsen, K. L. (1985a). Foraminiferal stratigraphy of Quaternary deposits in the Roar, Skjold and Dan fields, central North Sea. *Boreas* 14, 311–24.

Knudsen, K. L. (1985b). Foraminiferal faunas in Eemian deposits of the Oldenbüttel area near the Kiel Canal, Germany. *Geologisches Jahrbuch* A 86, 27–47.

Knudsen, K. L. (1985c). Correlation of Saalian, Eemian and Weichselian foraminiferal zones in North Jutland. *Bulletin of the Geological Society of Denmark* 33, 325–39.

Knudsen, K. L. (1986a). Quaternary foraminiferal stratigraphy: a review of recent Nordic research. *Striae* (Uppsala), 24, 35–8.

Knudsen, K. L. (1986b). Middle and Late Quaternary foraminiferal stratigraphy in the southern and central North Sea area. *Striae* (Uppsala), 24, 201–5.

Knudsen, K. L. (1987). Elsterian-Holsteinian foraminiferal stratigraphy in the North Jutland and Kattegat areas, Denmark. *Boreas* 16, 359–68.

Knudsen, K. L. (1988). Marine interglacial deposits in the Cuxhaven area, NW Germany: A comparison of Holsteinian, Eemian and Holocene foraminiferal faunas. *Eiszeitalter und Gegenwart* 38, 69–77.

Knudsen, K. L. (1989). Elsterian and Holsteinian faunas from the Wacken clay pit, Schleswig-Holstein, NW Germany. *Geologisches Jahrbuch* A 114, 3–14.

Knudsen, K. L., & Feyling-Hanssen, R. W. (1976). Ergebnisse der Foraminiferanalyse zur Quartärstratigrafie in Skandinavien. *Eiszeitalter und Gegenwart* 27, 82–92.

Knudsen, K. L., & A.-L. Lykke-Andersen. (1982). Foraminifera in Late Saalian, Early and Middle Weichselian of the Skaerumhede I boring. *Bulletin of the Geological Society of Denmark* 30, 97–109.

Knudsen, K. L., & Penney, D. N. (1987). Foraminifera and Ostracoda in Late Elsterian–Holsteinian deposits at Tornskov and adjacent areas in Jutland, Denmark. *Danmarks Geologiske Undersøgelse* B 10.

Koenigswald, W. von (1988). Paläoklimatische Aussage letztinterglazialer Säugetiere aus der nördliche Oberrheinebene. In *Zur Paläoklimatologie des letzten Interglazials im Nordteil der Oberrheinebene*, ed. W. von Koenigswald, pp. 205–314. Stuttgart: Gustav Fischer.

Kokkola, M. (1989). Is the till matrix transported – or is the way to its study wrong? *Geological Survey of Finland, Special Paper* 7, 55–8.

Kolp, O. (1986). Entwicklungsphasen des Ancylus-Sees. *Petermanns Geographische Mitteilungen*, 1986(2), 79–94.

Kolstrup, E. (1986). Reappraisal of the Upper Weichselian periglacial environment from Danish frost wedge casts. *Palaeogeography, Palaeoclimatology, Palaeoecology* 56, 237–49.

Kolstrup, E., Grün, R., Mejdahl, V., Packman, S. C., & Wintle, A. G. (1990). Stratigraphy and thermoluminescence dating of Late Glacial cover sand in Denmark. *Journal of Quaternary Science* 5(3), 207–24.

Kolstrup, E., & Jørgensen, J. B. (1982). Older and Younger coversand in southern Jutland (Denmark). *Bulletin of the Geological Society of Denmark* 30, 71–7.

Königsson, L.-K. (1977). Weathering in the fissure systems in southern Öland. *Geologiska Föreningens i Stockholm Förhandlingar* 99, 384–94.

Königsson, L.-K. (1980). The Quaternary history of the Baltic. The development of the Baltic during the Pleistocene. In *The Quaternary history of the Baltic*, ed. V. Gudelis & L.-K. Königsson, pp. 87–97. Acta Universitatis Upsaliensis, Symposia Universitatis Upsaliensis Annum Qungentesimum Celebrantis, 1, Uppsala.

Korpela, K. (1969). Die Weichsel-Eiszeit und ihr Interstadial in Peräpohjola (nördliches Nordfinnland) im Licht von submoränen Sedimenten. *Annales Academiae Scientiarum Fennicae* A III 99.

Kozarski, S. (1988). Time and dynamics of the last Scandinavian ice-sheet retreat from northwestern Poland. *Geographia Polonica* 55, 91–101.

Kristiansen, I. L., Mangerud, J., & Lømo, L. (1988). Late Weichselian / Early Holocene pollen- and lithostratigraphy in lakes in the Ålesund area, western Norway. *Review of Palaeobotany and Palynology* 53, 185–231.

Kristiansson, J. (1982). Varved sediments and the Swedish time scale. *Geologiska Föreningens i Stockholm Förhandlingar* 104, 273–5.

Kristiansson, J. (1986). The ice recession in the south-eastern part of Sweden. *University of Stockholm, Department of Quaternary Geology, Report* 7.

Krog, H. (1954). Pollen analytical investigation of a C[14]-dated Allerød section from Ruds Vedby. *Danmarks Geologiske Undersøgelse* II 80, 120–39.

Krog, H. (1965). On the post-glacial development of the Great Belt. *Baltica* 2, 47–60.

Krog, H. (1968). Late-glacial and Postglacial shoreline displacement in Denmark. In *Means of correlation of Quaternary successions*, ed. R. B. Morrison & H. E. Wright, Jr., pp. 421–35. Proceedings VII Congress International Association for Quaternary Research, 8. Salt Lake City: University of Utah Press.

Krog, H. (1973). The early Post-glacial development of the Store Belt as reflected in a former fresh water basin. *Danmarks Geologiske Undersøgelse, Årbog* 1972, 37–47.

Krog, H. (1978). The Late Weichselian freshwater beds at Nørre Lyngby: C-14 dates and pollen diagram. *Danmarks Geologiske Undersøgelse, Årbog* 1976, 29–43.

Krog, H. (1979). Late Pleistocene and Holocene shorelines in western Denmark. In *The Quaternary history of the North Sea*, ed. E. Oele, R. T. E. Schüttenhelm, & A. J. Wiggers, pp. 75–83. Acta Universitatis Upsaliensis, Symposia Universitatis Upsaliensis Annum Quingentesimum Celebrantis, 2, Uppsala.

Krumbein, W. C. (1937). Sediments and exponential curves. *Journal of Geology* 45, 577–601.

Krzywinski, K., & Stabell, B. (1984). Late Weichselian sea level changes at Sotra, Hordaland, western Norway. *Boreas* 13, 159–202.

Kuivamäki, A., & Vuorela, P. (1985). Käytetyn ydinpolttoaineen loppusijoitukseen vaikuttavat geologiset ilmiöt Suomen kallioperässä. *Geological Survey of Finland, Nuclear Waste Disposal Research, Report* YST-47.

Kujansuu, R. (1964). Nuorista siirroksista Lapissa. *Geologi, Suomen Geologinen Seura – Geologiska Sällskapet i Finland* 1964(3-4), 30–6.

Kujansuu, R. (1967). On the deglaciation of western Finnish Lapland. *Bulletin de la Commission géologique de Finlande* 232.

Kujansuu, R. (1972). Interstadiaalikerrostuma Vuotsossa. *Geologi, Suomen Geologinen Seura – Geologiska Sällskapet i Finland* 1972(5-6), 53–6.

Kukkamäki, T. J. (1968). Report on the work of the Fennoscandian Sub-Commission. *Third Symposium of the CRCM in Leningrad, May 22–29, 1968.*

Kukkamäki, T. J. (1971). Report on the work of the Fennoscandian Sub-Commission. *Symposium on Recent Movements of the Crust, Moscow, August 1971.*

Kulonpalo, M. (1969). Kärnäiittilohkareita keski ja etelä-Suomessa eli Suomen pisin lohkarevasta. *Geologi, Suomen Geologinen Seura – Geologiska Sällskapet i Finland* 1969(6), 80–1.

Kurimo, H. (1979). Deglaciation and early post-glacial hydrography in northern Kainuu and Peräpohjola, north-east Finland: A glacial morphological study. *Publications of the University of Joensuu* B II 10.

Kurkinen, I., & Niemelä, J. (1979). Rapautumahavainto Kauhajoella. *Geologi, Suomen Geologinen Seura – Geologiska Sällskapet i Finland* 1979(5), 79–81.

Kurkinen, I., Niemelä, J. & Tikkanen, J. (1989). Hienoainesmoreenia harjudeltan pohjalla. *Geologi, Suomen Geologinen Seura – Geologiska Sällskapet i Finland* 1989(3), 47–50.

Kurtén, B. (1966). A Late-Glacial find of Arctic fox (Alopex lagopus L.) from southwestern Finland. *Societas Scientiarum Fennica, Commentationes Biologicae* 29(6).

Kurtén, B. (1968). *Pleistocene mammals of Europe.* London: Weidenfeld and Nicolson.

Kurtén, B. (1986). Pleistocene mammals in Europe. *Striae* (Uppsala) 24, 47–9.

Kurtén, B. (1988). Fossil and subfossil mammals in Finland. *Memoranda Societatis pro Fauna et Flora Fennica* 64, 35–9.

Kvamme, T., Mangerud, J., Furnes, H., & Ruddiman, W. F. (1989). Geochemistry of Pleistocene ash zones in cores from the North Atlantic. *Norsk Geologisk Tidsskrift* 69, 251–72.

Kvasov, D. D. (1979). The Late-Quaternary history of large lakes and inland seas of Eastern Europe. *Annales Academiae Scientiarum Fennicae* A III 127.

Lagerbäck, R. (1988a). The Veiki moraines in northern Sweden – widespread evidence of an Early Weichselian deglaciation. *Boreas* 17, 469–86.

Lagerbäck, R. (1988b). Periglacial phenomena in the wooded areas of northern Sweden. *Boreas* 17, 487–99.

Lagerbäck, R. (1990). Late Quaternary faulting and paleoseismicity in northern Fennoscandia, with particular reference to the Lansjärv area, northern Sweden. *Geologiska Föreningens i Stockholm Förhandlingar* 112, 333–54.

Lagerbäck, R., & Robertsson, A.-M. (1988). Kettle holes – stratigraphical archives for Weichselian geology and palaeoenvironment in northernmost Sweden. *Boreas* 17, 439–68.

Lagerlund, E. (1983). The Pleistocene stratigraphy of Skåne, southern Sweden. In *Glacial deposits in North-West Europe*, ed. J. Ehlers, pp. 155–9. Rotterdam: Balkema.

Lahti, S. I. (1985). Porphyritic pyroxene-bearing granitoids – a strongly weathered rock group in central Finland. *Fennia* 163(2), 315–21.

Landvik, J. Y., & Hamborg, M. (1987). Weichselian glacial episodes in outer Sunnmøre, western Norway. *Norsk Geologisk Tidsskrift* 67, 107–23.

Lang, G. (1985). Palynological research in Switzerland 1925–1985. *Dissertationes Botanicae* 87, 11–82.

Lappalainen, V. (1962). The shore-line displacement on southern Lake Saimaa. *Acta Botanica Fennica* 64.

Larsen, E., Eide, F., Longva, O., & Mangerud, J. (1984). Allerød–Younger Dryas climatic inferences from cirque glaciers and vegetational development in the Nordfjord area, western Norway. *Arctic and Alpine Research* 16(2), 137–60.

Larsen, E., & Holtedahl, H. (1985). The Norwegian strandflat: A reconsideration of its age and origin. *Norsk Geologisk Tidsskrift* 65, 247–54.

Larsen, E., Longva, O., & Follestad, B. A. (1991). Formation of De Geer moraines and implications for deglaciation dynamics. *Journal of Quaternary Science* 6(4), 263–77.

Larsen, E., & Sejrup, H. P. (1990). Weichselian land-sea interactions: Western Norway–Norwegian Sea. *Quaternary Science Reviews* 9, 85–97.

Laurén, L., Lehtovaara, J., Boström, R., & Tynni, R. (1978). On the geology and the Cambrian sediments of the circular depression at Söderfjärden, western Finland. *Geological Survey of Finland Bulletin* 297.

Lavrova, M. (1961). Relationship between interglacial boreal transgression in the northern part of the USSR and Eemian in western Europe (in Russian). *Trudy Instituta Geologii Akademii Nauk, ESSR* 8, 65–88.

Lemdahl, G. (1988). Paleoclimatic and palaeoecological studies based on subfossil insects from Late Weichselian sediments in southern Sweden. *University of Lund, Department of Quaternary Geology, Thesis* 22.

Lepiksaar, J. (1986). The Holocene history of the theriofauna in Fennoscandia and Baltic countries. *Striae* (Uppsala) 24, 51–70.

Lepiksaar, J. (1992). Remarks on the Weichselian megafauna (*Mammuthus, Coelodonta* and *Bison*) of the "interglacial" area around the Baltic basin. *Annales Zoologicae Fennici* 28(3-4), 229–40.

Lidén, R. (1913). Geokronologiska studier öfver det finiglaciala skedet i Ångermanland. *Sveriges Geologiska Undersøkning* Ca 9.

Lidén, R. (1938). Den senkvartära strandförskjutningens förlopp och kronologi i Ångermanland. *Geologiska Föreningens i Stockholm Förhandlingar* 60, 397–404.

Lidmar-Bergström, K. (1988). Preglacial weathering and landform evolution in Fennoscandia. *Geografiska Annaler* 70 A, 273–83.

Liiva, A., Kessel, H., & Aaloe, A. (1979). Ilumetsa kraatrite vanus. *Eesti Loodus* 22, 762–4.

Liivrand, E. (1984). The interglacials of Estonia. *Annales Academiae Scientiarum Fennicae* A III 138.

Liivrand, E. (1991). Biostratigraphy of the Pleistocene deposits in Estonia and correlations in the Baltic region. *University of Stockholm, Department of Quaternary Geology, Report* 19.

Liljequist, G. H. (1974). Notes on the meteorological conditions in connection with continental land-ices in the Pleistocene. *Geologiska Föreningens i Stockholm Förhandlinger* 96, 293–8.

Liljegren, R. (1975). Subfossila vertebratfynd från Skåne. *University of Lund, Department of Quaternary Geology, Report* 8.

Liljegren, R. (1977). Jättehjorten och dess senglaciala miljö i Skåne. *Limhamniana* 1977.

Liljegren, R., & Lagerås, P. (1993). *Från mammutstäpp till kohage. Drjurens historia i Sverige.* Lund: Wallin & Dalholm.

Lindén, A. (1975). Till petrographical studies in an Archaean bedrock area in southern central Sweden. *Striae* (Uppsala) 1.

Lilliesköld, M. (1990). Lithology of some Swedish eskers. *University of Stockholm, Department of Quaternary Geology, Report* 17.

Lindroos, P. (1972). On the development of late-glacial and postglacial dunes in North Karelia, eastern Finland. *Geological Survey of Finland Bulletin* 254.

Linke, G. ed. (1986). Holstein – Symposium. *Guidebook to the Excursions, International Union of Quaternary Research, Subcommission on European Quaternary Stratigraphy,* Hamburg.

Linke, G. Katzenberger, O., & Grün, R. (1985). Description and ESR dating of the Holsteinian interglaciation. *Quaternary Science Reviews* 4, 319–31.

Ljungner, E. (1949). East-west balance of the Quaternary ice caps in Patagonia and Scandinavia. *Bulletin of the Geological Institutions of the University of Uppsala* 33, 11–96.

Long, D., & Morton, A. C. (1987). An ash fall within the Loch Lomond Stadial. *Journal of Quaternary Science* 2(2), 97–101.

Lukashov, A. D. (1982). *Guidebook for excursions A-4, C-4, Karelia, International Union for Quaternary Research, XI Congress, Moscow.*

Ludwig, A. O. (1980). The Quaternary history of the Baltic. The southern part. In *The Quaternary history of the Baltic,* ed. V. Gudelis & L.-K. Königsson, pp. 41–48. Acta Universitatis Upsaliensis, Symposia Universitatis Upsaliensis Annum Quingentesimum Celebrantis, 1, Uppsala.

Lundqvist, J. (1958). Beskrivning till jordartskarta över Värmlands län. *Sveriges Geologiska Undersökning* Ca 38.

Lundqvist, J. (1962). Patterned ground and related frost phenomena in Sweden. *Sveriges Geologiska Undersökning* C 583.

Lundqvist, J. (1965). The Quaternary in Sweden. In *The geologic systems: the Quaternary, Volume 1,* ed. K. Rankama, pp. 139–198. New York: Interscience Publishers.

Lundqvist, J. (1967). Submoräna sediment i Jämtlands län. *Sveriges Geologiska Undersökning* C 618.

Lundqvist, J. (1969a). Beskrivning till jordartskarta över Jämtlands län. *Sveriges Geologiska Undersökning* Ca 45.

Lundqvist, J. (1969b). Earth and ice mounds: a terminological discussion. In *The periglacial environment,* ed. T. L. Péwé, pp. 203–15. Arctic Institute of North America. Montreal: McGill-Queen's University Press.

Lundqvist, J. (1971). The interglacial deposit at Leveäniemi mine, Svappavaara, Swedish Lapland. *Sveriges Geologiska Undersökning* C 658.

Lundqvist, J. (1972). Ice-lake types and deglaciation pattern along the Scandinavian mountain range. *Boreas* 1, 27–54.

Lundqvist, J. (1973). Isavsmältningens förlopp i Jämtlands län. *Sveriges Geologiska Undersökning* C 681.

Lundqvist, J. (1974). Outlines of the Weichselian Glacial in Sweden. *Geologiska Föreningens i Stockholm Förhandlingar* 96, 327–39.

Lundqvist, J. (1977). Till in Sweden. *Boreas* 6, 73–85.

Lundqvist, J. (1978). New information about Early and Middle Weichselian interstadials in northern Sweden. *Sveriges Geologiska Undersökning* C 752.

Lundqvist, J. (1981). Weichselian in Sweden before 15,000 B.P. *Boreas* 10, 395–402.

Lundqvist, J. (1985). Deep-weathering in Sweden. *Fennia* 163(2), 287–92.

Lundqvist, J. (1986a). Stratigraphy of the central area of the Scandinavian glaciation. *Quaternary Science Reviews* 5, 251–68.

Lundqvist, J. (1986b). Late Weichselian glaciation and deglaciation in Scandinavia. *Quaternary Science Reviews* 5, 269–92.

Lundqvist, J. (1987a). Beskrivning till jordartskarta över Västernorrlands län och förutvarande fjällsjö k:n. *Sveriges Geologiska Undersökning* Ca 55.

Lundqvist, J. (1987b). Glaciodynamics of the Younger Dryas marginal zone in Scandinavia. *Geografiska Annaler* 69 A, 305–19.

Lundqvist, J. (1988a). Late glacial ice lobes and glacial landforms in Scandinavia. In *Genetic classification of glacigenic deposits,* ed. R. P. Goldthwait & C. L. Matsch, pp. 217–25. Rotterdam: Balkema.

Lundqvist, J. (1988b). Younger Dryas – Preboral moraines and deglaciation in southwestern Värmland, Sweden. *Boreas* 17, 301–16.

Lundqvist, J., & Lagerbäck, R. (1976). The Pärve fault: A lateglacial fault in the Precambrian of Swedish Lapland. *Geologiska Föreningens i Stockholm Förhandlingar* 98, 45–51.

Lundqvist, J., & Mejdahl, V. (1987). Thermoluminescence dating of eolian sediments in Central Sweden. *Geologiska Föreningens i Stockholm Förhandlingar* 109, 147–158.

Lundqvist, J., & Miller, U. (1992). Weichselian stratigraphy and glaciations in the Tåsjö-Hoting area, Central Sweden. *Research Papers, Sveriges Geologiska Undersökning* C 826.

Lundqvist, J., & Pleijel, C. (1976). Mammutbeten från Kånkback, Ragunda s:n, Jämtland. *Fauna och flora* (Stockholm) 71(3), 89–132.

Lykke-Andersen, A.-L. (1987). A Late Saalian, Eemian and Weichselian marine sequence at Nørre Lyngby, Vendsyssel, Denmark. *Boreas* 16, 345–57.

Magnusson, E. (1962). An interglacial deposit at Gallejaure, northern Sweden. *Geologiska Föreningens i Stockholm Förhandlingar* 84, 363–71.

Mäkinen, K. (1979). Interstadiaalinen turvekerrostuma Tervolan Kauvonkankaalla. *Geologi, Suomen Geologinen Seura – Geologiska Sällskapet i Finland* 1979(5), 82–87.

Mangerud, J. (1970). Late Weichselian vegetation and ice-front oscillations in the Bergen district, western Norway. *Norsk geografisk Tidsskrift* 24, 121–48.

Mangerud, J. (1972). Radiocarbon dating of marine shells, including a discussion of apparent age of recent shells from Norway. *Boreas* 1, 143–72.

Mangerud, J. (1973). Isfrie refugier i Norge under istidene. *Norges Geologiske Undersøkelse* 297.

Mangerud, J. (1977). Late Weichselian marine sediments containing shells, foraminifera, and pollen, at Ågotnes, western Norway. *Norsk Geologisk Tidsskrift* 57, 23–54.

Mangerud, J. (1980). Ice-front variations at different parts of the Scandinavian ice sheet, 13,000–10,000 years B.P. In *The Lateglacial of North-West Europe*, ed. J. J. Lowe, J. M. Gray, & T. E. Robinson, pp. 23–30. Oxford: Pergamon.

Mangerud, J. (1981). The Early and Middle Weichselian in Norway: a review. *Boreas* 10, 381–93.

Mangerud, J. (1983). The glacial history of Norway. In *Glacial deposits in north-west Europe*, ed. J. Ehlers, pp. 3–9. Rotterdam: Balkema.

Mangerud, J. (1987). The Allerød / Younger Dryas boundary. In *Abrupt climatic change*, ed. W. H. Berger & L. D. Labeyrie, pp. 163–71. Dordrecht: Reidel.

Mangerud, J. (1990). Paleoklimatologi. In *Drivhuseffekten og klimatutvicklingen*, pp. 102–51. Lillestrøm: Norsk Institutt for luftforskning, Rapport 21/90.

Mangerud, J., Andersen, S. T., Berglund, B. E., & Donner, J J. (1974). Quaternary stratigraphy of Norden, a proposal for terminology and classification. *Boreas* 3, 109–28.

Mangerud, J., & Gulliksen, S. (1975). Apparent radiocarbon ages of recent marine shells from Norway, Spitsbergen, and arctic Canada. *Quaternary Research* 5, 263–73.

Mangerud, J., Gulliksen, S., Larsen, E., Longva, O., Miller, G. H., Sejrup, H.-P., & Sønstergaard, E. (1981b). A Middle Weichselian ice-free period in western Norway: the Ålesund Interstadial. *Boreas* 10, 447–62.

Mangerud, J., Larsen, E., Longva, O., & Sønstergaard, E. (1979a). Glacial history of western Norway 15,000–10,000 B.P. *Boreas* 8, 179–87.

Mangerud, J., Lie, S. E., Furnes, H., Kristiansen, I. L., & Lømo, L. (1984). A Younger Dryas ash bed in western Norway, and its possible correlations with tephra in cores from the Norwegian Sea and the North Atlantic. *Quaternary Research* 21, 85–104.

Mangerud, J., Sønstergaard, E., & Sejrup, H.-P. (1979b). Correlation of the Eemian (interglacial) Stage and the deep-sea oxygen-isotope stratigraphy. *Nature* 277, 189–192.

Mangerud, J., Sønstergaard, E., Sejrup, H.-P., & Haldorsen, S. (1981a). A continuous Eemian–Early Weichselian sequence containing pollen and marine fossils at Fjøsanger, western Norway. *Boreas* 10, 137–208.

Mankinen, E. A., & Dalrymple, G. B. (1979). Revised geomagnetic polarity time scale for the interval 0–5 m.y. B.P. *Journal of Geophysical Research* 84 B 2, 615–26.

Marcussen, I. (1973). Stones in Danish tills as a stratigraphical tool. A review. *Bulletin of the Geological Institutions of the University of Uppsala, New Series* 5, 177–81.

Markuse, G., & Vesajoki, H. (1985). Periglasiaalisia ventifakteja Pohjois-Karjalasta. *Geologi, Suomen Geologinen Seura – Geologiska Sällskapet i Finland* 1985(4-5), 81–84.

Marthinussen, M. (1962). C 14-datings referring to shore lines, transgressions, and glacial substages in northern Norway. *Norges Geologiske Undersökelse* 215, 37–67.

Marthinussen, M. (1974). Contributions to the Quaternary geology of north-easternmost Norway and the closely adjoining foreign territories. *Norges Geologiske Undersøkelse* 315,

Martinson, D. G., Pisias, N. G., Hays, J. D., Imbrie, J., Moore, T. C. Jr., & Shackleton, N. J. (1987). Age dating and the orbital theory of the ice ages: development af a high-resolution 0 to 300,000-year chronostratigraphy. *Quaternary Research* 27, 1–29.

Martinsson, A. (1958). The submarine Palaeozoic of the Baltic area. *Bulletin of the Geological Institutions of the University of Uppsala* 38, 11–86.

Martinsson, A. (1980). The Quaternary history of the Baltic. The Pre-Quaternary substratum of the Baltic. In *The Quaternary history of the Baltic*, ed. V. Gudelis & L.-K. Königsson, pp. 77–86. Acta Universitatis Upsaliensis, Symposia Universitatis Upsaliensis Quingentesimum Celebrantis, 1, Uppsala.

Menke, B. (1968). Beiträge zur Biostratigraphie des Mittelpleistozäns in Norddeutschland. *Meyniana* 18, 15–42.

Menke, B. (1970). Ergebnisse der Pollenanalyse zur Pleistozän-Stratigraphie und zur Pliozän-Pleistozän-Grenze in Schleswig-Holstein. *Eiszeitalter und Gegenwart* 21, 5–21.

Menke, B., & Behre, K.-E. (1973). History of vegetation and biostratigraphy. *Eiszeitalter und Gegenwart* 23/24, 251–67.

Menke, B., & Tynni, R. (1984). Das Eeminterglazials und das Weichselfrühglazial von Rederstall/Dithmarschen und ihre Bedeutung für die mitteleuropäische Jungpleistozän-Gliederung. *Geologisches Jahrbuch* A 76, 3–120.

Meyer, K.-J. (1974). Pollenanalytische Untersuchungen und Jahresschichtenzählungen an der holstein-zeitlichen Kieselgur von Hetendorf. *Geologisches Jahrbuch* A 21, 87–105.

Miller, G. H., & Mangerud, J. (1985). Aminostratigraphy of European marine interglacial deposits. *Quaternary Science Reviews* 4, 215–78.

Miller, U. (1977). Pleistocene deposits of the Alnarp Valley, southern Sweden. Microfossils and their stratigraphical application. *University of Lund, Department of Quaternary Geology, Thesis* 4.

Miller, U. (1986). An outline of the Nordic natural landscape in the Late Pleistocene. *Striae* (Uppsala) 24, 9–14.

Miller, U., & Persson, C. (1973). A lump of clay embedded in glacial intermorainic sand. *Geologiska Föreningens i Stockholm Förhandlingar* 95, 342–6.

Miller, U., & Robertsson, A.-M. (1988). Late Weichselian and Holocene environmental changes in Bohuslän, southwestern Sweden. *Geographia Polonica* 55, 103–11.

Milthers, K. (1942). Ledeblokke og landskabsformer i Danmark. *Danmarks Geologiske Undersøgelse*, II 69.

Miskovsky, K. (1985). The Baltic Shield relief and its development. *Fennia* 163(2), 353–8.

Moe, D. (1970). The post-glacial immigration of Picea abies into Fennoscandia. *Botaniska Notiser* (Lund) 123, 61–6.

Moe, D. (1979). Tregrense-fluktuasjoner på Hardangervidda etter siste istid. In *Fortider i Søkelyset*, ed. R. Nydal, S. Westin, U. Hafsten, & S. Gulliksen, pp. 199–208. Trondheim: Laboratoriet for Radiologisk Datering.

Mohrén, E. (1938). Jättehjorthornets geologiska ålder. *Fauna och flora* 1938, 104–12.

Mölder, K., & Tynni, R. (1973). Über Finnlands rezente und subfossile Diatomeen VII. *Bulletin of the Geological Society of Finland* 45, 149–79.

Molodkov, A., & Raukas, A. (1988). The age of Upper Pleistocene marine deposits of the Boreal transgression on the basis of electron-spin-resonance (ESR) dating of subfossil mollusc shells. *Boreas* 17, 267–72.

Mook, W. G., & Waterbolk, H. T. (1985). Radiocarbon dating. *Handbook for Archaeologists*, 3, Strasbourg: European Science Foundation.

Müller, H. (1965). Eine pollenanalytische Neubearbeitung des Interglazialprofils von Bilshausen (Unter-Eichsfeld). *Geologisches Jahrbuch* 83, 327–52.

Müller, H. (1974a). Pollenanalytische Untersuchungen und Jahresschichtenzählungen an der holstein-zeitlichen Kieselgur von Munster-Breloh. *Geologisches Jahrbuch* A 21, 107–40.

Müller, H. (1974b). Pollenanalytische Untersuchungen und Jahresschichtenzählungen an der eem-zeitlichen Kieselgur von Bispingen/Luhe. *Geologisches Jahrbuch* A 21, 149–69.

Müller-Wille, L. L. (1969). Biometrical comparison of four populations of *Phoca hispida* Schreb. in the Baltic and White Seas and Lakes Ladoga and Saimaa. *Societas Scientiarum Fennica, Commentationes Biologicae* 31(3).

Munthe, H. (1892). Studier öfver baltiska hafvets kvartära historia. *Bihang till Kungliga Svenska Vetenskapsakademiens Handlingar* 18 II 1.

Munthe, H. (1910). Studies in the late-Quaternary history of southern Sweden. *Geologiska Föreningens i Stockholm Förhandlingar* 32, 1197–1293.

Nenonen, K. (1986). Orgaanisen aineksen merkitys moreenistratigrafiassa. *Geologi, Suomen Geologinen Seura – Geologiska Sällskapet i Finland* 1986(2), 41–4.

Nesje, A., & Sejrup, H. P. (1988). Late Weichselian / Devensian ice sheets in the North Sea and adjacent land areas. *Boreas* 17, 371–84.

Niemelä, J. (1971). Die quartäre Stratigrafie von Tonablagerungen und der Rückzug des Inlandeises zwischen Helsinki und Hämeenlinna in Südfinnland. *Geological Survey of Finland Bulletin* 253.

Niemelä, J. ed. (1979). Suomen sora- ja hiekkavarojen arviointiprojekti 1971–78. *Geological Survey of Finland, Report of Investigations* 42.

Niemelä, J., & Jungner, H. (1991). Thermoluminescence dating of Late Pleistocene sediments related to till-covered eskers from Ostrobothnia, Finland. *Geological Survey of Finland, Special Paper* 12, 135–8.

Niemelä, J., & Tynni, R. (1979). Interglacial and interstadial sediments in the Pohjanmaa region, Finland. *Geological Survey of Finland Bulletin* 302.

Niini, H. (1968). A study of rock fracturing in valleys of Precambrian bedrock. *Fennia* 97(8).

Nilsson, E. (1953). Om södra Sveriges senkvartära historia. *Geologiska Föreningens i Stockholm Förhandlangar* 75, 155–246.

Nilsson, E. (1968). Södra Sveriges senkvartära historia. Geokronologi, issjöar och landhöjning. *Kungliga Svenska Vetenskapsakademiens Handlingar* IV 12(1).

Nilsson, T. (1964a). Standardpollendiagramme und C 14 – Datierungen aus dem Ageröds Mosse im mittleren Schonen. *Lunds Universitets Årsskrift*, N.F. 2, 59(7).

Nilsson, T. (1964b). Entwicklungsgeschichtliche Studien im Ageröds Mosse, Schonen. *Lunds Universitets Årsskrift*, N.F. 2, 59(8).

Nilsson, T. (1983). *The Pleistocene: Geology and life in the Quaternary ice age*. Dordrecht: Reidel.

Nordhagen, R. (1933). De senkvartaere klimatvekslinger i Nordeuropa og deres betydning for kulturforskningen. *Instituttet for Sammenlignende Kulturforskning*. Oslo: Aschehoug.

Nordhagen, R. (1936). Skandinavien fjellflora og dens relasjoner til den siste istid. *Nordiska (19. skandinaviska) naturforskarmötet i Helsingfors 1936*, pp. 93–124. Helsingfors: Finska Litteratursällskapet.

Norrman, J. O. (1964a). Vätterbäckenets senkvartära strandlinjer. *Geologiska Föreningens i Stockholm Förhandlingar* 85, 391–413.

Norrman, J. O. (1964b). Lake Vättern. Investigations on shore and bottom morphology. *Geografiska Annaler* 1964(1-2).

Oele, E., & Schüttenhelm, R. T. E. (1979). Development of the North Sea after the Saalian glaciation. In *The Quaternary history of the North Sea*, ed. E. Oele, R. T. E. Schüttenhelm, & A. J. Wiggers, pp. 191–215. Acta Universitatis Upsaliensis, Symposia Universitatis Upsaliensis Annum Quingentesimum Celebrantis, 2, Uppsala.

Oele, E., Schüttenhelm, R. T. E., & Wiggers, A. J., ed. (1979). *The Quaternary history of the North Sea*. Acta Universitatis Upsaliensis, Symposia Universitatis Upsaliensis Annum Quingentesimum Celebrantis, 2, Uppsala.

Okko, M. (1962). On the development of the first Salpausselkä, west of Lahti. *Bulletin de la Commission géologique de Finlande* 202.

Okko, V. (1941). Über das Verhältnis der Gesteinszusammensetzung der Moräne zum Felsgrund in den Gebieten der Kartenblätter von Ylitornio und Rovaniemi im nördlichen Finnland. *Geologische Rundschau* 32(4/5), 627–43.

Okko, V. (1964). Maaperä. In *Suomen Geologia*, ed. K. Rankama, pp. 239–332. Helsinki: Kirjayhtymä.

Olausson, E., ed. (1982). The Pleistocene / Holocene boundary in south-western Sweden. *Sveriges Geologiska Undersökning* C 794.

Olsen, L. (1988). Stadials and interstadials during the Weichsel glaciation on Finnmarksvidda, northern Norway. *Boreas* 17, 517–39.

Olsen, L. (1990). Weichselian till stratigraphy and glacial history of Finnmarksvidda, North Norway. *Quaternary International* 3/4, 101–8.

Olsen, L., & Hamborg, M. (1983). Morenestratigrafi og isbevegelser fra Weichsel, sørvestra Finnmarksvidda, Nord-Norge. *Norges Geologiske Undersøkelse* 378, 93–113.

Olsen, L., & Hamborg, M. (1984). Weichselian till stratigraphy and ice movements, a model based mainly on clast fabric, Finnmarksvidda, northern Norway. *Striae* (Uppsala) 20, 69–73.

Olsson, I. U. (1979). A warning against radiocarbon dating of samples containing little carbon. *Boreas* 8, 203–207.

Overweel, C. J. (1977). Distribution and transport of Fennoscandian indicators. *Scripta Geologica*, Leiden, 43.

Papunen, H., & Gorbunov, G. I., ed. (1985). Nickel-copper deposits of the Baltic Shield and Scandinavian Caledonides. *Geological Survey of Finland Bulletin* 333.

Parsons, R. W., Prentice, I. C., & Saarnisto, M. (1980). Statistical studies on pollen representation in Finnish lake sediments in relation to forest inventory data. *Annales Botanici Fennici* 17, 379–93.

Påsse, T. (1983). Havsstrandens nivåförändringar i norra Halland under Holocen. *Geologiska Institutionen, Chalmers Tekniska Högskola – Göteborgs Universitet* A 45.

Påsse, T., Robertsson, A.-M., Miller, U., & Klingberg, F. (1988). A Late Pleistocene sequence at Margreteberg, southwestern Sweden. *Boreas* 17, 141–63.

Paterson, W. S. B. (1981). *The physics of glaciers*, 2nd ed. Oxford: Pergamon.

Paus, A. (1989). Late Weichselian vegetation, climate, and floral migration at Liastammen, North Rogaland, south-western Norway. *Journal of Quaternary Science* 4(3), 223–42.

Pedersen, S. A. S., Petersen, K. S., & Rasmussen, L. A. (1988). Observations on glaciodynamic structures at the Main Stationary Line in western Jutland, Denmark. In *Glaciotectonics: Forms and processes*, ed. D. G. Croot, pp. 177–83. Rotterdam: Balkema.

Peltoniemi, H., Eriksson, B., Grönlund, T., & Saarnisto, M. (1989). Marjamurto, an interstadial site in a till-covered esker area of central Ostrobothnia, western Finland. *Bulletin of the Geological Society of Finland* 61(2), 209–37.

Penttilä, S. (1963). The deglaciation of the Laanila area, Finnish Lapland. *Bulletin de la Commission gélogique de Finlande* 203.

Persson, C. (1966). Försök till tefrokronologisk datering av några svenska torvmossar. *Geologiska Föreningens i Stockholm Förhandlingar* 88, 361–94.

Persson, C. (1967a). Undersökning av tre sura asklager på Island. *Geologiska Föreningens i Stockholm Förhandlingar* 88, 500–19.

Persson, C. (1967b). Försök till tefrokronologisk datering i tre norska myrar. *Geologiska Föreningens i Stockholm Förhandlingar* 89, 181–97.

Persson, C. (1971). Tephrochronological investigation of peat deposits in Scandinavia and on the Faroe Islands. *Sveriges Geologiska Undersökning* C 656.

Persson, C. (1975). Speculations on the immigration of spruce into Sweden. *Geologiska Föreningens i Stockholm Förhandlingar* 97, 292–4.

Persson, C. (1983). Glacial deposits and the Central Swedish end moraine zone in eastern Sweden. In *Glacial deposits in north-west Europe*, ed. J. Ehlers, pp. 131–9. Rotterdam: Balkema.

Perttunen, M. (1977). The lithologic relation between till and bedrock in the region of Hämeenlinna, southern Finland. *Geological Survey of Finland Bulletin* 291.

Pesonen, L. J., Grieve, R. A. F., & Leino, M. A. H. (1991). Planeetta maan impaktikraattereista. *XV Geofysiikan Päivät, Geofysiikan Seura, Oulu*, 95–101.

Pesonen, L. J., Torsvik, T. H., Elming, S.-Å., & Bylund, G. (1989). Crustal evolution of Fennoscandia – palaeomagnetic constraints. *Tectonophysics* 162, 27–49.

Petersen, K. S. (1973). Tills in dislocated drift deposits on the Røsnaes Peninsula, Northwestern Sjaelland, Denmark. *Bulletin of the Geological Institutions of the University of Uppsala* 5, 41–9.

Petersen, K. S. (1978). Applications of glaciotectonic analysis in the geological mapping of Denmark. *Danmarks Geologiske Undersøgelse, Årbog* 1977, 53–61.

Petersen, K. S. (1984a). Stratigraphical position of Weichselian tills in Denmark. *Striae* (Uppsala) 20, 75–8.

Petersen, K. S. (1984b). Late Weichselian sea-levels and fauna communities in northern Vendsyssel, Jutland, Denmark. In *Climatic changes on a yearly to millennial basis*, ed. N.-A. Mörner & W. Karlén, pp. 63–8. Dordrecht: Reidel.

Petersen, K. S. (1985). The Late Quaternary history of Denmark. *Journal of Danish Archaeology* 4, 7–22.

Petersen, K. S., Odgaard, B., Knudsen, K. L., Rasmussen, L. A., & Pedersen, S. A. S. (1987). Excursion Guide to the Quaternary geology of the central part of Jutland. Middle and Late Pleistocene. *Danmarks Geologiske Undersøgelse, Internal report* 9.

Petersen, K. S., Rasmussen, K. L., Heinemeier, J., & Rud, N. (1992). Clams before Columbus? *Nature* 359, 679.

Peulvast, J.-P. (1985). In situ weathered rocks on plateaus, slopes and strandflat areas of the Lofoten-Vesterålen, North Norway. *Fennia* 163(2), 330–40.

Péwé, T. L., Church, R. E., & Andresen, M. J. (1969). Origin and palaeoclimatic significance of large-scale patterned ground in the Donnelly Dome area, Alaska. *Geological Society of America, Special Papers* 103.

Pirrus, E. & Tiirmaa, R. (1990). The meteorite craters in Estonia. In *Symposium, Fennoscandian impact structures, May 29–31. 1990. Programme and abstracts*, ed. L. J. Pesonen & H. Niemisara, pp. 51–2. Espoo: Geological Survey of Finland.

Post, L. von (1916). Om skogsträdpollen i sydsvenska torfmosslagerföljder. *Geologiska Föreningens i Stockholm Förhandlingar* 38, 384.

Post, L. von (1967). Forest tree pollen in south Swedish peat bog deposits (translation of lecture to the 16th convention of Scandinavian naturalists in Kristiania (Oslo) 1916). *Pollen et Spores* IX 3, 375–401.

Prentice, I. C. (1978). Modern pollen spectra from lake sediments in Finland and Finnmark, north Norway. *Boreas* 7, 131–53.

Prentice, H. C. (1981). A Late Weichselian and early Flandrian pollen diagram from Østervatnet, Varanger peninsula, NE Norway. *Boreas* 10, 53–70.

Prentice, H. C. (1982). Late Weichselian and early Flandrian vegetational history of Varanger peninsula, northeast Norway. *Boreas* 11, 187–208.

Rainio, H. (1985). Första Salpausselkä utgör randzonen för en landis som avancerat på nytt. *Geologi Suomen Geologinen Seura – Geologiska Sällskapet i Finland* 1985(4-5), 70–7.

Rainio, H. (1991). The Younger Dryas ice-marginal formations of southern Finland. In *Eastern Fennoscandian Younger Dryas end moraines*, ed. H. Rainio & M. Saarnisto, pp. 25–72. Espoo: Geological Survey of Finland, Guide 32.

Rainio, H., & Lahermo, P. (1976). Observations on dark grey basal till in Finland. *Bulletin of the Geological Society of Finland* 48, 137–52.

Rainio, H., & Lahermo, P. (1984). New aspects on the distribution and origin of the so-called dark till. *Striae* (Uppsala) 20, 45–7.

Ramsay, W. (1891). Über den Salpausselkä in östlichen Finnland. *Fennia* 4(2).

Ramsay, W. (1900). *Finlands geologiska utveckling ifrån istiderna intill våra dagar*, andra upplagan. Helsingfors: Lindstedts Antikvariska Bokhandel.

Ramsay, W. (1921). Strandlinjer i södra Finland. *Geologiska Föreningens i Stockholm Förhandlingar* 43, 495–7.

Ramsay, W. (1926). Nivåförändringar och stenåldersbosättning i det baltiska området. *Fennia* 47(4), 1–68.

Rankama, K., ed. (1965). *The geologic systems: the Quaternary, Volume 1*. New York: Interscience Publishers.

Rappol, M. (1986). Till in The Netherlands: A selective review. *Inqua Symposium on "Tills and endmoraines in The Netherlands and NW-Germany," University of Amsterdam, September 1986*.

Rasmussen, H. W. (1966). *Danmarks geologi*. Gjellerup.

Rasmussen, L. A. (1982). Weichselian stratigraphy in northern Jutland, Denmark. *Quaternary Studies in Poland* 3, 91–6.

Rasmussen, L. A., & Petersen, K. S. (1980). Resultater fra DGU's genoptagne kvartaergeologiske kortlaegning. *Dansk Geologisk Forening, Årsskrift* 1979, 47–54.

Raukas, A. (1965). Application of lithological methods in solving the stratigraphy and paleogeography of the Quaternary period (in Russian). *Academy of Sciences of the USSR, Siberian Branch, Geology and Geophysics.* Moscow: Nauka, 470–7.

Raukas, A. (1969). Composition and genesis of Estonian tills. *Zeszyty Naukowe, UAM Geografia* (Poznan) 8, 167–176.

Raukas, A. (1977). Ice-marginal formations and the main regularities of the deglaciation in Estonia. *Zeitschrift für Geomorphologie, Supplement* 27, 68–78.

Raukas, A. (1978). *Pleistocene deposits of the Estonian SSR* (in Russian). Tallinn: Academy of Sciences of the Estonian SSR.

Raukas, A., ed. (1985). Methods of investigation of glacial and aqueoglacial deposits for genetical and practical purposes. *Institute of Geology, Academy of Sciences of the Estonian SSR, INQUA Soviet Section, Inqua Commission on the Genesis and Lithology of Quaternary Deposits. Invitation, programme and excursion guide.*

Raukas, A. (1988). *Eestimaa viimastel aastimiljonitel.* Tallinn: Valgus.

Raukas, A., & Hyvärinen, H., eds. (1992). *Geology of the Gulf of Finland* (in Russian). Tallinn: Institute of Geology, Estonian Academy of Sciences.

Raukas, A., & Karukäpp, R. (1979). Eesti liustikutekkeliste akumulatiivsete saarkorgustike ehitys ja kujunemine. In *Eesti NSV saarkõrgustike ja järvenõgude kujunemine,* ed. A. Raukas, pp. 9–28. Tallinn: Valgus.

Raukas, A., Rähni, E., & Miidel, A. (1971). *Marginal glacial formations in North Estonia* (in Russian). Tallinn: Institute of Geology, Estonian Academy of Sciences.

Repo, R. (1957). Untersuchungen über die Bewegungen des Inlandeises in Nordkarelien. *Bulletin de la Commission géologique de Finlande* 179,

Repo, R., & Tynni, R. (1967). Zur spät- und postglazialen Entwicklung im Ostteil des ersten Salpausselkä. *Comptes Rendus de la Société géologique de Finlande* 39, 133–59.

Repo, R., & Tynni, R. (1971). Observations on the Quaternary geology of an area between the 2nd Salpausselkä and the ice-marginal formation of Central Finland. *Bulletin of the Geological Society of Finland* 43, 185–202.

Richmond, G. M. (1990). The INQUA-approved provisional Lower/Middle Pleistocene boundary. *INQUA Sub-Commission on European Quaternary Stratigraphy, Cromer Symposium, Norwich, Abstracts,* 27–8.

Richmond, G. M., & Fullerton, D. S. (1986). Introduction to Quaternary glaciations in the United States of America. *Quaternary Science Reviews* 5, 3–10.

Richter, E., Baudenbacher, R., & Eissmann, L. (1986). Die Eiszeitgeschichte in der Umgebung von Leipzig. *Altenburger Naturwissenschaftliche Forschungen* 3.

Ringberg, B. (1971). Glacialgeologi och isavsmältning i östra Blekinge. *Sveriges Geologiska Undersökning* C 661.

Ringberg, B. (1979). Varve chronology of glacial sediments in Blekinge and northeastern Skåne, southeastern Sweden. *Boreas* 8, 209–15.

Ringberg, B. (1988). Late Weichselian geology of southeastern Sweden. *Boreas* 17, 243–63.

Ringberg, B. (1989). Upper Late Weichselian lithostratigraphy in western Skåne, southernmost Sweden. *Geologiska Föreningens i Stockholm Förhandlingar* 111, 319–37.

Ringberg, B. (1991). Late Weichselian clay varve chronology and glaciolacustrine environment during deglaciation in southeastern Sweden. *Sveriges Geologiska Undersökning* Ca 79.

Ringberg, B., & Rudmark, L. (1985). Varve chronology based upon glacial sediments in the area between Karlskrona and Kalmar, southeastern Sweden. *Boreas* 14, 107–10.

Risberg, J., Miller, U., & Brunnberg, L. (1991). Deglaciation, Holocene shore displacement and coastal settlements in eastern Svealand, Sweden. *Quaternary International,* 9, 33–7.

Ristaniemi, O. (1984). Ancylusjärven aikainen rannansiirtyminen Salpausselkävyöhykkeessä Karjalohjan-Kiskon alueella Lounais-Suomessa. *Publications of the Department of Quaternary Geology, University of Turku* 53.

Ristaniemi, O., & Glückert, G. (1987). The Ancylus transgression in the area of Espoo – the first Salpausselkä, southern Finland. *Bulletin of the Geological Society of Finland* 59(1), 45–69.

Ristaniemi, O., & Glückert, G. (1988). Ancylus- ja litorinatransgressiot Lounais-Suomessa. *Annales Universitatis Turkuensis* C 67, 129–45.

Ristiluoma, S. (1974). Fossiilisia jääkiiloja Torniojokilaaksossa. *Terra* 86(1), 3–6.

Roaldset, E., Pettersen, E., Longva, O., & Mangerud, J. (1982). Remnants of preglacial weathering in western Norway. *Norsk Geologisk Tidsskrift* 62, 169–78.

Robertsson, A.-M. (1988). Biostratigraphical studies of interglacial and interstadial deposits in Sweden. *University of Stockholm, Department of Quaternary Geology, Report* 10.

Robertsson, A.-M., & Garcia Ambrosiani, K. (1988). Late Pleistocene stratigraphy at Boliden, northern Sweden. *Boreas* 17, 1–14.

Robertsson, A.-M., & Rodhe, L. (1988). A Late Pleistocene sequence at Seitevare, Swedish Lapland, *Boreas* 17, 501–9.

Robin, G. de Q. (1964). Glaciology. *Endeavour* XXIII 89, 102–7.

Rokoengen, K., Bugge, T., & Løfaldi, M. (1979). Quaternary geology and deglaciation of the continental shelf off Troms, north Norway. *Boreas* 8, 217–27.

Ronnert, L., Svedhage, K., & Wedel, P. O. (1992). Luminescence dates of pre-Late Weichselian reddish-brown clay in and south of the Middle Swedish end-moraine zone. *Geologiska Föreningens i Stockholm Förhandlingar* 114, 323–5.

Rosberg, J. E. (1892). Ytbildningar i ryska och finska Karelen med särskild hänsyn till de karelska randmoränerna. *Fennia* 7(2).

Rosberg, J. E. (1925). Jättegrytor i södra Finland. *Fennia* 46(1).

Rose, J. (1989). Stadial type sections in the British Quaternary. In *Quaternary type sections: Imagination or reality?,* ed. J. Rose & C. Schlüchter, pp. 45–67. Rotterdam: Balkema.

Roth, E., & Poty, B. (1985). *Méthodes de datation par les phénomènes nucléaires naturels.* Collection du Commissariat à l'énergie atomique, Serie Scientifique. Paris: Masson.

Rózycki, S. Z. (1961). *Guidebook for excursion from the Baltic to the Tatras, Part II, Volume I, Middle Poland.* International Union for Quaternary Research, VI Congress, Poland.

Ruuhijärvi, R. (1960). Über die regionale Einteilung der nordfinnischen Moore. *Annales Botanici Societatis Zoologicae Botanicae Fennicae Vanamo* 31(1).

Ruuhijärvi, R. (1962). Palsasoista ja niiden morfologiasta siitepölyanalyysin valossa. *Terra* 74(2), 58–68.

Rzechowski, K. (1986). Pleistocene till stratigraphy in Poland. *Quaternary Science Reviews* 5, 365–372.

Saarnisto, M. (1968). The Flandrian history of Lake Höytiäinen, eastern Finland. *Bulletin of the Geological Society of Finland* 40, 71–98.

Saarnisto, M. (1970). The Late Weichselian and Flandrian history of the Saimaa lake complex. *Societas Scientiarum Fennica, Commentationes Physico-Mathematicae* 37.

Saarnisto, M. (1971a). The upper limit of the Flandrian transgression of Lake Päijänne. *Societas Scientiarum Fennica, Commentationes Physico-Mathematicae* 41, 149–70.

Saarnisto, M. (1971b). The history of Finnish lakes and Lake Ladoga. *Societas Scientiarum Fennica, Commentationes Physico-Mathematicae* 41, 371–88.

Saarnisto, M. (1977). Deformational structures in the first Salpausselkä end moraine, Joutsenonkangas, south-eastern Finland. *Bulletin of the Geological Society of Finland* 49(1), 65–72.

Saarnisto, M. (1981). Holocene emergence history and stratigraphy in the area north of the Gulf of Bothnia. *Annales Academiae Scientiarum Fennicae* A III 130.

Saarnisto, M. (1985). Long varve series in Finland. *Boreas* 14, 133–7.

Saarnisto, M. (1988). Time-scales and dating. In *Vegetation history*, ed. B. Huntley & T. Webb, III, pp. 77–112. Norwell, Mass.: Kluwer.

Saarnisto, M., Huttunen, P., & Tolonen, K. (1977). Annual lamination of sediments in Lake Lovojärvi, southern Finland, during the past 600 years. *Annales Botanici Fennici* 14, 35–45.

Saarnisto, M., & Peltoniemi, H. (1984). Glacial stratigraphy and compositional properties of till in Kainuu, eastern Finland. *Fennia* 162(2), 163–99.

Saarnisto, M., & Siiriäinen, A. (1970). Laatokan transgressioraja. *Suomen Museo* 1970, 10–22.

Saarse, L., Rajamäe, R., Heinsalu, A., & Vassiljev, J. (1990). Formation of the meteor crater Lake Kaali (Island Saaremaa, Estonia). In *Symposium, Fennoscandian impact structures, May 29–31, 1990. Programme and abstracts*, ed. L. J. Pesonen & H. Niemisara, p. 55. Espoo: Geological Survey of Finland.

Salmi, M. (1948). The Hekla ashfalls in Finland A.D. 1947. *Comptes Rendus de la Société géologique de Finlande* 21, 87–96.

Salmi, M. (1968). Development of palsas in Finnish Lapland. *Third International Peat Congress, Quebec, Canada 1968*, Ottawa, 182–9.

Salomaa, R., & Matiskainen, H. (1985). New data on shoreline displacement and archeological chronology in southern Ostrobothnia and northern Satakunta. *Suomen Muinaismuistoyhdistys, Iskos* (Helsinki) 5, 141–55.

Salonen, V.-P. (1986). Glacial transport distance distributions of surface boulders in Finland. *Geological Survey of Finland Bulletin* 338.

Salonen, V.-P. (1987). Observations on boulder transport in Finland. *Geological Survey of Finland, Special Paper* 3, 103–10.

Salonen, V.-P. (1991). Glacial dispersal of Jotnian sandstone fragments in southwestern Finland. *Geological Survey of Finland, Special Paper* 12, 127–30.

Sauramo, M. (1918). Geochronologische Studien über die spätglaziale Zeit in Südfinnland. *Fennia* 41(1).

Sauramo, M. (1923). Studies on the Quaternary varve sediments in southern Finland. *Bulletin de la Commission géologique de Finlande* 60.

Sauramo, M. (1924). Lehti B 2, Tampere, Maalajikartan selitys. *Suomen Geologinen Yleiskartta, Suomen Geologinen Komissioni – Geologiska Kommissionen i Finland.*

Sauramo, M. (1925a). Über die Bändertone in den ostbaltischen

Ländern vom geochronologischen Standpunkt. *Fennia* 45(6).

Sauramo, M. (1925b). Geochronologische Studien in Russland. *Geologiska Föreningens i Stockholm Förhandlingar* 47, 521–2.

Sauramo, M. (1929). The Quaternary geology of Finland. *Bulletin de la Commission géologique de Finlande* 86.

Sauramo, M. (1934). Zur spätquartären Geschichte der Ostsee. *Comptes Rendus de la Société géologique de Finlande* 8, 28–87.

Sauramo, M. (1940). *Suomen luonnon kehitys jääkaudesta nykyaikaan*. Porvoo: Werner Söderström Oy.

Sauramo, M. (1949). Das dritte Scharnier der fennoskandischen Landhebung. *Societas Scientiarum Fennica, Årsbok-Vuosikirja* 27(4).

Sauramo, M. (1958). Die Geschichte der Ostsee. *Annales Academiae Scientiarum Fennicae* A III 51.

Sauramo, M., & Auer, V. (1928). On the development of Lake Höytiäinen in Carelia and its ancient flora. *Communicationes Instituti Forestalis Fennicae* 13.

Schütrumpf, R. (1967). Die Profile von Loopstedt und Geesthacht in Schleswig-Holstein. Ein Beitrag zur vegetationsgeschichtlichen Gliederung des jüngeren Pleistozänes. *Fundamenta* B 2, 136–67.

Sederholm, J. J. (1911). Extension du glacier continental dans l'Europe septentrionale et transportation de blocs erratiques fennoscandiens (Atlas de Finland 1910, Carte No. 5). *Fennia* 30, 56–61.

Sederholm, J. J. (1913). Weitere Mitteilungen über Bruchspalten, mit besonderer Beziehung zur Geomorphologie von Fennoskandia. *Fennia* 34(4).

Segerstråle, S. G. (1927). Skalmärgelfyndigheterna i Finland. *Fennia* 47(8).

Segerstråle, S. G. (1956). The distribution of glacial relicts in Finland and adjacent Russian areas. *Societas Scientiarum Fennica, Commentationes Biologicae* 15(18).

Segerstråle, S. G. (1957). On the immigration of the glacial relicts of northern Europe, with remarks on their prehistory. *Societas Scientiarum Fennica, Commentationes Biologicae* 16(16).

Segerstråle, S. G. (1976). Immigration of glacial relicts into northern Europe. *Boreas* 5, 1–5.

Segerstråle, S. G. (1982). The immigration of glacial relicts into northern Europe in the light of recent geological research. *Fennia* 160(2), 303–12.

Sejrup, H. P. (1987). Molluscan and foraminiferal biostratigraphy of an Eemian–Early Weichselian section on Karmøy, southwestern Norway. *Boreas* 16, 27–42.

Sejrup, H. P., Miller, G. H., Brigham-Grette, J., Løvlie, R., & Hopkins, D. (1984). Amino acid epimerization implies rapid sedimentation in Arctic Ocean cores. *Nature* 310, 772–5.

Seppälä, M. (1971). Evolution of eolian relief of the Kaamasjoki-Kiellajoki river basin in Finnish Lapland. *Fennia* 104.

Serebryanni, L. R., & Raukas, A. (1966). Transbaltic correlations of the late Pleistocene ice-marginal features (in Russian). In *Verchnij plejstocen, stratigrafija i absoljutnaja geochronologija*, pp. 12–28. Moscow: Academy of Sciences of the USSR.

Serebryanni, L. R., & Raukas, A. (1967). Correlation of Gothiglacial ice marginal belts in the Baltic Sea depression and the neighboring countries. *Baltica* 3, 235–49.

Serebryanni, L. R., & Raukas, (1970). Über die eiszeitliche

Geschichte der Russischen Ebene im oberen Pleistozän. *Petermanns Geographische Mitteilungen* 1970(3), 161–72.

Serebryanni, L. R., Raukas, A., & Punning, J.-M. (1970). Fragments of the natural history of the Russian plain during the Late Pleistocene with special reference to radiocarbon datings of fossil organic matter from the Baltic region. *Baltica* 4, 351–66.

Sernander, R. (1908). On the evidence of Postglacial changes of climate furnished by peat-mosses of northern Europe. *Geologiska Föreningens i Stockholm Förhandlingar* 30, 465–73.

Shackleton, N. J. (1969). The last interglacial in the marine and terrestrial records. *Proceedings of the Royal Society of London* B 174, 135–54.

Shackleton, N. J., Backman, J., Zimmerman, H., Kent, D. V., Hall, M. A., Roberts, D. G., Schnitker, D., Baldauf, J. G., Despraires, A., Homrighausen, R., Huddlestun, P., Keene, J. B., Kaltenback, A. J., Krumsiek, K. A. O., Morton, A. C., Murray, J. W., & Westberg-Smith, J. (1984). Oxygen isotope calibration of the onset of ice-rafting and history of glaciation in the North Atlantic region. *Nature* 307, 620–3.

Shackleton, N. J., Berger, A., & Peltier, W. (1990). An alternative astronomical calibration of the lower Pleistocene timescale based on ODP Site 677. *Transactions of the Royal Society of Edinburgh, Earth Sciences* 81, 251–61.

Shackleton, N. J., & Matthews, R. K. (1977). Oxygen isotope stratigraphy of Late Pleistocene coral terraces in Barbados. *Nature* 268, 618–20.

Shackleton, N. J., & Opdyke, N. D. (1973). Oxygen isotope and palaeomagnetic stratigraphy of Equatorial Pacific Core V28–238: Oxygen isotope temperatures and ice volumes on a 10^5 year and 10^6 year scale. *Quaternary Research* 3, 39–55.

Shackleton, N. J., & Turner, C. (1967). Correlation between marine and terrestrial Pleistocene successions. *Nature* 216, 1079–82.

Shotton, F. W. (1973). General principles governing the subdivision of the Quaternary System. In *A correlation of Quaternary deposits in the British Isles,* ed. G. F. Mitchell, L. F. Penny, F. W. Shotton, & R. G. West, pp. 1–7. Geological Society of London, Special Report, 4. Edinburgh: Scottish Academic Press.

Shotton, F. W. (1986). Glaciations in the United Kingdom. *Quaternary Science Reviews* 5, 293–97.

Sibrava, V. (1986). Correlation of European glaciations and their relation to the deep-sea record. *Quaternary Science Reviews* 5, 433–41.

Simola, L. K. (1963). Über die postglazialen Verhältnisse von Vanajavesi, Leteensuo und Lehijärvi sowie die Entwicklung ihrer Flora. *Annales Academiae Scientiarum Fennicae* A III 70.

Simonen, A. (1980). The Precambrian in Finland. *Geological Survey of Finland Bulletin* 304,

Sirén, G. (1961). Skogsgränstallen som indikator för klimatfluktuationerna i norra Fennoskandien under historisk tid. *Communicationes Instituti Forestalii Fennicae* 54(2).

Sjørring, S. (1974). Über spätpleistozäne Glazialdynamik und -stratigraphie in Ost-Dänemark. *Eiszeitalter und Gegenwart* 25, 208–9.

Sjørring, S. (1981). Pre-Weichselian till stratigraphy in western Jutland, Denmark. *Mededelingen van de Rijks Geologische Dienst,* 34-10, 62–8.

Sjørring, S. (1982). The Weichselian till stratigraphy in the southern part of Denmark. *Quaternary Studies in Poland* 3, 103–9.

Smart, P. L. (1991). Electron spin resonance (ESR) dating. In *Quaternary dating methods – a user's guide,* ed. P. L. Smart & P. D. Frances, pp. 128–60. Quaternary Research Association, Technical Guide, 4. Nottingham: MI Press.

Söderman, G. (1980). Slope processes in cold environments of northern Finland. *Fennia* 158(2), 83–152.

Sokolova, L. F., Malyasova, E. S., Vishnevskaya, E. M., & Lavrova, M. A. (1972). A new find of Mga interglacial deposits in the central parts of the Karelian Isthmus (in Russian). *Bulletin of the University of Leningrad* 12, 124–31.

Sollid, J. L., Andersen, S., Hamre, N., Kjeldsen, O., Salvigsen, O., Sturød, S., Tveitå, T., & Wilhelmsen, A. (1973). Deglaciation of Finnmark, north Norway. *Norsk geografisk Tidsskrift* 27, 233–325.

Sollid, J. L., & Sørbel, L. (1974). Palsa bogs at Haugtjørnin, Dovrefjell, south Norway. *Norsk geografisk Tidsskrift* 28, 53–60.

Sollid, J. L., & Sørbel, L. (1979). Deglaciation of western central Norway. *Boreas* 8, 233–9.

Sørensen, R. (1979). Late Weichselian deglaciation in the Oslofjord area, south Norway. *Boreas* 8, 241–6.

Spjeldnaes, N. (1973). Moraine stratigraphy, with examples from the basal Cambrian ("Eocambrian") and Ordovician glaciations. *Bulletin of the Geological Institutions of Uppsala, New Series* 5, 165–71.

Stoker, M. S., Harland, R., Morton, A. C., & Graham, D. K. (1989). Late Quaternary stratigraphy of the northern Rockall Trough and Faeroe-Shetland Channel, northeast Atlantic Ocean. *Journal of Quaternary Science* 4(3), 211–22.

Strahler, A. N. (1965). *The earth sciences.* New York: Harper & Row.

Strömberg, B. (1985a). Revision of the lateglacial Swedish varve chronology. *Boreas* 14, 101–5.

Strömberg, B. (1985b). New varve measurements in Västergötland. *Boreas* 14, 111–115.

Strömberg, B. (1989). Late Weichselian deglaciation and clay varve chronology in east-central Sweden. *Sveriges Geologiska Unhdersökning* Ca 73.

Strömberg, B. (1990). A connection between the clay varve chronologies in Sweden and Finland. *Annales Academiae Scientiarum Fennicae* A III 154.

Stuart, A. J. (1991). Mammalian extinctions in the late Pleistocene of northern Eurasia and North America. *Biological Reviews of the Cambridge Philosophical Society* 66, 453–562.

Stuiver, M., Kromer, B., Becker, B., & Ferguson, C. W. (1986). Radiocarbon age calibration back to 13 300 years BP and the ^{14}C age matching of the German oak and US Bristlecone pine chronologies. *Radiocarbon* 28(2B), 969–79.

Stuiver, M., Robinson, S. W., & Yang, I. C. (1979). ^{14}C dating to 60,000 years B.P. with proportional counters. In *Radiocarbon dating,* ed. R. Berger & H. E. Suess, pp. 202–15. Berkeley: University of California Press.

Sue, J.-P., & Zagwijn, W. H. (1983). Plio-Pleistocene correlations between the northwestern Mediterranean region and northwestern Europe according to recent biostratigraphic and palaeoclimatic data. *Boreas* 12, 153–66.

Sugden, D. E., & John, B. S. (1976). *Glaciers and landscape.* London: Edward Arnold.

Sundius, N., & Sandegren, R. (1948). Interglacialfyndet vid Långsele. *Sveriges Geologiska Undersökning* C 495.

Suutarinen, O. (1983). Recomputation of land uplift values in Finland. *Reports of the Finnish Geodetic Institute* 83(1).

Svendsen, J. I., & Mangerud, J. (1987). Late Weichselian and Holocene sea-level history for a cross-section of western Norway. *Journal of Quaternary Science* 2, 113–32.

Svendsen, J. I., & Mangerud, J. (1990). Sea-level and pollen stratigraphy on the outer coast of Sunnmøre, western Norway. *Norsk Geologisk Tidsskrift* 70, 111–34.

Svensson, H. (1962). Några iakttagelser från palsområden. Flygbildanalys och fältstudier i nordnorska frostmarksområden. *Norsk geografisk Tidsskrift* 18, 212–27.

Svensson, H. (1964). Fossil tundramark på Laholmsslätten. *Sveriges Geologiska Undersökning* C 598.

Svensson, H. (1965). Aerial photographs for tracing and investigating fossil tundra ground in Scandinavia. *Biuletyn peryglacjalny* (Lodz) 21, 321–5.

Svensson, H., Källander, H., Maack, A., & Öhrngren, S. (1967). Polygonal ground and solifluction features. *Lund Studies in Geography* A 40.

Svensson, N.-O. (1989). Late Weichselian and Early Holocene shore displacement in the central Baltic, based on stratigraphical and morphological records from eastern Småland and Gotland, Sweden. *University of Lund, Department of Quaternary Geology, Thesis* 25.

Svensson, M.-O. (1991). Late Weichselian and Early Holocene shore displacement in the central Baltic area. *Quaternary International* 9, 7–26.

Sykes, G. (1991). Amino acid dating. In *Quaternary dating methods – a user's guide*, ed. P. L. Smart & P. D. Frances, pp. 161–76. Quaternary Research Association, Technical Guide, 4. Nottingham: MI Press.

Synge, F. M. (1969). The raised shorelines and deglaciation chronology of Inari, Finland, and south Varanger, Norway. *Geografiska Annaler* 51 A, 193–206.

Synge, F. M. (1980). A morphometric comparison of raised shorelines in Fennoscandia, Scotland and Ireland. *Geologiska Föreningens i Stockholm Förhandlingar* 102, 235–249.

Tallantire, P. A. (1972). The regional spread of spruce (Picea abies (L.) Krast.) within Fennoscandia: a reassessment. *Norwegian Journal of Botany* 19(1).

Tallantire, P. A. (1974). The palaeohistory of the grey alder (Alnus incana (L.) Moench.) and black alder (A. glutinosa (L.) Gaertn.) in Fennoscandia. *New Phytologist* 73, 529–46.

Tallantire, P. A. (1981). Some reflections on hazel (Corylus avellana L.) on its boundary in Fennoscandia during the Post-glacial. *Acta Palaeobotanica* 21, 161–71.

Tanner, V. (1914). Studier öfver kvartärsystemet i Fennoskandias nordliga delar. III. *Bulletin de la Commission géologique de Finlande* 38.

Tanner, V. (1930). Studier över kvartärsystemet i Fennoskandias nordliga delar. IV. *Bulletin de la Commission géologique de Finlande* 88.

Tanner, V. (1938). Die Oberflächengestaltung Finnlands. *Bidrag till Kännedom om Finlands Natur och Folk, Utgifna av Finska Vetenskaps-Societeten* 86.

Tauber, H. (1970). The Scandinavian varve chronology and C 14 dating. In *Nobel Symposium 12, Radiocarbon variations and absolute chronology*, ed. I. U. Olsson, pp. 173–96. Stockholm: Almqvist & Wiksell.

Tavast, E. (1987). Inherited character of modern drainage net in north and central Baltic. In *Palaeohydrology of the temperate zone. I. Rivers and lakes*, ed. A Raukas & L. Saarse, pp. 70–83. Tallinn: Valgus.

Tavast, E. (1992). The bedrock topography on the southern slope of the Fennoscandian Shield and in the transitional zone to the platform. *Dissertationes Geologicae Universitatis Tartuensis* 2.

Thompson, R. (1991). Palaeomagnetic dating. In *Quaternary dating methods – a user's guide*, ed. P. L. Smart & P. D. Frances, pp. 177–98. Quaternary Research Association, Technical Guide, 4. Nottingham: MI Press.

Thomsen, E., & Vorren, T. O. (1986). Macrofaunal palaeoecology and stratigraphy in Late Quaternary shelf sediments off northern Norway. *Palaeogeography, Palaeoclimatology, Palaeoecology* 56, 103–50.

Thomson, P. W. (1941). Die Klima- und Waldentwicklung des von K. Orviku entdeckten Interglazials von Ringen bei Dorpat / Estland. *Zeitschrift der Deutschen Geologischen Gesellschaft* 93(6), 274–82.

Thoresen, M., & Bergersen, O. F. (1983). Submorene sedimenter i Folldal, Hedmark, Sørøst-Norge. *Norges Geologiske Undersøkelse* 389, 37–55.

Tolonen, K. (1966). Stratigraphic and rhizopod analyses on an old raised bog, Varrassuo, in Hollola, south Finland. *Annales Botanici Fennici* 3, 147–66.

Tolonen, K., & Ruuhijärvi, R. (1976). Standard pollen diagrams from the Salpausselkä region of southern Finland. *Annales Botanici Fennici* 13, 155–96.

Tolonen, K., Siiriänen, A., & Thompson, R. (1975). Prehistoric field erosion sediment in Lake Lojärvi, S. Finland and its palaeomagnetic dating. *Annales Botanici Fennici* 12, 161–4.

Tolonen, M. (1978). Palaeoecology of annually laminated sediments in Lake Ahvenainen, S. Finland. II. Comparison of dating methods. *Annales Botanici Fennici* 15, 209–22.

Tolvanen, V. (1922). Der Alt-Päijänne. *Fennia* 43(5).

Treter, U. (1984). *Die Baumgrenzen Skandinaviens*. Wissenschaftliche Paperbacks, Geographie. Wiesbaden: Franz Steiner Verlag.

Tulkki, P. (1977). The bottom of the Bothnian Bay, geomorphology and sediments. *Merentutkimuslaitoksen Julkaisuja, Havsforskningsinstitutets Skrifter* (Helsinki) 241.

Turner, C. (1975). The correlation and duration of Middle Pleistocene interglacial periods in Northwest Europe. In *After the Australopithecines*, ed. K. W. Butzer & G. L. Isaac, pp. 259–308. The Hague: Mouton.

Turner, C., & West, R. G. (1968). The subdivision and zonation of interglacial periods. *Eiszeitalter und Gegenwart* 19, 93–101.

Tynni, R. (1960). Ostseestadium während der Allerödzeit in Askola, Ost-Uusimaa (Südfinnland). *Comptes Rendus de la Société géologique de Finlande* 32, 149–57.

Tynni, R. (1966). Über spät- und postglaziale Uferverschiebung in der Gegend von Askola, Südfinnland. *Bulletin de la Commission géologique de Finlande* 223.

Usinger, H. (1978). Bölling-Interstadial und Laacher Bimstuff in einem neuen Spätglazial-Profil aus dem Vallensgård Mose / Bornholm: Mit pollen-grössenstatistischer Trennen der Birken. *Danmarks Geologiske Undersøgelse, Årbog* 1977, 5–29.

Vartanyan, S. L., Garutt, V. E., & Sher, A. V. (1993). Holocene dwarf mammoths from Wrangel Island in the Siberian Arctic. *Nature* 362, 337–40.

Velichko, A. A., Isayeva, L. L., Makeyev, V. M., Matishov, G. G., & Faustova, M. A. (1984). Late Pleistocene glaciation of the arctic shelf, and the reconstruction of Eurasian ice sheets. In *Late Quaternary environments of the Soviet Union*, ed. A. A. Velichko, pp. 35–41. London: Longman.

Vesajoki, H. (1985). Fossiilisten jääkiilojen kehityssarja Ilomantsissa. *Geologi, Suomen Geologinen Seura – Geologiska Sällskapet i Finland* 1985(9-10), 176–80.

Viiding, H. (1976). Andmete kogumisest Eesti suurte ränrahnude kohta. In *Eesti NSV maapõne kaitsest,* ed. H. Viiding, pp. 148–61. Tallinn: Valgus.

Viiding, H. (1981). *The boulders of Lahemaa.* Tallinn: Valgus.

Viiding, H., Gaigalas, A., Gudelis, V., Raukas, A., & Tarvydas, R. (1971). *Crystalline indicator boulders in the east Baltic area* (in Russian). Vilnius: Mintis.

Virkkala, K. (1955). On glaciofluvial erosion and accumulation in the Tankavaara area, Finnish Lapland. *Acta Geographica* 14, 393–412.

Virkkala, K. (1958). Stone counts in the esker of Hämeenlinna, southern Finland. *Comptes Rendus de la Société géologique de Finlande* 30, 87–103.

Virkkala, K. (1962). Lehti 2123, Tampere, Maalajikartan selitys. *Explanation to the map of superficial deposits,* Geological Survey of Finland.

Virkkala, K. (1963). On ice-marginal features in southwestern Finland. *Bulletin de la Commission géologique de Finlande* 210.

Virkkala, K. (1969). On the lithology and provenance of the till of a gabbro area in Finland. In *Etudes sur le Quaternaire dans le Monde,* ed. M. Ters, pp. 711–714. Union Internationale pour l'Etude du Quaternaire, VIII Congrès. Paris.

Vorren, K.-D. (1972). Stratigraphical investigations of a palsa bog in northern Norway. *Astarte, Journal of Arctic Biology* 5, 39–71.

Vorren, K.-D. (1978). Late and Middle Weichselian stratigraphy of Andøya, north Norway. *Boreas* 7, 19–38.

Vorren, K.-D. (1979). Recent palsa datings, a brief survey. *Norsk geografisk Tidsskrift* 33, 217–20.

Vorren, T. O. (1979). Weichselian ice movements, sediments and stratigraphy on Hardangervidda, South Norway. *Norges Geologiske Undersøkelse* 350.

Vorren, T. O. (1982). Weichselian stratigraphy in Troms, North Norway. *Quaternary Studies in Poland* 3, 119–28.

Vorren, T. O., Hald, M., & Lebesbye, E. (1988). Late Cenozoic environments in the Barents Sea. *Paleoceanography* 3(5), 601–12.

Vorren, T. O., & Kristoffersen, Y. (1986). Late Quaternary glaciation in the south-western Barents Sea. *Boreas* 15, 51–59.

Vorren, T. O., & Roaldset, E. (1977). Stratigraphy and lithology of Late Pleistocene sediments at Møsvatn, Hardangervidda, South Norway. *Boreas* 6, 53–69.

Vorren, T. O., Rønnevik, H. C., & Reiersen, J. E. (1980). Kontinentalsokkelen utafor Norge i nord. *Ottar, Universitetet i Tromsø* 118.

Vorren, T. O., & Vorren, K.-D. (1979). Siste istids utvikling i Troms, på Andøya og på sokkelen utenfor. In *Fortiden i Søkelyset,* ed. R. Nydal, S. Westin, U. Hafsten, & S. Gulliksen, pp. 271–83. Trondheim: Laboratoriet for Radiologisk Datering.

Vorren, T. O., Vorren, K.-D., Alm, T., Gulliksen, S., & Løvlie, R. (1988). The last deglaciation (20,000 to 11,000 B.P.) on Andøya, northern Norway. *Boreas* 17, 41–77.

Watson, R. A., & Wright, H. E. Jr. (1980). The end of the Pleistocene: a general critique of chronstratigraphic classification. *Boreas* 9, 153–63.

Wegmüller, S., & Welten, M. (1973). Spätglaziale Bimstufflagen des Laacher Vulkanismus im Gebiet der westlichen Schweiz und der Dauphiné (F). *Eclogae Geologicae Helvetiae* 66. 533–41.

Welten, M. (1981). Verdrängung und Vernichtung der anspruchsvollen Gehölze am Beginn der letzten Eiszeit und die Korrelation der Frühwürm-Interstadiale in Mittel- und Nordeuropa. *Eiszeitalter und Gegenwart* 31, 187–202.

West, R. G. (1968). *Pleistocene geology and biology.* London: Longmans.

West, R. G. (1971). Studying the past by pollen analysis. In *Oxford biology readers,* ed. J. J. Head & O. E. Lowenstein, pp. 3–16. Oxford: Oxford University Press.

West, R. G. (1977). *Pleistocene geology and biology,* 2nd ed. London: Longman.

West, R. G. (1984). Interglacial, interstadial and oxygen isotope stages. *Dissertationes Botanicae* 72, 345–57.

West, R. G. (1988). The record of the cold stages. *Philosophical Transactions of the Royal Society of London* B 318, 505–22.

West, R. G. (1989). The use of type localities and type sections in the Quaternary, with especial reference to East Anglia. In *Quaternary type sections: Imagination or reality?,* ed. J. Rose & S. Schlüchter, pp. 3–10. Rotterdam: Balkema.

West, R. G., Dickson, C. A., Catt, J. A., Weir, A. H., & Sparks, B. W. (1974). Late Pleistocene deposits at Wretton, Norfolk. II: Devensian deposits. *Philosophical Transactions of the Royal Society of London* B 267, 337–420.

Winn, K., Averdieck, F.-R., Erlenkeuser, H., & Werner, F. (1986). Holocene sea level rise in the western Baltic and the question of isostatic subsidence. *Meyniana* 38, 61–80.

Winterhalter, B., Flodén, T., Ignatius, H., Axberg, S., & Niemistö, L. (1981). Geology of the Baltic Sea. In *The Baltic Sea,* ed. A. Voipio, pp. 1–121. Amsterdam: Elsevier Oceanography Series, 30.

Wintle, A. G. (1991). Luminescence dating. In *Quaternary dating methods – a user's guide,* ed. P. L. Smart & P. D. Frances, pp, 108–27. Quaternary Research Association, Technical Guide, 4. Nottingham: MI Press.

Woldstedt, P. (1958). *Das Eiszeitalter, Grundlinien einer Geologie des Quartär, Zweiter Band.* Stuttgart: Ferdinand Enke Verlag.

Woldstedt, P. (1969). *Handbuch der Stratigraphischen Geologie, II: Quartär.* Stuttgart: Ferdinand Enke Verlag.

Wright, H. E. Jr. (1984). Sensitivity and response time of natural systems to climatic change in the late Quaternary. *Quaternary Science Reviews* 3, 91–131.

Wright, W. B. (1937). *The Quaternary ice age,* 2nd ed. London: Macmillan.

Zagwijn, W. H. (1957). Vegetation, climate and time-correlations in the Early Pleistocene of Europe. *Geologie en Mijnbouw* 19, 233–44.

Zagwijn, W. H. (1961). Vegetation, climate and radiocarbon datings in the Late Pleistocene of the Netherlands. Part I: Eemian and Early Weichselian. *Memoirs of the Geological Foundation in the Netherlands* 14, 15–46.

Zagwijn, W. H. (1963). Pleistocene stratigraphy in the Netherlands, based on changes in vegetation and climate. *Verhandelingen van het Koninklijke Nederlands Geologisch Mijnbouwkundig Genootschap, Geologische Serie* 21(2), 173–96.

Zagwijn, W. H. (1973). Pollenanalytic studies of Holsteinian and Saalian beds in the northern Netherlands. *Medelingen van de Rijks Geologische Dienst* 24, 139–56.

Zagwijn, W. H. (1975). Variations in climate as shown by pollen analysis, especially in the Lower Pleistocene of Europe. In *Ice ages: Ancient and modern,* ed. A. E. Wright & F. Moseley, pp. 137–152. *Geological Journal,* Special Issue 6.

Zagwijn, W. H. (1979). Early and Middle Pleistocene coastlines in the southern North Sea basin. In *The Quaternary history of the North Sea,* ed. E. Oele, R. T. E. Schüttenhelm, & A. J. Wiggers, pp. 31–42. Acta Universitatis Upsaliensis, Symposia Universitatis Upsaliensis Annum Quingentesimum Celebrantis, 2, Uppsala.

Zagwijn, W. H. (1983). Sea-level changes in the Netherlands during the Eemian. *Geologie en Mijnbouw* 62, 437–50.

Zagwijn, W. H. (1985). An outline of the Quaternary stratigraphy of the Netherlands. *Geologie en Mijnbouw* 64, 17–24.

Zagwijn, W. H. (1986). The Pleistocene of The Netherlands with special reference to glaciation and terrace formation. *Quaternary Science Reviews* 5, 341–5.

Zagwijn, W. H. (1989). The Netherlands during the Tertiary and the Quaternary: A case history of lowland evolution. *Geologie en Mijnbouw* 68, 107–20.

Zagwijn, W. H. (1990). Vegetation and climate during warmer intervals in the Late Pleistocene of western and central Europe. *Quaternary International* 3/4, 57–67.

Zagwijn, W. H., van Montfrans, H. M., & Zandstra, J. G. (1971). Subdivision of the "Cromerian" in the Netherlands; pollen-analysis, palaeomagnetism and sedimentary petrology. *Geologie en Mijnbouw* 50, 41–58.

Zagwijn, W. H., & Paepe, R. (1968). Die Stratigraphie der weichselzeitlichen Ablagerungen der Niederlands und Belgiens. *Eiszeitalter und Gegenwart* 19, 129–46.

Zandstra, J. G. (1986). Explanation to the distribution map with the composition of the Fennoscandian crystalline erratic indicator pebbles (Saalian, The Netherlands). *Inqua Symposium on "Tills and endmoraines in The Netherlands and NW-Germany," University of Amsterdam, September 1986.*

Ziegler, P. A., & Louwerens, C. J. (1979). Tectonics of the North Sea. In *The Quaternary history of the North Sea,* ed. E. Oele, R. T. E. Schüttenhelm, & A. J. Wiggers, pp. 7–22. Acta Universitatis Upsaliensis, Symposia Universitatis Upsaliensis Quingentesimum Celebrantis, 2, Uppsala.

Zilliacus, H. (1987). De Geer moraines in Finland and the annual moraine problem. *Fennia* 165(2), 145–239.

Index